Microstructure, Characteristics, and Failure Behaviors of Cementitious Composites

Microstructure, Characteristics, and Failure Behaviors of Cementitious Composites

Editors

Yuan Gao
Junlin Lin
Xupei Yao

Basel • Beijing • Wuhan • Barcelona • Belgrade • Novi Sad • Cluj • Manchester

Editors

Yuan Gao
Nantong University
Nantong
China

Junlin Lin
Southeast University
Nanjing
China

Xupei Yao
Zhengzhou University
Zhengzhou
China

Editorial Office
MDPI AG
Grosspeteranlage 5
4052 Basel, Switzerland

This is a reprint of articles from the Special Issue published online in the open access journal *Materials* (ISSN 1996-1944) (available at: https://www.mdpi.com/journal/materials/special_issues/IZ4WVJEM12).

For citation purposes, cite each article independently as indicated on the article page online and as indicated below:

Lastname, A.A.; Lastname, B.B. Article Title. *Journal Name* **Year**, *Volume Number*, Page Range.

ISBN 978-3-7258-2511-0 (Hbk)
ISBN 978-3-7258-2512-7 (PDF)
doi.org/10.3390/books978-3-7258-2512-7

© 2024 by the authors. Articles in this book are Open Access and distributed under the Creative Commons Attribution (CC BY) license. The book as a whole is distributed by MDPI under the terms and conditions of the Creative Commons Attribution-NonCommercial-NoDerivs (CC BY-NC-ND) license.

Contents

About the Editors vii

Preface ix

Zofia Szweda, Justyna Kuziak, Liwia Sozańska-Jędrasik and Dominik Czachura
Analysis of the Effect of Protective Properties of Concretes with Similar Composition on the Corrosion Rate of Reinforcing Steel Induced by Chloride Ions
Reprinted from: *Materials* **2023**, *16*, 3889, doi:10.3390/ma16103889 1

Baozhen Jiang, Kotaro Doi and Koichi Tsuchiya
The Constituent Phases and Micromechanical Properties of Steel Corrosion Layers Generated by Hyperbaric-Oxygen Accelerated Corrosion Test
Reprinted from: *Materials* **2023**, *16*, 4521, doi:10.3390/ma16134521 29

Zhangjianing Cheng, Junying Wang, Junxiang Hu, Shuaijie Lu, Yuan Gao, Jun Zhang and Siyao Wang
Influence of the Graphene Oxide on the Pore-Throat Connection of Cement Waste Rock Backfill
Reprinted from: *Materials* **2023**, *16*, 4953, doi:10.3390/ma16144953 40

Shaofeng Zhang, Ronggui Liu, Chunhua Lu, Junqing Hong, Chunhong Chen and Jiajing Xu
Influence of Nano-SiO$_2$ Content on Cement Paste and the Interfacial Transition Zone
Reprinted from: *Materials* **2023**, *16*, 6310, doi:10.3390/ma16186310 54

Siyao Wang, Jingtao Hu, Zhiyuan Sun, Yuan Gao, Xiao Yan and Xiang Xue
Efficiency and Mechanism of Surface Reinforcement for Recycled Coarse Aggregates via Magnesium Phosphate Cement
Reprinted from: *Materials* **2024**, *17*, 122, doi:10.3390/ma17010122 67

Liqun Lu, Yingze Li, Yuncheng Wang, Fengjuan Wang, Zeyu Lu, Zhiyong Liu and Jinyang Jiang
Prediction of Hydration Heat for Diverse Cementitious Composites through a Machine Learning-Based Approach
Reprinted from: *Materials* **2024**, *17*, 715, doi:10.3390/ma17030715 85

Michał Pyzalski, Karol Durczak, Agnieszka Sujak, Michał Juszczyk, Tomasz Brylewski and Mateusz Stasiak
Synthesis and Investigation of the Hydration Degree of CA$_2$ Phase Modified with Boron and Fluorine Compounds
Reprinted from: *Materials* **2024**, *17*, 2030, doi:10.3390/ma17092030 98

Zhibin Qin, Jiandong Wu, Zhenhao Hei, Liguo Wang, Dongyi Lei, Kai Liu and Ying Li
Study on the Effect of Citric Acid-Modified Chitosan on the Mechanical Properties, Shrinkage Properties, and Durability of Concrete
Reprinted from: *Materials* **2024**, *17*, 2053, doi:10.3390/ma17092053 121

Nilam Adsul, Jun-Woo Lee and Su-Tae Kang
Investigating the Impact of Superabsorbent Polymer Sizes on Absorption and Cement Paste Rheology
Reprinted from: *Materials* **2024**, *17*, 3115, doi:10.3390/ma17133115 139

Muawia Dafalla, Ahmed M. Al-Mahbashi and Ahmed Alnuaim
Tension Capacity of Crushed Limestone–Cement Grout
Reprinted from: *Materials* **2024**, *17*, 3860, doi:10.3390/ma17153860 158

About the Editors

Yuan Gao

Yuan Gao is a Lecturer at the Nantong University. His research focuses on carbon nanomaterials engineered cement. Yuan has published more than 50 SCI papers concerning mechanical properties, impermeability, fractal characteristics of pore structure, material simulation optimization, and application research of carbon nanocomposite cement-based materials. He was a visiting scholar at Monash University from Oct 2019 to Sep 2021. Yuan completed his PhD in Civil Engineering at the China University of Mining and Technology in 2021.

Junlin Lin

Junlin Lin is a Lecturer at Southeast University. He obtained his PhD degree in civil engineering from Monash University. His research focuses on low-carbon cementitious composites, nanoengineering, characterization of cementitious composites, machine learning, and intelligent composites design.

Xupei Yao

Xupei Yao is a researcher at Zhengzhou University. He graduated from Monash University with a PhD degree in 2021. His research focuses on designing advanced materials for civil engineering with artificial intelligence. Xupei has published over 30 SCI papers on developing high-performance daytime radiative cooling materials and strong, durable cementitious nanocomposites.

Preface

This special issue publication was supported by the General Program of the National Natural Science Foundation of China (No. 52350004), Natural Science Foundation of China (No. 52408271), Natural Science Foundation of Jiangsu Province (No. BK20230615), China Postdoctoral Science Foundation (Nos. 2023TQ0299, 2023M743218) and Scientific and technological project in Henan Province (242102230135).

Yuan Gao, Junlin Lin, and Xupei Yao
Editors

Article

Analysis of the Effect of Protective Properties of Concretes with Similar Composition on the Corrosion Rate of Reinforcing Steel Induced by Chloride Ions

Zofia Szweda [1,*], Justyna Kuziak [2], Liwia Sozańska-Jędrasik [3] and Dominik Czachura [4]

[1] Department of Building Structures, Faculty of Civil Engineering, Silesian University of Technology, 44-100 Gliwice, Poland
[2] Institute of Building Engineering, Department of Building Materials Engineering, Faculty of Civil Engineering, Warsaw University of Technology, 00-637 Warsaw, Poland; justyna.kuziak@pw.edu.pl
[3] Łukasiewicz Research Network–Upper Silesian Institute of Technology, Centre of Welding, 44-100 Gliwice, Poland; liwia.sozanska-jedrasik@imz.lukasiewicz.gov.pl
[4] Smart Solutions, Measurement Center 3D, 03-046 Warsaw, Poland; dominik.czachura@smart-solutions.pl
* Correspondence: zofia.szweda@polsl.pl

Abstract: This study presents a comparison of the protective properties of three concretes of similar composition on the effect of chloride ions. To determine these properties, the values of the diffusion and migration coefficients of chloride ions in concrete were determined using both standard methods and the thermodynamic ion migration model. We tested a comprehensive method for checking the protective properties of concrete against chlorides. This method can not only be used in various concretes, even those with only small differences in composition, but also in concretes with various types of admixtures and additives, such as PVA fibers. The research was carried out to address the needs of a manufacturer of prefabricated concrete foundations. The aim was to find a cheap and effective method of sealing the concrete produced by the manufacturer in order to carry out projects in coastal areas. Earlier diffusion studies showed good performance when replacing ordinary CEM I cement with metallurgical cement. The corrosion rates of the reinforcing steel in these concretes were also compared using the following electrochemical methods: linear polarization and impedance spectroscopy. The porosities of these concretes, determined using X-ray computed tomography for pore-related characterization, were also compared. Changes in the phase composition of corrosion products occurring in the steel–concrete contact zone were compared using scanning electron microscopy with a micro-area chemical analysis capability, in addition to X-ray microdiffraction, to study the microstructure changes. Concrete with CEM III cement was the most resistant to chloride ingress and therefore provided the longest period of protection against chloride-initiated corrosion. The least resistant was concrete with CEM I, for which, after two 7-day cycles of chloride migration in the electric field, steel corrosion started. The additional use of a sealing admixture can cause a local increase in the volume of pores in the concrete, and at the same time, a local weakening of the concrete structure. Concrete with CEM I was characterized as having the highest porosity at 140.537 pores, whereas concrete with CEM III (characterized by lower porosity) had 123.015 pores. Concrete with sealing admixture, with the same open porosity, had the highest number of pores, at 174.880. According to the findings of this study, and using a computed tomography method, concrete with CEM III showed the most uniform distribution of pores of different volumes, and had the lowest total number of pores.

Keywords: composition of concrete; corrosion rate; impedance spectroscopy; linear polarization; microdiffraction; porosity of concretes

1. Introduction

Depending on their location and function, building structures are exposed to many harmful factors. The most common causes of reinforcement corrosion and concrete damage are carbonation and chloride penetration [1]. The faster the carbon dioxide or chlorides penetrate the concrete, the faster the passive layer on the rebar is destroyed and the corrosion process begins [2]. A very important issue is the corrosion of the reinforcement in foundations that are exposed to the harmful effects of chloride ions present in groundwater, especially in coastal environments, but also in industrial areas [3]. The use of traditional protective measures, consisting of coating the surface of concrete with various insulating layers, is expensive and time-consuming. At the same time, often after a short period of use of the building, these layers decompose in an aggressive and humid ground environment, ceasing to fulfill their protective role. In addition, the products resulting from the decomposition of these coatings are a source of environmental contamination [4]. Therefore, it is important to use such a quality of concrete that, with a properly selected thickness of concrete lagging, can provide a sufficiently good protective barrier to delay the penetration of chloride ions, thus ensuring the safe use of the building structure. The rate of penetration of chloride ions into concrete is characterized by the value of the diffusion coefficient. Currently, there are many methods for determining the value of this coefficient. The first group of methods consists of long-term diffusion tests [5], while another group of methods is based on electric field-accelerated migration tests [6]. Unfortunately, the values of coefficients obtained on the basis of tests conducted according to different standards and methods will ultimately differ from each other [7]. In one study [8], a number of various methods for determining diffusion and migration coefficients that can be carried out on different types of concrete were analyzed, and compared against the method of determining the diffusion coefficient conducted in accordance with the thermodynamic migration model [9]. This analysis showed that this coefficient accurately describes the natural process of chloride ion penetration into concrete.

A factor affecting the rate of chloride ion penetration into concrete is undoubtedly its porosity. However, the relationship between pore diameter and chloride diffusivity is reported to be linear by some researchers, such as Moon et al. [10] and Schutter [11], while Sherman et al. [12] found that the correlation between chloride ion migration coefficient and saturation was very low. Similarly, in one study [13], it has been shown that there is no clear correlation between chloride migration rate and pore size. There is also a hypothesis that in cementitious materials there are both ink-bottle pore and closed pore types, through which the penetration of chloride ions is strongly limited. However, these are detectable using a modified mercury porosimetry method [14,15], in addition to a gravimetric method. It was proven that the dominant factor affecting the value of the diffusion coefficient can often be the calcium aluminate content of the cement contained in the concrete [16]. It is difficult to find a correlation between porosity and diffusion coefficient values in such a complex material as concrete due to the fact that both the methods for determining porosity and the methods of determining the diffusion coefficient are dependent on many variables, and it thus is difficult to make these experiments repeatable. Therefore, a comprehensive analysis of the properties of concrete is necessary to assess its protective properties.

In recent years, the X-ray computed microtomography method (XRD) has been used to determine the porosity of concrete. This method, although promising, is not without some disadvantages. The main problem is the relationship between sample size and spatial resolution. In order to obtain a sufficiently high spatial resolution, it is important to use a sufficiently small sample. However, such a small section of the tested material may not represent the properties of the whole object [17,18]. Despite this, the method is being used with increasing frequency to determine both the porosity of cementitious materials and to visualize the corrosion processes occurring in reinforced concrete materials [19–21].

In addition to evaluating the diffusion and porosity properties, it is necessary to assess the rate of development in corrosion processes against the reinforcement in concrete. To evaluate the rate of corrosion development, nondestructive polarization methods are used: the Linear Polarization Resistance (LPR) method [22], and electrochemical impedance spectroscopy (EIS) [23]. Since the diffusion processes are long-lasting, the action of an electric field was used to accelerate the penetration of chloride ions into the concrete, similar to the work [24,25], while the corrosion processes occurred naturally.

The type of corrosion products of reinforcement (i.e., the type of oxides formed) depends on the environment in which the corroding elements are located. The initial stages of atmospheric corrosion of carbon steel in both rural and urban atmospheres leads to the formation of lepidocrocite (γ-FeOOH) and goethite (α-FeOOH), which are the main phases present in corrosion products, regardless of the environment. In chloride-containing coastal environments, akaganeite (β-FeO(OH,Cl)), magnetite (Fe_3O_4), and siderite ($FeCO_3$) are common. However, under simulated coastal environments, ferrihydrite ($Fe(OH)_3$) is also found [26,27]. Corrosion products accumulating on the surface of reinforcing steel in concrete, while increasing their volume, cause tensile stresses that induce cracking in the concrete lagging. Depending on the type of corrosion products, different levels of increased volume are observed. The largest, almost a sixfold increase, is observed when the corrosion products consist of 100% $Fe(OH)_3 \cdot 3H_2O$. In contrast, $Fe(OH)_2$ and $Fe(OH)_3$ hydroxides increase in volume by 2.17 and 1.76 times more than noncorroded steel, respectively. Determining the type of corrosion products is important in the process of modeling the cracking of concrete lagging [28].

The purpose of this study was to test the effectiveness of Improving the protective properties of the plain concrete used in the production of precast foundations. Both the modification of the concrete mixture by replacing Portland cement CEM I 42.5 R with metallurgical cement CEM III/A 42.5 N-LH/HSR/NA, in addition to using a sealing admixture in another mixture, were proposed. A number of tests were carried out to evaluate the efficacy of the proposed modifications on the mechanical and protective properties of the steel reinforcements of building foundations against corrosion processes. The novelty in the work is in using a comprehensive method for determining the resistance of concrete to the action of chloride ions. This method can be used in a variety of concretes, not only in those with only slight differences in composition, but also in concretes with various types of admixtures and additives, such as PVA fibers or the addition of other postindustrial waste. In order to determine the protective properties of concrete, the value of chloride ion diffusion coefficients in concrete was determined using the standard methods according to: ASTM1220, NT BUILD 443, NT BUILD 449, and ASTM1556, as well as using the thermodynamic ion migration model. The corrosion rate of reinforcing steel in these concretes was also compared using electrochemical methods: linear polarization LPR and impedance spectroscopy EIS. The porosities of these concretes (determined using the weight method, in addition to using X-ray computed tomography for pore-related characterization) were also compared. Using scanning electron microscopy (SEM) with microarea chemical analysis capabilities, as well as X-ray microdiffraction (XRD) to study microstructural changes, changes in the phase composition of corrosion products occurring in the steel–concrete contact zone were compared. It turns out that a slight modification of the composition of the concrete used can result in a huge improvement in its protective properties, ensuring the safety and durability of the reinforced concrete structures made from it.

2. Materials

2.1. Type of Material Tested

The tests were performed on three types of concrete mix. All of the concretes tested were designed to be ordinary concretes (used for precast foundation elements). Three types of natural rounded aggregate (from Lafarge Basalt Mine in Lubień, compliant with the standard PN-EN 12620 [29]) were utilized for mix creation: sand, $0 \div 2$ mm (722 kg/m^3); gravel, $2 \div 8$ mm (512 kg/m^3); and gravel, $8 \div 16$ mm (681 kg/m^3). The first tested concrete (B1) contained CEM I 42.5 R cement in its composition. The next concrete (B2) differed from the first only in the use of CEM III/A 42.5 N-LH/HSR/NA cement. The last concrete (B3) was further modified using a refining admixture in the amount recommended by the manufacturer (0.5% of the weight of the cement). The detailed compositions and properties of mixes are presented in Table 1.

Table 1. Composition, properties, and compressive strength of concrete mixtures.

Constituent	B	B2	B3
Cement	368		
Aggregate	1915		
Water	147		
w/c	0.4		
Plasticizer (0.5% m.c.)			
Sealing admixture (0.8% m. c.)			
Compressive strength f_{cm} MPa	62.4	63.9	56.9
Volume weight γ_b kg/m^3	2271	2241	2269

2.2. Type of Samples Used in the Tests

The compressive strength of the concrete was determined using cubic samples possessing 150 mm sides, which were stored and made in accordance with PN-EN12350-3:2011 [30].

3. Test Method

Next, a series of tests was conducted to determine the protective properties of the tested concretes against the effects of corrosion on the concrete's reinforcements by chloride ions. In order to perform these tests, 12 samples of each type of concrete were prepared.

To determine the depth of penetration by the pressurized water, cubic samples with 150 mm sides were used. In order to perform these tests, 3 samples of each type of concrete were prepared.

To determine the open porosity of hardened concrete, 1 sample for every type of concrete (each with a volume of about 10 cm^3) was used.

The X-ray computed microtomography tests used concrete cores with a diameter of about 18 mm and a height of 50 mm, cut with a diamond core drill from cylindrical specimens with a diameter of 100 mm and a height of 50 mm (Figure 1).

Figure 1. Concrete cores prepared for testing using X-ray computed microtomography.

For examination of the rate of penetration of chloride ions into concrete, concrete disks that were 50 mm thick and 100 mm in diameter were used. In order to perform these tests, 21 samples for each type of concrete were prepared.

The non-destructive polarization tests (LPR and EIS) were carried out each time on 2 test pieces in the shape of cylinders that were 100 mm in diameter and 60 mm high, made of each of the three types of concrete. A reinforcing bar was placed in each cylinder. A reinforcement cover 20 mm in thickness was used (Figure 2).

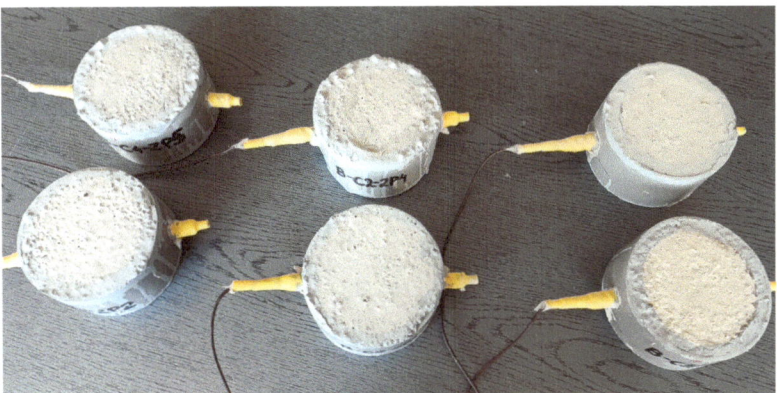

Figure 2. Concrete samples prepared for testing using non-destructive polarization tests (LPR and EIS).

The research for phase identification and quantitative analysis (using the Rietveld method) of corrosion products was conducted on three rebar specimens in concrete lagging made of 3 types of concrete, following simulations of the deterioration process under the formation of corrosion hazards caused by the presence of chloride ions. Figure 3 shows two smaller cylindrical test specimens made of B1 and B2 concrete; these had been cut out of the larger cylindrical test specimens tested, as outlined in Section 3.6, using a diamond-core drill so as to minimize rebar fragments. The samples disintegrated during cutting.

Figure 3. Examples of two smaller cylindrical test tubes that were cut from larger cylindrical test tubes previously subjected to chloride ion migration and corrosion measurements: (**a**) concrete sample B1; (**b**) concrete sample B2.

3.1. Compressive Strength Test after 1, 2, 7, and 28 Days of Maturation

The test was conducted after 1, 2, 7, and 28 days of concrete curing, which had been performed with the use of machine for testing the compressive strength of concrete. The results are shown in Tables A1–A3 of the Appendix A.

3.2. Determination of the Depth of Penetration of Pressurized Water

Testing the depth of water penetration under pressure was performed according to EN 12390-8, Testing of concrete: Part 8: Depth of water penetration under pressure [31]. A water pressure of 0.5 MPa was maintained for 72 h. After this time, the sample was split, to determine the moisture level.

3.3. Tests Open Porosity of Hardened Concrete

Open porosity was defined as the ratio of the differences between m_{SD} (the weight of the specimens completely saturated with water) and m_{OD} (the weight of the specimens dried at 60 °C); in both cases, the weight of the specimens was checked daily until the change in the weight of the specimens was less than 0.1%) against the differences between m_{SD} (the weight of the specimens completely saturated with water) and m_{FD} (the floating weight of the specimens with the use of hydrostatic balance), calculated according to the following equation

$$\varepsilon = \frac{m_{SD} - m_{OD}}{m_{SD} - m_{FD}} \cdot 100\% \tag{1}$$

3.4. Use of X-ray Computed Microtomography to Determine the Porosity of Concrete

A Nikon Corporation Industrial Metrology Business Unit, Tokyo, Japan XT H 225 ST industrial microtomograph equipped with a reflection lamp was used for the tomographic study; at the maximum approximation of small specimens, it proposed a resolution of 3 μ The optimal resolution used in this case was 10 μ, which made it possible to obtain reliable images of the scanned sample. The resulting images were then reconstructed using Nikon's CTPro 3D 6th generation software. After reconstruction, data preparation and all analyses were carried out using the appropriate tools available in VGStudio MAX 3.1 software. Prior to defectoscopy, an area 13 mm in diameter and 15 mm in height, slightly smaller than the scanned area, was taken from each scanned core. Clipping was necessary due to artifacts and imperfections in the tomographic image created during scanning. The data prepared in this way was analyzed for porosity.

3.5. Examination of the Rate of Penetration of Chloride Ions into Concrete

3.5.1. Testing the Diffusion of Chloride Ions through a Concrete Sample, in Accordance with the Norwegian Standard NT BUILD443 [32] and the American Standard ASTM1556 [33]

The test was conducted in accordance with the standards NT BUILD 443 [32] and ASTM C 1556-03 [33]. The test was carried out on three specimens per type of concrete in each case. The exact test method is described in the paper [8].

The value of the D coefficient is determined by adjusting the chloride concentration graph. This graph is obtained by calculating the distribution of chloride ion concentration expressed to the cement mass according to the solution of the diffusion equation, with the concentrations of these ions determined in the test:

$$C_{cal}^1(x,t) = C_{0,cal}^1 \left(1 - erf \frac{x}{2\sqrt{Dt}}\right) \tag{2}$$

To determine the best fit calculation of the diffusion coefficient complying with the results of the experiment, the mean squared error was calculated based on the following formula:

$$s = \sqrt{\frac{\sum_{i=1}^{n}[c_{cal}(x,t) - c(x,t)]^2}{n-1}} \tag{3}$$

where $c(x,t)$ is the chloride ion concentration that was measured during the test at a distance x (mm) from the edge of the element (%), t is time (s), and n is the number of concrete layers in which the chloride concentration was determined. The determined values of the diffusion coefficient are presented in Section 4.5.

3.5.2. Testing the Permeability of Chloride Ions through a Concrete Sample, in Accordance with American Standards AASHTO T 277 [34] and ASTM C1202-97 [35]

The study was conducted in accordance with the procedure described in standards AASHTO T 277 [34] and ASTM C1202-97 [35], each time being performed on three samples made of each concrete type.

Based on the determined charge, the value of the diffusion coefficient was also calculated using the Nernst–Einstein equation

$$D_{NE} = \frac{RT}{z^2 F^2} \frac{t_i}{C_i \gamma_i \rho_{BR}}, \quad \rho_{BR} = \frac{100}{\sigma}, \quad \sigma = \frac{QL}{VtA} \quad (4)$$

where D_{NE} is the diffusion coefficient (m^2/s), R is the universal gas constant (J/Kmol), T is absolute temperature (K), z is the valence of ions (-), F is the Faraday constant (C/mol), t_i is 1 transport number of chloride ions (-), γ_i is 1 activity coefficient of chloride ions (-), C_i is the concentration of chloride ions (mol/m^3), ρ_{BR} is volumetric resistivity (Ωm), σ is conductivity (Ωm^{-1}), L is sample thickness (m), V is electrical potential (V), A is the cross-sectional area of a sample (m^2), and t is time (s).

3.5.3. Testing the Permeability of Chloride Ions through a Concrete Sample, in Accordance with the Norwegian Standard NT BUILD492 [36]

The tests were carried out in accordance with the procedure described in NT BUILD492 [36], being performed each time on three samples per concrete type. The exact method of testing is described in the paper [8].

In accordance with the standard NT BUILD492 [36], the migration coefficient was calculated on the basis of the penetration depth, which was determined with the colorimetric method using a 0.1 M silver nitrate AgNO$_3$ solution, according to the formula:

$$\alpha = 2\sqrt{\frac{RTL}{zFU}} erf^{-1}\left(1 - \frac{2c_d}{c_0}\right), \quad D_T = \frac{RTL}{zFU} \frac{x_d - \alpha\sqrt{x_d}}{t_d} \quad (5)$$

where c_0 is the concentration of chlorides in the source chamber and the concrete, respectively, to the depth x_d; t_d is test duration (hours); L is element thickness, (mm); U is the value of the applied voltage; c_d is the chloride concentration at which the color changes (0.07 M for OPC concrete); c_0 is the chloride concentration in the cathode solution (2 M); and erf^{-1} is the inverse of the error function.

The depth of chloride penetration was determined in one of the three test samples made from each concrete.

In turn, the other two samples were used to determine the distribution of chloride ion concentration in concrete using the method described in the papers [8,9,37]. In this method, a device called a Profile Grinding Kit from Germann Instruments was used to obtain concrete dust from ten layers, each 2 mm thick. Then, after obtaining solutions by modeling the pore liquid, chloride ion steepness was measured with a CX-701 multimeter from Elmetron, using an ion-selective electrode. The value of the diffusion coefficient was determined using Equations (2) and (3). The determined values of the diffusion coefficient are presented in Section 4.5.

3.5.4. Migration Studies and Determination of Diffusion Coefficient Values Based on the Thermodynamic Migration Model

Migration tests were conducted using the method described in the papers [8,9], and [37]. Six cylindrical specimens, 100 mm in diameter and 50 mm in height, were made from each concrete type, and examined. Chloride migration tests were conducted at two time points: t_1 = 24 and t_2 = 48 h. At the end of migration, the level of chloride ion concentration in the water extracts obtained from stratified stripped concrete was determined using a Profile Grinding Kid device with both a diamond drill and an attachment that allowed the extraction of 2 mm thick concrete layers, to a depth of 20 mm.

Based on measurements of the mass density distribution ρ_1 of chloride ions migrating in concrete under the influence of an electric field, the authoritative value of the diffusion coefficient was determined using the relationship [8,9,37]

$$D^1 = \frac{\overline{j^1}(a)a\Delta t}{\frac{z^1FUg}{RTh}\left[\overline{\rho_1^1}+\overline{\rho_2^1}+\ldots+\overline{\rho_n^1}\right]\Delta t - B}, \quad B \cong \omega \frac{z^1FUg}{RTh}\left(\overline{\rho_1^1}+\overline{\rho_2^1}+\ldots+\overline{\rho_n^1}\right)\Delta t. \quad (6)$$

In this expression, $\overline{j^1}(a)$ is the value of the mass flow of chloride ions passing through the plane situated at "a" distance $x = a$; $\overline{\rho_1^1}, \overline{\rho_2^1}$, and $\overline{\rho_n^1}$ are the averaged mass densities of ion Cl$^-$ at the midpoints of consecutive intervals $(0, g), (g, 2g), \ldots, [(n-1)g, a]$ in time Δt. The first component of the denominator defines the stationary part of the chloride ion flow, while the second component (B) defines the nonstationary part (in this study, the value $B = 0$). Furthermore, z^1 is the ion valence, $R = 8.317$ J/mol·K is the universal gas constant, $F = 96\,487$ C/mol is the Faraday constant, U is the voltage between the electrodes, and h is the specimen height. The determined values of the diffusion coefficient are presented in Section 4.5.

3.5.5. Diffusion Tests

Diffusion tests were carried out using the method described in the papers [8,9,37]. Six cylindrical specimens for each concrete, each 100 mm in diameter and 50 mm in height, were examined each time. Chloride migration tests were conducted at two time points: $t_3 = 30$ and $t_4 = 60$ days. After diffusion, the value of chloride ion concentration in the water extracts was determined in the same way as after migration tests.

3.6. Measurements of Linear Polarization Resistance LPR

Polarization studies of the corrosion rate in the reinforcements using a polarization resistance measurement method (LPR) were carried out using a potentiostat (Gamry Reference 600, made by Gamry Instruments, Warminster, United States of America), coupled with a computer set. The measurement was recorded in a three-electrode system in which the working electrode was the sample reinforcement, the reference electrode was a silver chloride electrode, and the counterelectrode was a stainless-steel circle with a diameter close to that of the concrete samples. In order to accelerate the very long timescale for adequate chloride diffusion in the concrete, accelerated electromigration of chloride ions was used, similar to the works [24,25], while the corrosion processes induced due to the threshold chloride content of concrete occurred naturally. The corrosion current i_{corr} could be calculated via polarization resistance R_p obtained using an LPR measurement, according to the Stern–Geary equation

$$R_p = \left.\frac{dE}{di}\right|_{i\to 0,\, E\to E_{corr}}, i_{corr} = \frac{b_a b_c}{2.303 R_p (b_a + b_c)}, \quad (7)$$

where b_a is constant of anodic reactions, and b_c is constant of cathodic reactions.

The corrosion current density clearly determines the corrosion intensity of steel, and this is because, according to Faraday's law, the mass of losses (m(mg)) is proportional to the flowing current (I_{corr} (μA/cm^2))

$$\Delta m = kI_{corr}t, I_{corr} = \frac{i_{corr}}{A}, \quad (8)$$

where k is electrochemical equivalent, and t is time. The above relationship shows the correlation of the corrosion current density, with the linear corrosion rate (V_r (mm/year) expressed as follows:

$$V_r = 0.0116\, i_{corr} \quad (9)$$

Corrosion rate (V_r (mm/year) is determined from the average cross-section loss around the bar circumference, in mm, per 1 operational year of the structure. The detailed results from the analysis with the calculated densities for corrosion current are shown in Tables A4–A6 of the Appendix B.

3.7. EIS Measurements for Steel in Concrete

Measurements of EIS spectra for steel in concrete were carried out before the LPR measurements in the same measurement set-up, and on the same samples as the LPR measurements described in Section 3.6. EIS spectra were recorded at the corrosion potential in the frequency range of 100 kHz to 50 mHz. The amplitude of the perturbation signal was 10 mV, and 10 points per decade of frequency were recorded.

The impedance spectra obtained were analyzed using the electrical equivalent circuit shown in Figure 4. The part of the spectra recorded at high frequencies describes the properties of the concrete and corresponds to the left part of the electrical equivalent circuit, which contained the following elements: the electrolyte resistance (R_e), concrete resistance (R_c), and constant phase element describing concrete cover (CPE_c). The impedance spectra at medium and high frequencies characterizes steel, and corresponds to the right-hand part of the electrical equivalent circuit, which contains the ohmic resistance in pits or defects of the passive layer (R_{pit}), the constant phase element for passive surface (CPE_{pas}), the charge transfer resistance (R_{ct}), constant phase element describing double layer on the steel (CPE_{dl}), and the Warburg (diffusional) impedance (W).

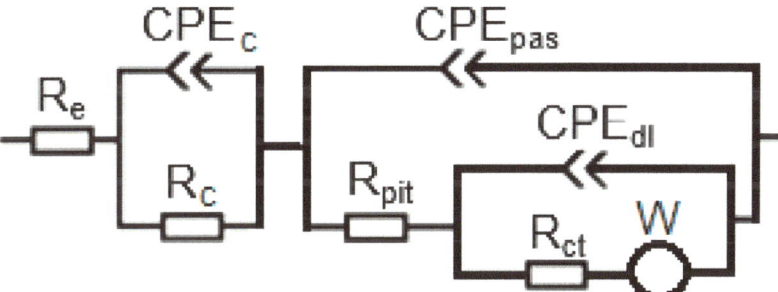

Figure 4. Equivalent circuit for steel in concrete (R_e is the electrolyte resistance; R_c is concrete resistance; CPE_c is the constant phase element describing concrete cover; R_{pit} is the ohmic resistance in pits or defects of passive layer; CPE_{pas} is the constant phase element for passive surface; R_{ct} is the charge transfer resistance; CPE_{dl} is the constant phase element describing double layer on the steel; and W is the Warburg (diffusional) impedance.

3.8. Phase Identification and Quantitative Analysis of Corrosion Products, Using the Rietveld Method

Measurements were performed with an Empyrean X-ray diffractometer acquired from PANalytical, using filtered iron radiation in a pixel detector configuration. Identification of phase composition was carried out in accordance with the M1-RTG accredited procedure, entitled *"Phase Identification"*, in addition to the International Centre for Diffraction Data's PDF-4+ database. Quantitative analysis was performed using the Rietveld method, in accordance with the accredited procedure M2-RTG, entitled *"Quantitative Phase Analysis"*. Observations were carried out using a JSM 7200F scanning electron microscope from JOEL, with a BSE backscattered electron detector and an EDS detector, to perform microarea chemical analysis.

4. Results

4.1. Compressive Strength Test Results after 1, 2, 7, and 28 Days of Maturation

Figure 5 shows the compressive strength comparison of the tested concretes.

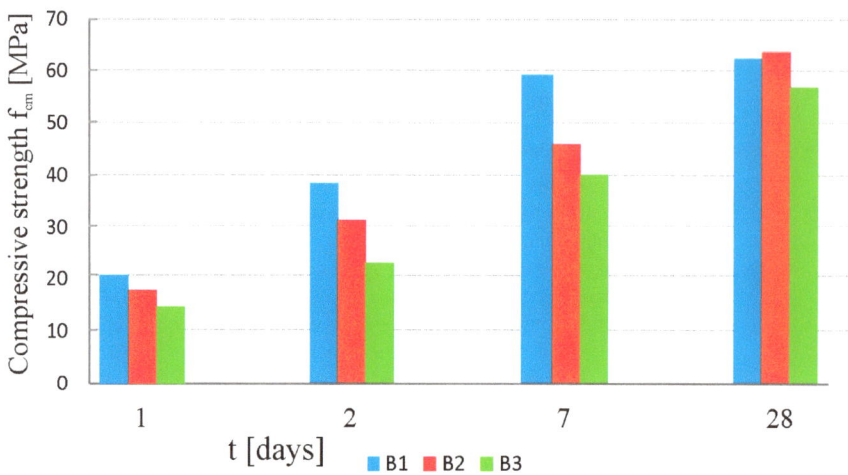

Figure 5. Compressive strength of concretes B1, B2, and B3 after 1, 2, 7, and 28 days of maturation.

Changing the cement from Portland cement (for B1 concrete) to metallurgical cement (for B2 and B3 concretes) caused a reduction in compressive strength in the first 7 days of maturation. This reduction in compressive strength could also have been caused by the sealer admixture (B3 concrete). After the 28th day of maturation, the strength of concrete with metallurgical cement was the highest, and a clear upward trend was also visible. The properties of blast furnace slag, which is a component of metallurgical cement, allow us to hypothesize that the strength of B2 and B3 concretes will be higher than that of B1 concrete after a longer maturation time.

4.2. Results of Pressurized Water Penetration Depth Tests

Table 2 summarizes the depth of penetration of water under pressure, as determined using the method outlined in the standard PN-EN 12390-8 [31], in three samples made from each concrete.

Table 2. Depth of penetration of pressurized water.

Concrete	B1.1	B1.2	B1.3	B2.1	B2.2	B2.3	B3.1	B3.2	B3.3
Depth of penetration [mm]	23	20	25	14	6	14	18	18	22
Average depth [mm]		23			11			19	

Figure 6 shows cross-sections of the test samples, cut after the pressure water penetration test.

Figure 6. Cross-sections of the test samples obtained after splitting them immediately after moistening with water at a pressure of 0.5 MPa maintained for 72 h: (**a**) sample B1.1; (**b**) sample B1.2; (**c**) sample B1.3; (**d**) sample B2.1; (**e**) sample B2.2; (**f**) sample B2.3; (**g**) sample B3.1; (**h**) sample B3.2; and (**i**) sample B3.3.

The use of metallurgical cement significantly reduced the depth of pressurized water penetration. However, the addition of a sealing admixture did not significantly improve the achieved effect in B3 concrete. It should be noted that the penetration of water deep into each sample was of a local nature. The magnitude of penetration forming about 20 mm was small and appropriate for concretes with a good quality of cement matrix.

4.3. Results of Testing the Open Porosity of Hardened Concrete

Table 3 presents the values of the weight of the sample determined using the gravimetric method with a hydrostatic balance, and the porosity calculated according to the formula (1).

Table 3. Values of the open porosity and mass of a sample made of concrete B1, B2, and B3, determined using the gravimetric method with a hydrostatic balance.

Concrete	Dry Mass g	Water Saturation Mass g	Floating Mass g	Open Porosity %
B1	216.85	225.3	131.4	9
B2	121.36	126.2	72.8	9
B3	148.37	153.4	87.9	8

As can be seen from the results, the tested concretes did not differ substantially in terms of open porosity as determined using the gravimetric method. Concrete B3 (with sealing admixture) had slightly better properties than the other concretes.

4.4. Results of Porosity Analysis of Concrete Cores using X-ray Computed Microtomography

Figure 7 shows cross sections made in B1, B2, and B3 concrete, subjected to image analysis in VGStudio MAX 3.1 software.

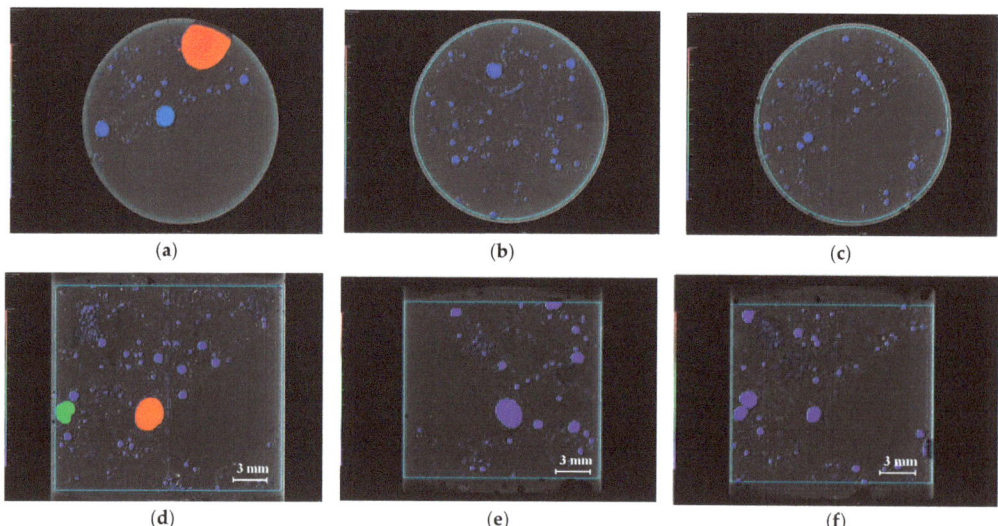

Figure 7. Selected cross sections taken during the tomographic examination, together with the image analysis performed in VGStudio MAX 3.1 software: cross section of the cylinder (**a**) concrete B1, (**b**) concrete B2, and (**c**) concrete B3; longitudinal cross-section of the cylinder (**d**) concrete B1, (**e**) concrete B2, and (**f**) concrete B3.

These were based on tomographic studies of a smaller cylindrical sample, measuring 18 mm in diameter and 21 mm in height.

On the basis of the analysis performed, the porosity of the concretes was calculated, expressed as the percentage of the sum of all air voids (imaged using the tomograph) against the volume of the fragment of the cylindrical sample with a diameter of 18 mm and a height of 21 mm, amounting to $p_1 = 14\%$, $p_2 = 6\%$, and $p_3 = 6\%$ for concretes B1, B2, and B3, respectively. According to these tomograph tests, it can be said that the replacement of the ordinary cement (CEM I) with metallurgical cement (CEM III) resulted in a reduction in the porosity of the concrete by almost half. On the other hand, the use of a sealing agent had practically no effect on the porosity.

Figure 8 shows graphs of the frequency of pores of a given volume occurring in the analyzed concrete samples. Concrete B1 was characterized by the highest porosity, and had 140.537 pores; concrete B2, characterized by having half the porosity of B1, had 123.015 pores, while concrete B3, with the same porosity as B2, had the highest number of pores, at 174.880. Considering the above graphs, it can be concluded that concrete B3 has both the most pores and the highest volume (Figure 8f). Overall, B2 concrete is characterized by the smallest number of pores. Concrete B1 had 1.1 times more pores than concrete B2, while concrete B3 had 1.4 times as many pores. In summary, it can be seen that B2, while having a low porosity compared to concrete B1, simultaneously had fewer pores with a large volume compared to concrete B3.

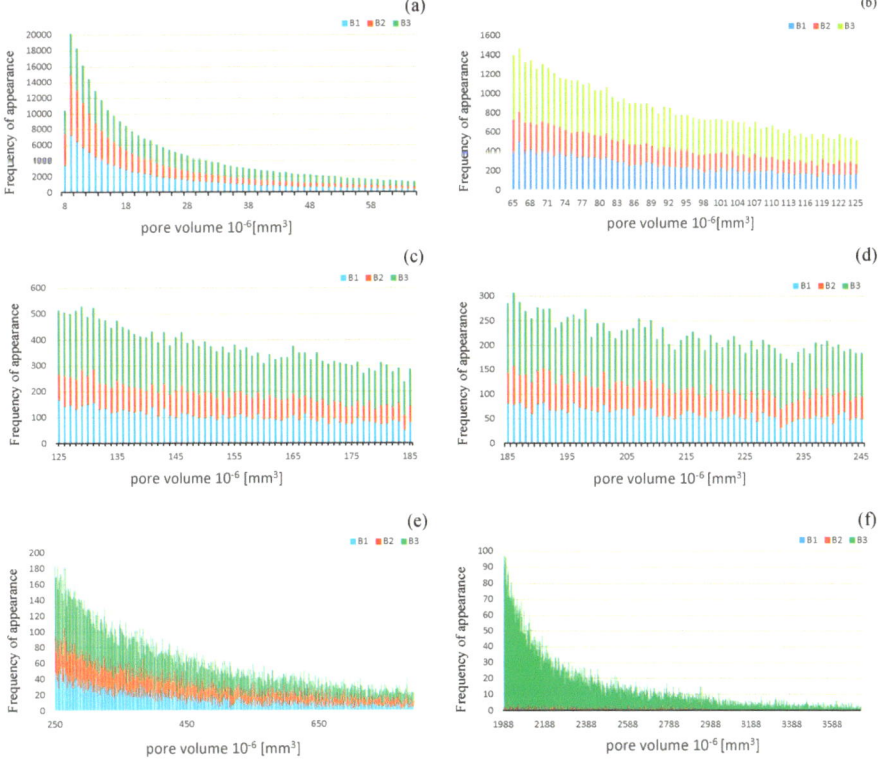

Figure 8. Frequency of pores of a given volume occurring in analyzed concrete samples B1, B2, B3: (**a**) volume range $(8-64) \times 10^{-6}$ mm^3; (**b**) volume range $(65-124) \times 10^{-6}$ mm^3; (**c**) volume range $(125-184) \times 10^{-6}$ mm^3; (**d**) volume range $(185-249) \times 10^{-6}$ mm^3; (**e**) volume range $(250-1987) \times 10^{-6}$ mm^3; (**f**) volume range $(1987-3788) \times 10^{-6}$ mm^3.

4.5. Determination of the Rate of Penetration of Chloride Ions into Water-Saturated Concretes

All three concretes were tested using both standard methods and the thermodynamic method. A detailed discussion of the methods, including a comparison with the results obtained in different concrete mixtures, are presented in the paper [8]. Table 4 summarizes the diffusion and migration coefficient values determined using various methods. The table includes the values of the diffusion coefficient D_{NE} determined from the charge measured during the migration test, conducted according to the rules of the standard ASTM C1202-997 [35] after applying the Nernst–Einstein Equation (4). Another coefficient specified in the standard (D_T) was calculated on the basis of migration tests, in accordance with the rules contained in the standard NT BUILD 492 [36], based on the depth of penetration of chloride ions, determined with the colorimetric method using the Equation (2). Subsequent values of the migration rate D^1_{migr} were determined by fitting the concentration curve that had been plotted according to Equation (2) to the results obtained in a migration test conducted in accordance with the standard NT BUILD 492 [36]. On the basis of this coefficient, using Equation (2), which in a simplified manner takes into account a certain multiplier linking the diffusion flow to the migration of chloride ions, the value of the diffusion coefficient D^1_{dif} was determined. Another diffusion coefficient (D) was determined from diffusion tests conducted in accordance with the standard NT BUILD 443 [32]; this was determined using Equation (2) based on the lowest value of the mean–square error (3). The next two values of the diffusion coefficient (D^{t1} and D^{t2}) were determined by fitting a concentration curve that had been determined numerically from Equations (4) and (2) based on diffusion tests. Another value of the diffusion coefficient (D^1) was determined on the basis of both migration studies and Equation (6), which was the solution of the thermodynamic migration model.

Table 4. Diffusion and migration coefficient values determined using different methods [8].

Mix ID	D_{N-E} (σ) **	D_T (∂) ***	$D^1_{migr} 10^{-3}$ ($s(x)$) *	D^1_{dyf} ($s(x)$) *	D; ($s(x)$) *	D^{t1}; ($s(x)$) *	D^{t2} ($s(x)$) *	D^1
B1	3.67; (0.28)	0.48; (±0.06)	12.5; (0.69)	12.5; (0.69)	1.20; (0.85)	4.84; (0.34)	2.42; (0.61)	4.84
B2	1.34; (0.05)	2.0; (±0.06)	16; (0.74)	16; (0.74)	2.96; (0.21)	3.84; (0.25)	1.92; (0.62)	1.48
B3	1.86; (0.05)	1.41; (±0.06)	130; (0.48)	130; (0.48)	2.32; (0.67)	5.88; (0.54)	4.52; (0.33)	2.27

* ($s(x)$)—the value of mean squared error in brackets; ** (σ)—the value of standard deviation in brackets; and *** (∂)—the value of error due to the length of the scale in brackets.

According to an analysis of various methods of determining the diffusion coefficient in different types of concrete, each of which having been carried out in the paper [8], the value of diffusion coefficient that best represents the natural diffusion process was the coefficient D^1, determined from the thermodynamic migration model. The value of this coefficient was used to predict the durability time of a structure made of one of the three concretes considered.

It was assumed that the threat of corrosion of the reinforcing bars causes, according to the standard PN-EN 1992-1-1 Eurokod 2 [38], chloride ion concentration with a critical value of C_K = 0.4% of cement mass. The calculations of chloride concentration changes at the reinforcement–concrete interface of the lagging (x = 15, x = 25, x = 35, and x = 40 mm), which were ascertained according to the known solution of the diffusion Equation (2), assuming the D^1 value of the diffusion coefficient for the calculations, determined using a thermodynamic migration model. At the edge of the concrete, the chloride concentration was assumed to equal C_0 = 0.8% of cement mass. Computer-determined changes in time *t* of chloride concentration at the reinforcement–concrete interface of the lagging (i.e., 15 mm, 25 mm, 35 mm, and 40 mm, respectively) are shown in Figure 9.

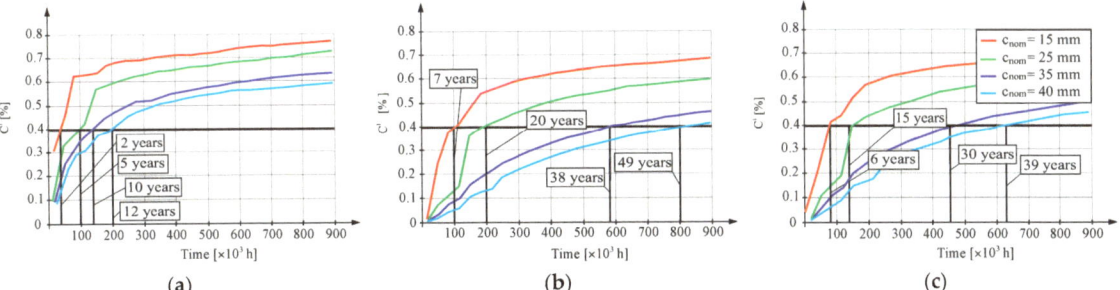

Figure 9. Time-dependent change in chloride concentration at the surface of the reinforcement at different lagging thicknesses (15, 25, 35, and 40 mm): (**a**) concrete B1, (**b**) concrete B2, and (**c**) concrete B3.

The corrosion hazard of the reinforcement in B1 concrete will occur after just 2 years of using the structure with the applied lagging thickness of 15 mm, while with the application of a lagging thickness of 25 mm the corrosion hazard will occur after 5 years. Application of the minimum thickness of lagging around the reinforcement required by the standard EN-PN 1992-1-1 [38] in construction class S2 and exposure class XD2/D3 (amounting to 35 and 40 mm, respectively) would avoid corrosion risks to the reinforcement for a period of 10 and 12 years, respectively, at the lagging thicknesses considered. The corrosion hazard of reinforcement in B2 concrete will occur after just 7 years of operation in structures with the applied lagging thickness of 15 mm; by contrast, with the application of lagging thickness of 25 mm, the corrosion hazard will occur after 20 years. Application of the minimum thickness of lagging of reinforcement required by the standard EN-PN 1992-1-1 [38] for construction class S2 and exposure class XD2/D3 (amounting to 35 and 40 mm, respectively) would avoid the corrosion risk of reinforcement for a period of 38 and 49 years, respectively, at the lagging thicknesses considered. The corrosion hazard of reinforcement in B3 concrete will occur already after 6 years of operation in structures with the applied lagging thickness of 15 mm; by contrast, with the application of lagging thickness of 25 mm, the corrosion hazard will occur after 15 years. Application of the minimum thickness of lagging of reinforcement required by the standard EN-PN 1992-1-1 for construction class S2 and exposure class XD2/D3 (amounting to 35 and 40 mm, respectively) would avoid the corrosion risk of reinforcement for a period of 30 and 39 years, respectively, at the lagging thicknesses considered.

4.6. Determination of Corrosion Rate of Reinforcing Steel Using the Linear Polarization Method (LPR)

The specimens, for which a number of measurements were taken during the whole research process, were analyzed. They were the B1_P1, B1_P2 and B2_P1, B2_P2 and B3_P1, and B3_P2 specimens. The first measurement was a reference, prior to the chloride migration in the concrete. The second measurement was taken after 7 days of chloride migration onto concrete. After waiting a further 7 days, the third measurement was taken after 14 days of chloride migration into concrete. After waiting 7 days again, the fourth measurement was taken, 21 days after chloride migration into the concrete. Likewise, the fifth measurement, taken after 28 days of chloride migration into the concrete and after waiting 7 days. Three seven-day cycles of chloride ion recharge with an electric field were performed for B1 concrete. For B2 concrete, five seven-day cycles of chloride ion recharge were carried out using an electric field. For B3 concrete, four seven-day cycles of recharging chloride ions with an electric field were carried out. Each time the seven-day migration process was completed, post measurements of the corrosion current density were made after waiting seven days from the electric system being turned off. Charging cycles were applied until the corrosion current density value that would indicate advanced corrosion was obtained. Exemplary shapes for the six selected measuring elements are illustrated in Figure 10.

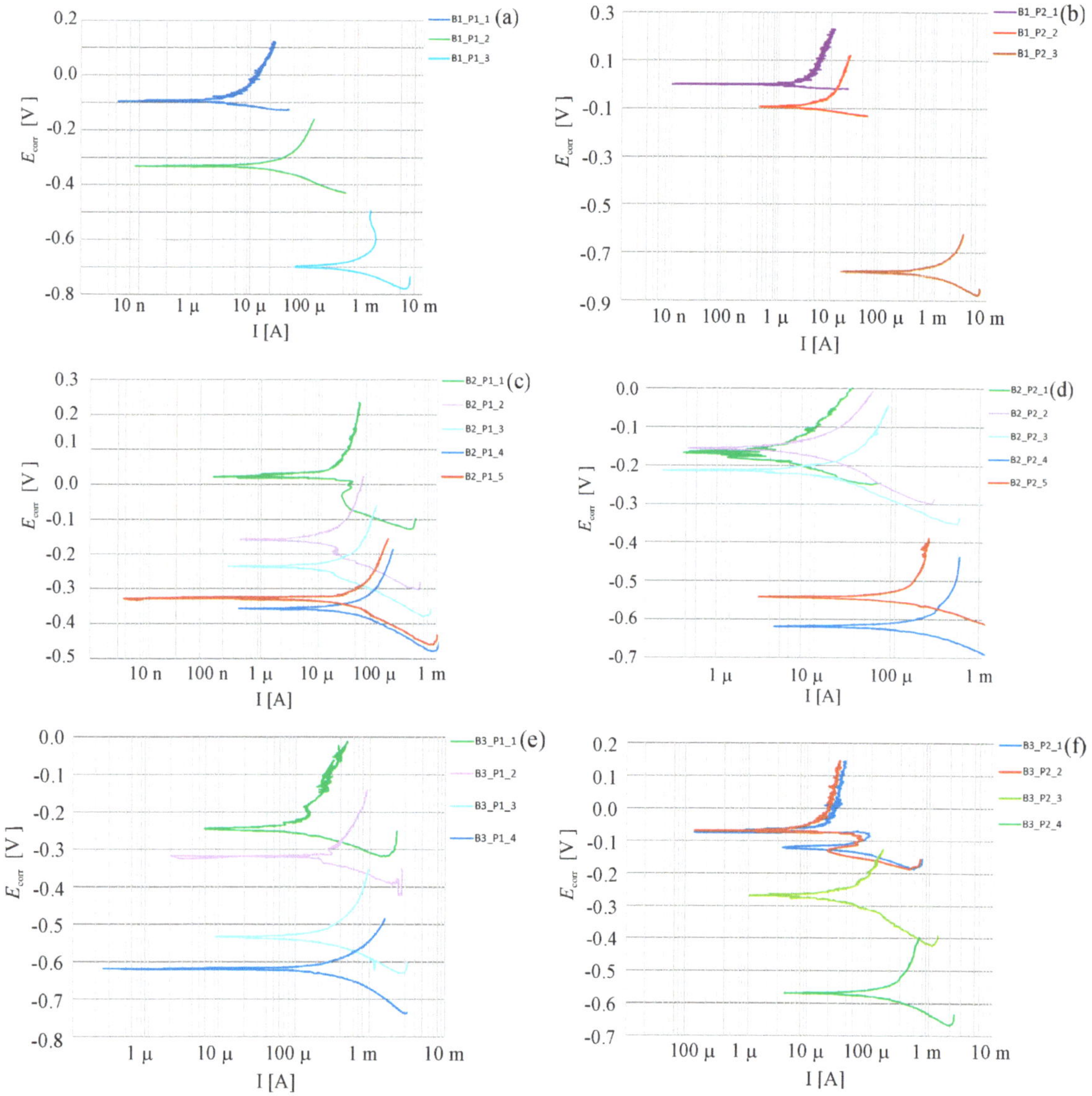

Figure 10. Potentiodynamic polarization curves for steel reinforcement in concrete obtained for the selected specimens: (**a**) B1_P1; (**b**) B1_P2; (**c**) B2_P1; (**d**) B2_P2; (**e**) B3_P1; and (**f**) B3_P2, after successive cycles of chloride ion migration to concrete.

A similar change in the distribution of polarization curves over time was observed for the specimens B1_P1 and B1_P2 (Figure 9a,b) after the first reference measurement of corrosion potential, taken prior to migration ($E_{corr} = -95$(B1_P1 _1); -2(B1_P2 _1) mV. Then, in both cases, after the first migratory recharge, a decrease in the corrosion potentials were observed, with values: $\Delta E_{corr} = 236$ mV (B1_P1 _2), and $\Delta E_{corr} = 90$ mV (B1_P2 _2), which can indicate the beginning of the corrosion process. After another recharge, corrosion potential values were observed in both probes, indicating a very high probability of corrosion

(E_{corr} = −694(B1_P1 _3); −781(B1_P1 _3) mV. In B2 concrete, a decrease in corrosion potential value was observed, from the initial value of (E_{corr} = −141(B2_P1 _1); −170(B2_P1 _2) mV to the value at the last measurement of (E_{corr} = −328(B2_P1 _5); −429(B2_P2 _5) mV. In sample B2_P2, there was a slight increase in the corrosion potential by a value of ΔE_{corr} = 15 mV. After the second charge, the values of the corrosion potential in both samples reached a value above the 200 mV (E_{corr} = −237(B2_P1 _2); −214(B2_P2 _2) mV, which may indicate a 5% corrosion potential. After the third charge, corrosion potential values obtained in both samples were of the value (E_{corr} = −357(B2_P1 _3); −614(B2_P2 _3) mV, indicating a high probability of corrosion.

In B3 concrete, a decrease in corrosion potential was also observed during all test cycles, but the initial values in one of the samples (E_{corr} = −242(B3_P1 _1); −320(B3_P1 _2) mV could already indicate a 50% possibility of corrosion. However, in sample B3_P2, in the first two measurements, the value of corrosion potential (E_{corr} = −23(B3_P1 _1); −49(B3_P1 _2) mV indicated the absence of corrosion. After a further two recharges, corrosion potential values obtained in both samples (E_{corr} = −618(B3_P1 _1); −568(B3_P1 _2) mV were indicative of a 95% possibility of corrosion occurrence [39].

Figure 11a presents a comparison of the results from six measurements of corrosion current density (i_{corr}) of the steel reinforcement in concrete from two chosen test elements made of tested concretes. Figure 11b shows a comparison of results from six measurements of corrosion potential (E_{corr}) of the steel reinforcement in concrete from two chosen test elements made of tested concretes.

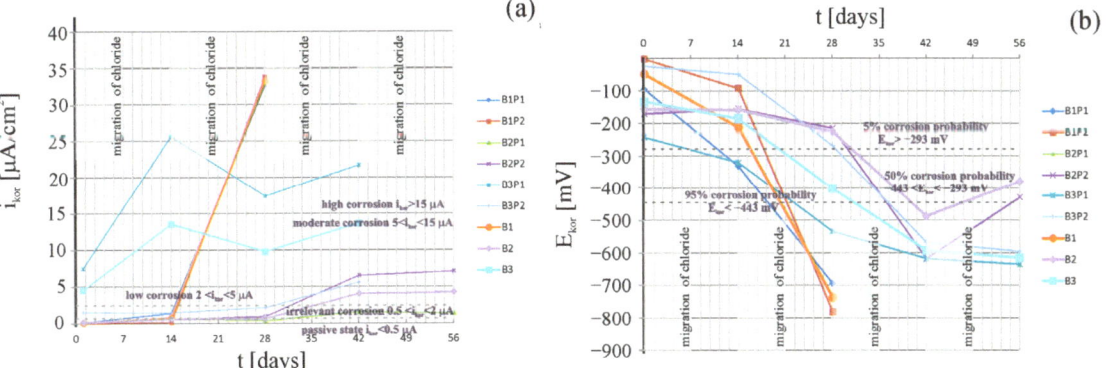

Figure 11. Distributions of (**a**) the corrosion current densities, and (**b**) the corrosion potential, obtained for selected specimens B1P1, B1P2, B2P1, B2P2, B3P1, and B3P2: 1-before chloride migration; 2-after 7 days, 3-after 14 days, 4-after 21, and 5-after 28 days of migration.

The analysis was made on the basis of assumptions made in the work [40]. The first reference measurement taken prior to migration indicated that the average value of corrosion current intensity (B1 ($\bar{i}_{corr}$ = 0.1 μA); B2 ($\bar{i}_{corr}$ = 0.18 μA)) suggested the irrelevant corrosion in B1 and B2 concretes. In contrast, concrete B3 ($\bar{i}_{corr}$ = 4.52 μA) suggested a low corrosion. Another measurement taken boht after 7 days of chloride ions migration under the accelerated action of the electric field and 7 days after switching off the system indicated that the increase in the average corrosion current rate was fastest (an 88% increase) in B1 concrete ($\overline{\Delta i}_{corr}$ = 0.7 μA), which was also followed by a large increase of 72% in B2 concrete ($\overline{\Delta i}_{corr}$ = 0.45 μA), and a marginally smaller one (a 67% increase) in B3 concrete ($\overline{\Delta i}_{corr}$ = 9.06 μA). After another 7-day charging with chloride ions, a massive increase was found for concrete B1 ($\overline{\Delta i}_{corr}$ = 32.37 μA), reaching the value of high corrosion ($\bar{i}_{corr}$ = 32.37 μA). On the other hand, in B2 concrete we saw a slight (11%) increase ($\overline{\Delta i}_{corr}$ = 0.08 μA), and in B2 concrete there was even a decrease of 38%. Due to the high values of corrosion current in concrete B1, measurements of corrosion current velocity were terminated after the second recharge. In the other two concretes, another 7-day recharge of chloride ions was applied. After this third migration cycle, the measured value of the average corrosion current in concrete B2 showed an increase of $\overline{\Delta i}_{corr}$ = 3.37 μA, yielding an average value of $\overline{i}_{corr} = 4.08$ μA, which indicated low corrosion levels. In contrast, the B3 concrete showed an increase in the average corrosion current value of $\overline{\Delta i}_{corr}$ = 3.86 μA. The obtained average value of the corrosion current rate was $\overline{i}_{corr} = 13.68$ μA, which indicated very high corrosion. Thus, the recharge process was stopped. The value of chloride ion concentration at the surface of the reinforcing steel far exceeded the critical value of C_K = 0.4% for the norm [38]. In B2 concrete, another stage of chloride ion migration was applied, after which the average value $\overline{i}_{corr}$ = 4.26 μA was obtained, indicating low corrosion. However, due to the fact that the B2P2 sample yielded a value of $\overline{i}_{corr} = 7.12$ μA, it ultimately indicated high corrosion. However, in sample B2P1, the increase in corrosion rate values was higher than 20 times, so the measurements were stopped.

4.7. Analysis of EIS Spectra for Steel in Concrete

Before the injection of chloride ions into the concrete, fragments of large arcs were visible on the Nyquist spectra for steel (Figures 12–14). This is a characteristic course of the impedance spectra for steels in the passive state. On the Bode diagrams, high impedance values at low frequencies were observed, indicating a low corrosion rate of the steel.

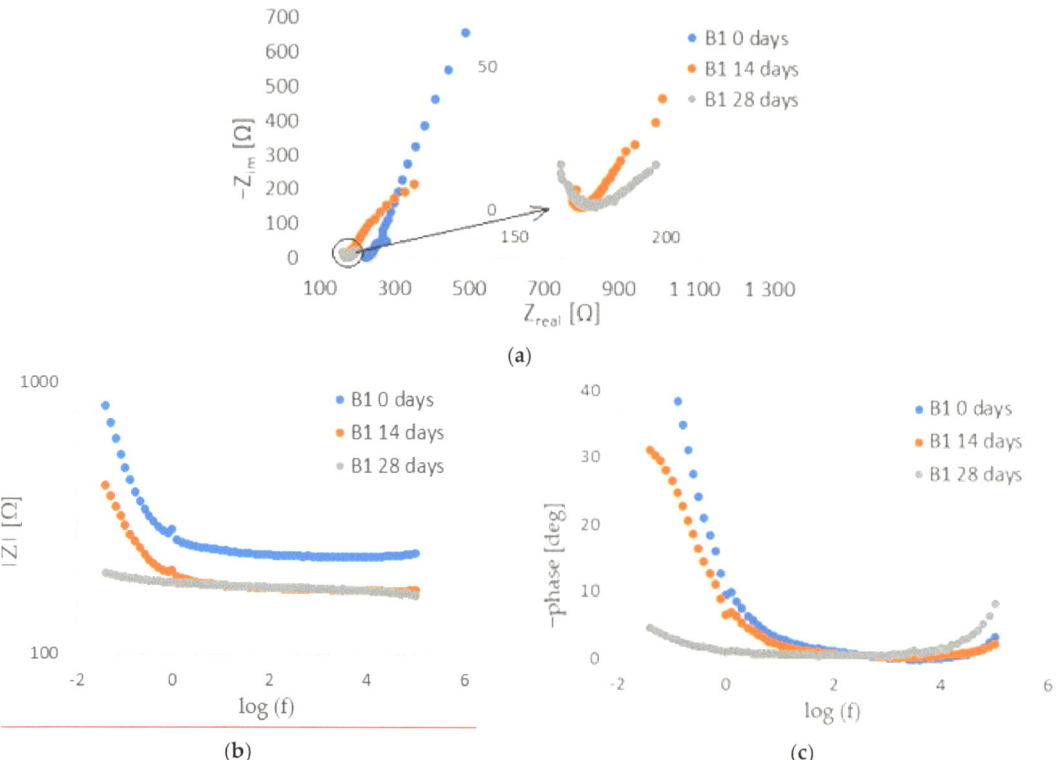

Figure 12. Examples of impedance spectra ((**a**) Nyquist plot, (**b**) impedance, and (**c**) phase angle shift) for steel in B1 concrete during tests.

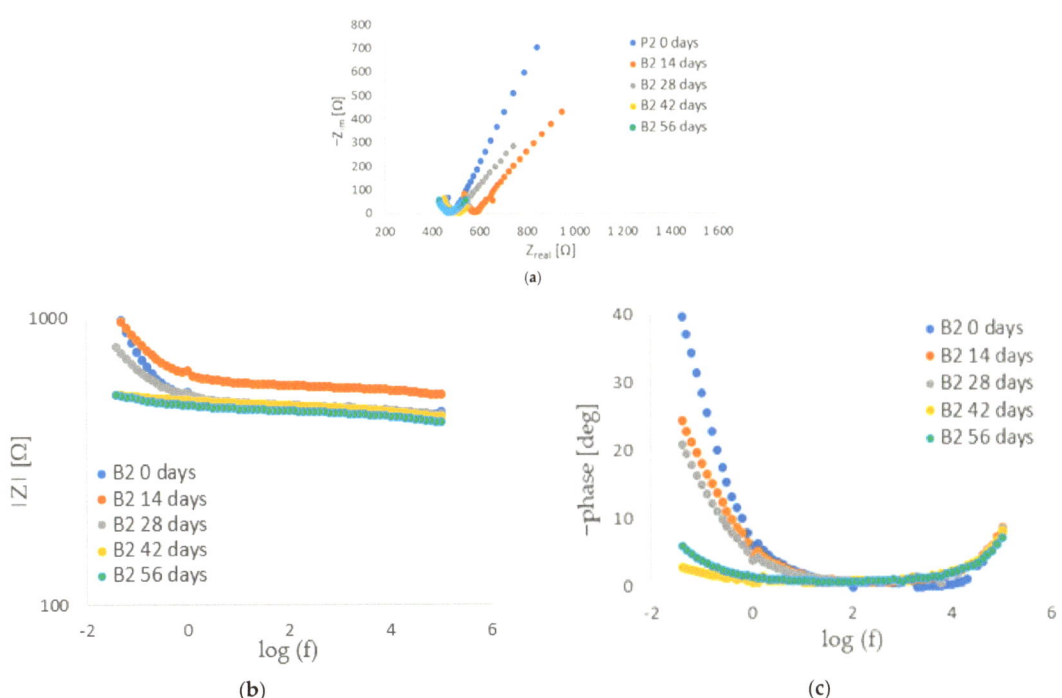

Figure 13. Examples of impedance spectra ((**a**) Nyquist plot, (**b**) impedance, and (**c**) phase angle shift) for steel in B2 concrete during tests.

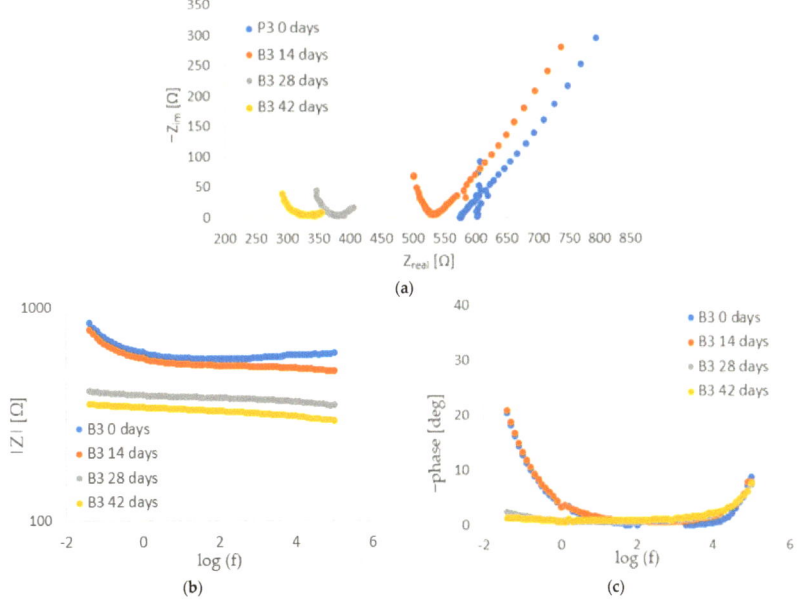

Figure 14. Examples of impedance spectra ((**a**) Nyquist plot, (**b**) impedance, and (**c**) phase angle shift) for steel in B3 concrete during tests.

The introduction of chlorides into the concrete resulted in a decrease in the impedance values at high frequencies (corresponding to the sum of the electrolyte and concrete resistance $R_e + R_c$) associated with an increase in the conductivity of the concrete due to an increase in the ion concentration in the concrete pore liquid. The sum of $R_e + R_c$ determined from the analysis of the spectra for concretes B2 and B3 was greater than it was for concrete B1 (Figure 15, Tables A7–A9 in Appendix B), indicating a tighter concrete cover for concretes B2 and B3. This is consistent with the results of the concrete porosity tests (concretes B2 and B3 had lower porosity than B1).

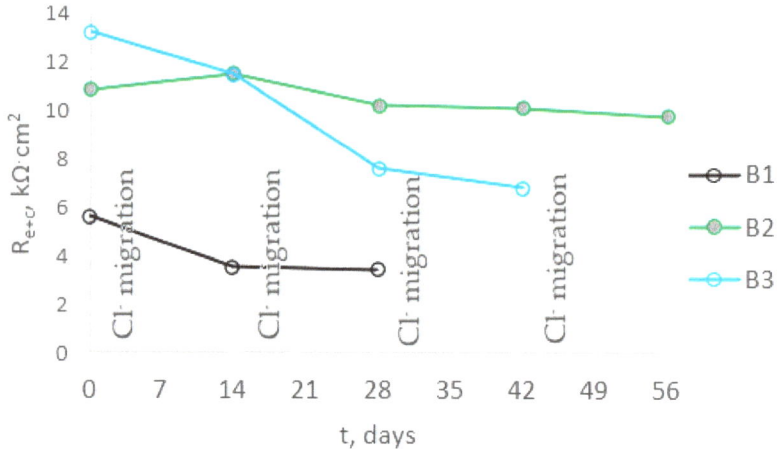

Figure 15. The dependence of the mean values of the sum of the resistance of electrolyte and concrete for B1, B2, and B3 concrete on the migration time of chlorides.

Successive cycles of chloride migration into the concrete resulted in changes in the course of the impedance spectra; a shift in the phase shift angle towards lower frequencies, as well as a decrease in the impedance and phase values at low frequencies, were observed, indicating the development of corrosion processes in the steel. It can be seen that, before the introduction of chlorides, the impedance spectra for steel in B3 concrete were characterized by lower values of impedance and phase shift angle. It can therefore be concluded that the steel in B3 concrete was not as highly corrosion resistant as the steel in B1 and B2 concrete. This may indicate an adverse effect of the use of the admixture on the corrosion of the reinforcement. After two chloride injection cycles, a significant acceleration of corrosion processes is evident.

The determined polarization resistance R_p (R_p = Rpit + Rct) decreased with the duration of the tests (Figure 16). The value of Rp was inversely proportional to the corrosion rate, so the observed changes in this parameter indicated an increase in the corrosion rate following the introduction of chlorides into the concrete. The rate of change in Rp depended on the type of concrete. The earliest large decrease in Rp, indicating a significant acceleration of steel corrosion, was recorded for B1 concrete. In this concrete, the steel started to corrode intensively after two cycles of chloride migration (2 × 7 days). The Rp value for steel in B3 concrete also decreased significantly after two cycles of chloride migration, but was on average four times higher than the value for B1 concrete, indicating slower corrosion of steel in B3 concrete. The slowest decrease in Rp value was observed for steel in B2 concrete. After two cycles of chloride migration, the values of this parameter were still high, and indicated passive state of the steel. The Rp value for steel in B3 concrete also decreased significantly after two cycles of chloride migration, but was on average four times higher than the value for B1 concrete, indicating slower corrosion of steel in B3 concrete. The slowest decrease in R_p value was observed for steel in B2 concrete; after two cycles of chloride migration, the values of this parameter were still high and indicated passivation of the steel. However, after three cycles of chloride migration, the R_p values for the steel in B3 concrete decreased significantly, indicating a significant corrosion rate of the steel. The results obtained were consistent with the determined diffusion coefficients of the concrete. The steel in B1 concrete (i.e., in the concrete with the highest diffusion coefficient) started to corrode after the shortest time following chloride introduction into the concrete. The R_p values determined from the analysis of the impedance spectra of steel in concrete confirmed the results obtained from the analysis of the steel polarization curves.

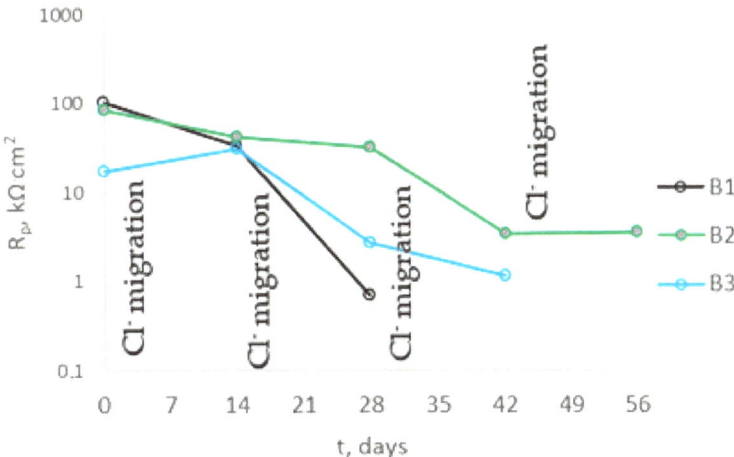

Figure 16. Dependence of mean polarization resistance values for steel in B1, B2, and B3 concrete on chloride migration time.

With increasing chloride migration time, changes in the parameters describing the constant phase element (CPE_{dl}) were also observed (Figure 17). The value of the Y_{03} parameter increased with the increase in chloride migration time. This indicated an increase in the capacity of the double layer, associated with an increase in thickness. An increase in parameter Y_{03} was accompanied by a decrease in parameter n. This indicated a deterioration in the homogeneity of the double layer on the steel. Such changes are observed due to a deterioration of the protective properties of the passive layer on the steel and the development of pits on the steel surface.

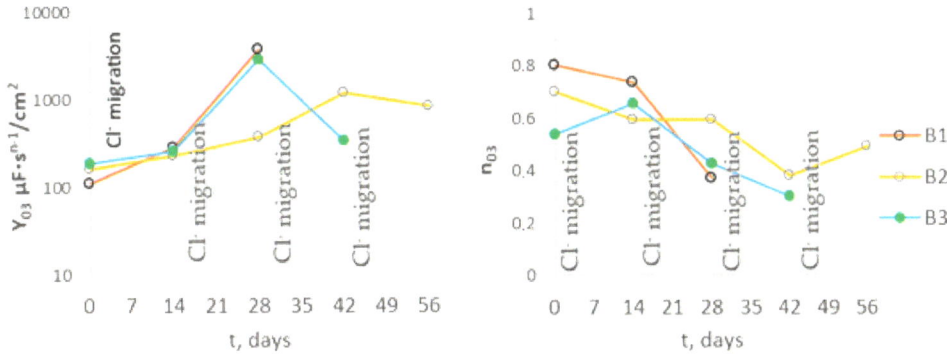

Figure 17. Dependence of the mean values of the parameters of constant phase element describing the double layer on steel in B1, B2, and B3 concrete on the migration time of chlorides.

4.8. Results of Phase Identification and Quantitative Analysis of Corrosion Products on Reinforcing Steel, Using the Rietveld Method

Scanning electron microscopy (SEM), equipped with microarea chemical analysis capabilities and X-ray microdiffraction (μXRD), were used to study microstructural changes, in addition to changes in the phase composition of corrosion products occurring in the steel–concrete contact zone. These changes were observed during the development of corrosion processes caused by an increase in the concentration of chloride ions in the concrete. Examples of tests performed using X-ray microdiffraction (XRD) and surface micrography are shown in Figure 18a–c.

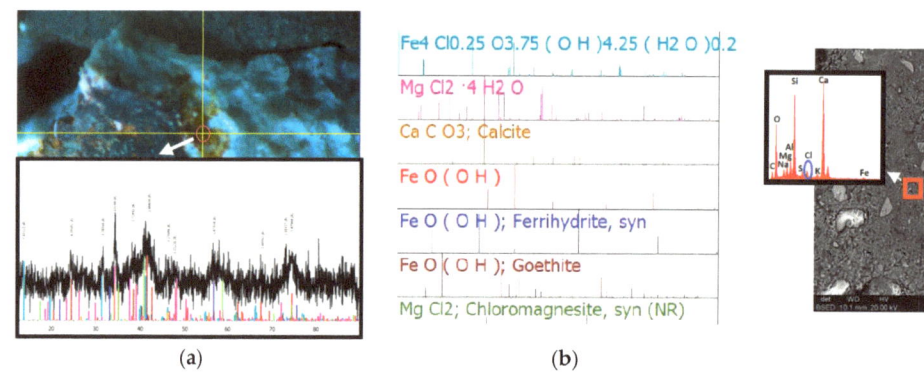
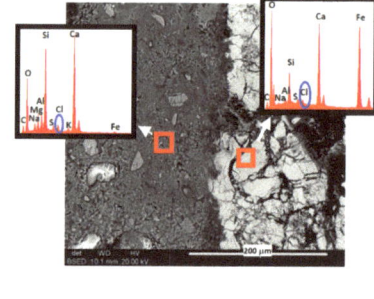

(a)   (b)   (c)

Figure 18. Surface of concrete adjacent to the surface of the bar: (**a**) image from the camera in the X-ray diffractometer showing the area examined using X-ray diffraction (XRD, highlighted in red, with a diameter of 400 μm) and the measured diffractogram at bottom left; (**b**) phase identification made on the basis of the measured diffractogram, which is also included; (**c**) SEM micrographs under backscattered electron (BSE) light and the results of chemical composition analysis from the areas marked on the micrographs (X-ray emission spectra (EDS)).

The use of phase identification showed that there are differences in the iron compounds formed depending on the type of sample. Lepidocrocite and akaganeite are present in all areas tested, while the presence of δ-type iron oxide was only confirmed in the sample from B2 concrete; likewise, goethite was identified only in the sample from B3 concrete. The only iron compound containing chlorine is akaganeite, though chlorine–calcium compounds also appeared in the sample from concrete B3. In addition, the presence of hydrated iron–potassium oxide was possible. No reflections of chlorine, potassium, or sodium compounds were registered in any of the samples.

Table 5 shows the results of quantitative phase analysis of corrosion products.

Table 5. Results of phase analysis of corrosion products.

Identified Phase Components Share [mass %]	B1	B2	B3
Magnetite Fe_3O_4	34.7 ± 2.1	-	-
Lepidocrocite γ-FeO(OH)	36.2 ± 1.8	10.0 ± 1.1	35.9 ± 1.7
Akaganeite $Fe_4Cl_{0.44}O_{3.55}(OH)_{4.44}(H_2O)_{0.2}$	28.7 ± 2.0	24.0 ± 1.7	10.9 ± 0.7
Bassanite $CaSO_4(H_2O)_{0.5}$	<1.0	-	-
Feroxyhyte δ-FeOOH	-	59.0 ± 2.1	-
Goethite α-FeO(OH)	-	-	48.7 ± 2.0
$CaCl_2$	-	-	3.3 ± 0.8

Only a small amount of bassanite was identified in the B1 concrete sample. Aqueous conversion of bassanite (Ca-SO4 0.5H$_2$O) to gypsum (CaSO$_4$ 2H$_2$O) controls the setting process of gypsum plaster [41], hence it can be assumed that there is a danger of gypsum nucleation in this concrete.

Quartz was not included in the quantitative calculations. In the B1 concrete sample, the proportions of iron compounds were at a similar level; magnetite and lepidocrocite in the amount of about 35% of the mass and akaganeite, i.e., about 29% of the mass. In the sample from concrete B2, the dominant component was iron oxyhydroxide of the type δ, comprising 59% of the mass; the share of lepidocrocite and magnetite was again in similar amounts (10% of the mass and 7% of the mass, respectively), while akaganeite was 24% of the mass. In the B3 concrete sample, goethite was the most abundant at 48.7% of the mass; lepidocrocite was at the same level as it was in the B1 concrete sample, i.e., 35.9% of the mass. In this sample, the share of akaganeite was the lowest, at about 10% of the mass. This was also the lowest share of this mineral in all the samples tested.

It can be concluded that a common feature of the tested samples was the presence of lepidocrocite and akaganeite, although their proportions varied from sample to sample. In each of the samples, a different phase was dominant; in the B1 concrete sample, all three iron compounds were found in similar amounts, while in the B2 concrete sample, half of the tested material was formed of iron oxide type δ, and in sample No. 3, the main phase was goethite. Since lepidocrocite was the result of hydration of goethite (and there was the least amount of it in the B2 concrete sample), it can be assumed that corrosion processes in the B2 concrete sample occurred more slowly than in the other samples.

5. Conclusions

The type of cement influences the rate of chloride diffusion in the concrete, and thus the duration of corrosion initiation of the reinforcement by chlorides penetrating into the concrete from the environment.

Of the concretes tested, concrete B2 with CEM III cement was the most resistant to chloride ingress, and therefore provided the longest period of protection for the reinforcement against chloride-initiated corrosion. The fastest, occurring after only two 7-day cycles of chloride migration in the electric field, was the steel corrosion that started in concrete with CEM I cement (concrete B1).

The results of the concrete diffusion and porosity tests correlated with the time of the onset of steel corrosion in the tested concretes due to chloride penetration into the concrete.

Similarly, studies of both the microstructure and changes in the phase composition of corrosion products occurring in the steel –concrete contact zone confirmed these observations as having the smallest amount of lepidocrocite, which was the result of goethite hydration in the smallest amount, occurring in the sample made of B2 concrete.

The additional use of a sealing admixture did not seem to be the most effective strategy, as it can cause a local increase in the volume of pores in the concrete, and, at the same time, a local weakening of the concrete structure, as can be seen from the distribution of the number of pores in a given volume found using computed tomography. According to the study conducted with this method, B2 concrete shows the most uniform distribution of pores of different volumes, with the lowest total number of pores.

The test results obtained were consistent with the results obtained in the work [42], where the use of metallurgical cement in the samples (together with the simultaneous addition of an air-entraining agent) showed better protective properties, as determined using the galvanostatic pulse method.

Author Contributions: Conceptualization, Z.S.; methodology, Z.S., J.K., L.S.-J. and D.C.; software, Z.S.; validation, Z.S., J.K., L.S.-J. and D.C.; formal analysis, Z.S., J.K. and L.S.-J.; investigation, Z.S., J.K. and L.S.-J.; resources, Z.S.; data curation, Z.S.; writing—original draft preparation, Z.S.; writing—review and editing, Z.S.; visualization, Z.S.; supervision, Z.S.; project administration, Z.S.; funding acquisition, Z.S. All authors have read and agreed to the published version of the manuscript.

Funding: The research was financed by Silesian University of Technology (Poland) 03/020/RGH_20/0096 and in part within the project BK-208/RB-2/2023 (03/020/BK_23/0147). This research was supported in part by the Faculty of Civil Engineering, Warsaw University of Technology.

Institutional Review Board Statement: Not applicable.

Informed Consent Statement: Not applicable.

Data Availability Statement: The data presented in this study are available on request from the corresponding author.

Acknowledgments: The tests were carried out in the Laboratory of the Faculty of Civil Engineering, and in the Institute for Ferrous Metallurgy, contributing to the Łukasiewicz Research Network, and in Smart Solutions, Measurement Center 3D, with the use of funds obtained thanks to the Statutory work of the Department of Building Structures of the Silesian University of Technology in Poland.

Conflicts of Interest: The authors declare no conflict of interest.

Appendix A

Table A1. Compressive strength results of B1 concrete after 1, 2, 7 and 28 days of curing.

Test Date [Day]	Compressive Force [kN]	Compressive Strength [MPa]	Average Compressive Strength [Mpa]
1	456.3	20.3	20.3
	441.7	19.6	
	470.2	20.9	
2	834.7	37.1	38.6
	896.54	39.8	
	875.23	38.9	
7	1303.7	57,9	59.1
	1328.2	59,0	
	1356.4	60,3	
28	1452.2	64.5	62.4
	1422.7	63.2	
	1339.5	59.5	

Table A2. Compressive strength results of B2 concrete after 1, 2, 7 and 28 days of curing.

Test Date [Day]	Compressive Force [kN]	Compressive Strength [Mpa]	Average Compressive Strength [Mpa]
1	384.0	17.1	17.8
	397.4	17.7	
	418.2	18.6	
2	713.0	31.7	31.4
	719.4	32.0	
	683.9	30.4	
7	1008.3	44.8	46.0
	1027.5	45.7	
	1069.0	47.5	
28	1413.7	62.8	63.9
	1454.0	64.6	
	1445.3	64.2	

Table A3. Compressive strength results of B3 concrete after 1, 2, 7 and 28 days of curing.

Test Date [Day]	Compressive Force [kN]	Compressive Strength [Mpa]	Average Compressive Strength [Mpa]
1	335.8 327.1 330.7	14.9 14.5 14.7	14.7
2	523.7 514.6 522.0	23.3 22.9 23.2	38.6
7	861.7 983.2 861.3	38.3 43.7 38.3	59.1
28	1228.5 1304.7 1309.4	54.6 58.0 58.2	62.4

Appendix B

Table A4. Comparison of results from analyzing polarization curves obtained for two selected specimens (B1.1 and B1.2), measuring before and after 14, 28, 42 and 56 days of chloride migration.

Measure No.	Time Days	E_{corr} mV	b_a mV	b_c mV	R_p kΩ	$R_p A$ kΩcm²	I_{corr} µA/cm²	V_r mm/year
B1.1	0	−95	399	25	3.01	68.33	0.15	0.002
B1.2	0	−2	345	12	4.79	108.73	0.05	0.001
B1.1	14	−333	394	72	0.82	18.70	1.41	0.016
B1.2	14	−92	622	30	2.93	66.51	0.19	0.002
B1.1	28	−694	61	710	0,03	0.75	32.56	0.378
B1.2	28	−781	447	49	0.03	0.57	33,79	0.392

Table A5. Comparison of results from analyzing polarization curves obtained for two selected specimens (B2.1 and B2.2), measuring before and after 14, 28, 42, and 56 days of chloride migration.

Measure No.	Time Days	E_{corr} mV	b_a mV	b_c mV	R_p kΩ	$R_p A$ kΩcm²	I_{corr} µA/cm²	V_r mm/year
B2.1	0	−10	701	54	4.57	103.74	0.21	0.002
B2.2	0	−170	239	52	5.71	129.59	0.14	0.002
B2.1	14	−161	500	59	1.52	34.50	0.66	0.008
B2.2	14	−155	265	109	2.52	57.11	0.59	0.007
B2.1	28	−237	480	23	1.01	22.93	0.42	0.005
B2.2	28	−214	440	82	1.33	30.21	0.99	0.012
B2.1	42	−357	78	92	0.51	11.49	1.60	0.019
B2.2	42	−619	85	70	0.11	2.54	6.56	0.076
B2.1	56	−328	96	84	0.61	13.87	1.40	0.016
B2.2	56	−429	368	69	0.16	3.54	7.12	0.083

Table A6. Comparison of results from analyzing polarization curves obtained for two selected specimens (B3.1 and B3.2), measuring before and after 14, 28, 42, and 56 days of chloride migration.

Measure No.	Time Days	E_{corr} mV	b_a mV	b_c mV	R_p kΩ	$R_p A$ kΩcm²	I_{corr} µA/cm²	V_r mm/Year
B3.1	0	−242	172	94	0.16	3.52	7.50	0.087
B3.2	0	−23	200	119	0.93	21.11	1.53	0.018
B3.1	14	−320	210	127	0.06	1.34	25.66	0.298
B3.2	14	−49	566	90	1.00	22.70	1.49	0.017
B3.1	28	−534	579	80	0.08	1.75	17.46	0.203
B3.2	28	−268	315	107	0.70	15.89	2.18	0.025
B3.1	42	−618	476	78	0.06	1.34	21.73	0.252
B3.2	42	−568	590	34	0.11	2.47	5.64	0.065

Table A7. Parameters of the electrical equivalent circuit elements describing the EIS spectra of steel in B1 concrete.

Sample	t Days	$R_e + R_c$ $k\Omega/cm^2$	CPE$_c$ Y_{01} $nF \cdot s^{n-1}/cm^2$	n_1	CPE$_{pas}$ Y_{02} $\mu F \cdot s^{n-1}/cm^2$	n_2	R_{pit} $k\Omega/cm^2$	CPE$_{ct}$ Y_{03} $\mu F \cdot s^{n-1}/cm^2$	n_3	R_{ct} $k\Omega/cm^2$	W $\mu F \cdot s^{n-1}/cm^2$
B1.1	0	5.2	0.019	1.00	54	0.79	0.71	131	0.77	60.5	0.000
B1.2	0	6.2	0.383	0.82	56	0.80	1.97	91	0.83	146	0.002
B1.1	14	3.7	0.014	1.00	114	0.32	0.31	262	0.76	17.1	0.000
B1.2	14	3.5	0.393	0.80	91	0.53	0.20	312	0.72	49.1	0.221
B1.1	28	3.6	0.120	0.96	80	0.30	0.42	3216	0.43	0.48	0.099
B1.2	28	3.4	0.048	1.00	2	0.73	0.16	1164	0.31	0.33	7040

Table A8. Parameters of the electrical equivalent circuit elements describing the EIS spectra of steel in B2 concrete.

Sample	t Days	$R_e + R_c$ $k\Omega/cm^2$	CPE$_c$ Y_{01} $nF \cdot s^{n-1}/cm^2$	n_1	CPE$_{pas}$ Y_{02} $\mu F \cdot s^{n-1}/cm^2$	n_2	R_{pit} $k\Omega/cm^2$	CPE$_{ct}$ Y_{03} $\mu F \cdot s^{n-1}/cm^2$	n_3	R_{ct} $k\Omega/cm^2$	W $\mu F \cdot s^{n-1}/cm^2$
B2.1	0	11.1	0.023	1.00	44	0.49	0.5	198	0.69	84.1	0.001
B2.2	0	10.8	0.082	0.90	25	0.64	0.6	125	0.72	84.2	0.006
B2.1	14	12.2	0.073	0.91	31	0.30	1.4	136	0.63	48.8	1.759
B2.1	28	11.0	0.098	0.88	12	0.40	1.2	335	0.56	34.5	0.060
B2.2	28	10.3	0.087	0.90	34	0.30	1.3	216	0.60	30.6	0.000
B2.2	42	10.2	0.186	0.84	28	0.32	1.4	1189	0.38	2.04	647
B2.2	56	9.8	0.205	0.83	31	0.32	1.2	846	0.50	2.27	0.000

Table A9. Parameters of the electrical equivalent circuit elements describing the EIS spectra of steel in B3 concrete.

Sample	t Days	$R_e + R_c$ $k\Omega/cm^2$	CPE$_c$ Y_{01} $nF \cdot s^{n-1}/cm^2$	n_1	CPE$_{pas}$ Y_{02} $\mu F \cdot s^{n-1}/cm^2$	n_2	R_{pit} $k\Omega/cm^2$	CPE$_{ct}$ Y_{03} $\mu F \cdot s^{n-1}/cm^2$	n_3	R_{ct} $k\Omega/cm^2$	W $\mu F \cdot s^{n-1}/cm^2$
B3.1	0	13.3	0.017	1.00	215	0.74	2.6	70	0.55	12.18	56.00
B3.2	0	13.2	0.094	0.86	43	0.70	0.6	313	0.63	18.7	0.000
B3.1	14	11.6	0.049	0.94	17	0.46	0.7	264	0.59	42.15	7.181
B3.2	14	11.5	0.301	0.80	81	0.61	0.9	250	0.73	17.3	0.000
B3.1	28	7.7	0.069	0.92	9	0.42	0.9	2194	0.33	1.15	0.000
B3.2	28	7.6	0.323	0.82	193	0.32	1.1	2043	0.53	2.23	0.097
B3.1	42	6.9	0.243	0.85	0.4	0.81	0.3	346	0.30	0.88	7546

References

1. Fuhaid, A.F.A.; Niaz, A. Carbonation and Corrosion Problems in Reinforced Concrete Structures. *Buildings* **2022**, *12*, 586. [CrossRef]
2. Cheng, Y.; Zhang, Y.; Wu, C.; Jiao, Y. Experimental and Simulation Study on Diffusion Behavior of Chloride Ion in Cracking Concrete and Reinforcement Corrosion. *Adv. Mater. Sci. Eng.* **2018**, *2018*, 8475384. [CrossRef]
3. Rqlü, R.; Dylgrylü, G.; Savi, J. Damage of Concrete and Reinforcement of Reinforced-Concrete Foundations Caused by Environmental Effects. *Procedia Eng.* **2015**, *117*, 411–418. [CrossRef]
4. Millán Ramírez, G.P.; Byliński, H.; Niedostatkiewicz, M. Deterioration and Protection of Concrete Elements Embedded in Contaminated Soil: A Review. *Materials* **2021**, *14*, 3253. [CrossRef] [PubMed]
5. Szyszkiewicz, K.; Jasielec, J.J.; Królikowska, A.; Filipek, R. Determination of Chloride Diffusion Coefficient in Cement-Based Materials—A Review of Experimental and Modeling Methods: Part I—Diffusion Methods. *Cem. Wapno Beton* **2017**, *2017*, 52–66.
6. Jasielec, J.; Szyszkiewicz, K.; Filipek, R. Determination of Chloride Diffusion Coefficient in Cement-Based Materials—A Review of Experimental and Modeling Methods: Part II—Migration Methods. *Cem. Wapno Beton* **2017**, *2017*, 154–167.
7. Roberto, J.; Junior, H.; Balestra, C.E.T.; Medeiros-junior, R.A. Comparison of test methods to determine resistance to chloride penetration in concrete: Sensitivity to the effect of fly ash. *Constr. Build. Mater.* **2021**, *277*, 122265. [CrossRef]
8. Szweda, Z.; Gołaszewski, J.; Ghosh, P.; Lehner, P.; Konečný, P. Comparison of Standardized Methods for Determining the Diffusion Coefficient of Chloride in Concrete with Thermodynamic Model of Migration. *Materials* **2023**, *16*, 637. [CrossRef]
9. Szweda, Z.; Zybura, A. Theoretical model and experimental tests on chloride diffusion and migration processes in concrete. *Procedia Eng.* **2013**, *57*, 1121–1130. [CrossRef]
10. Moon, H.Y.; Kim, H.S.; Choi, D.S. Relationship between average pore diameter and chloride diffusivity in various concretes. *Constr. Build. Mater.* **2006**, *20*, 725–732. [CrossRef]
11. De Schutter, G.; Audenaert, K. Evaluation of water absorption of concrete as a measure for resistance against carbonation and chloride migration. *Mat. Struct.* **2004**, *37*, 591–596. [CrossRef]

12. Sherman, M.R.; Mcdonald, D.B.; Pfeifer, D.W. Durability aspects of precast prestressed concrete. Part 2: Chloride permeability study. *PCI J.* **1996**, *41*, 76–95. [CrossRef]
13. Ruixing, C.; Song, M.; Jiaping, L. Relationship between chloride migration coefficient and pore structures of long-term water curing concrete. *Constr. Build. Mater.* **2022**, *341*, 127741. [CrossRef]
14. Kaufmann, J.; Loser, R.; Leemann, A. Analysis of cement-bonded materials by multi-cycle mercury intrusion and nitrogen sorption. *J. Colloid Interface Sci.* **2009**, *336*, 730–737. [CrossRef]
15. Zhang, Y.; Yang, B.; Yang, Z.; Ye, G. Ink-bottle Effect and Pore Size Distribution of Cementitious Materials Identified by Pressurization–Depressurization Cycling Mercury Intrusion Porosimetry. *Materials* **2019**, *12*, 1454. [CrossRef]
16. Ribeiro, D.V.; Pinto, S.A.; Amorim, N.S.; Júnior, A.J.S.; Neto, S.I.H.L.; Marques, S.L.; França, M.J.S. Effects of binders characteristics and concrete dosing parameters on the chloride diffusion coefficient. *Cem. Concr. Compos.* **2021**, *122*, 104114. [CrossRef]
17. Batista, Í. X-ray Computed Microtomography technique applied for cementitious materials: A review. *Micron* **2018**, *107*, 1–8. [CrossRef]
18. Estadual, U.; Cruz, D.S.; Sanchez, J.; De Assis, J.T. Obtaining Porosity of Concrete Using X-ray Microtomography or Digital Scanner. *J. Chem. Chem. Eng.* **2014**, *8*, 371–377. [CrossRef]
19. Dong, B.; Shi, G.; Dong, P.; Ding, W.; Teng, X.; Qin, S.; Liu, Y.; Xing, F.; Hong, S. Visualized tracing of rebar corrosion evolution in concrete with x-ray micro- computed tomography method. *Cem. Concr. Compos.* **2018**, *92*, 102–109. [CrossRef]
20. Dong, B.; Fang, G.; Liu, Y.; Dong, P.; Zhang, J.; Xing, F.; Hong, S. Monitoring reinforcement corrosion and corrosion-induced cracking by X- ray microcomputed tomography method. *Cem. Concr. Compos.* **2017**, *100*, 311–321. [CrossRef]
21. Itty, P.; Serdar, M.; Meral, C.; Parkinson, D.; Macdowell, A.A.; Monteiro, P.J.M. In situ 3D monitoring of corrosion on carbon steel and ferritic stainless steel embedded in cement paste. *Corros. Sci.* **2014**, *83*, 409–418. [CrossRef]
22. Güneyisi, E.; Gesoğlu, M.; Karaboğa, F.; Mermerdaş, K. Corrosion behavior of reinforcing steel embedded in chloride contaminated concretes with and without metakaolin. *Compos. Part B Eng.* **2013**, *45*, 1288–1295. [CrossRef]
23. Ribeiro, D.V.; Abrantes, J.C.C. Application of electrochemical impedance spectroscopy (EIS) to monitor the corrosion of reinforced concrete: A new approach. *Constr. Build Mater.* **2016**, *111*, 98–104. [CrossRef]
24. Szweda, Z.; Jaśniok, T.; Jaśniok, M. Evaluation of the effectiveness of electrochemical chloride extraction from concrete on the basis of testing reinforcement polarization and chloride concentration. *Ochr. Przed. Korozją* **2018**, *61*, 3–9. [CrossRef]
25. Szweda, Z. Evaluating the Impact of Concrete Design on the Effectiveness of the Electrochemical Chloride Extraction Process. *Materials* **2023**, *16*, 666. [CrossRef]
26. Antunes, R.A.; Ichikawa, R.U.; Martinez, L.G.; Costa, I. Characterization of corrosion products on carbon steel exposed to natural weathering and to accelerated corrosion tests. *Int. J. Corros.* **2014**, *2014*, 419570. [CrossRef]
27. Vera, R.; Villarroel, M.; Carvajal, A.M.; Vera, E.; Ortiz, C. Corrosion products of reinforcement in concrete in marine and industrial environments. *Mater. Chem. Phys.* **2009**, *114*, 467–474. [CrossRef]
28. Pan, T.; Wang, L. Finite-Element Analysis of Chemical Transport and Reinforcement Corrosion-Induced Cracking in Variably Saturated Heterogeneous Concrete. *J. Eng. Mech.* **2011**, *137*, 334–345. [CrossRef]
29. PN EN 12620:2013-08; Aggregates for Concrete. Polish Committee for Standardization: Warszawa, Poland, 2013.
30. PN EN 12390-3:2019-07; Concrete Research—Part 3: Compressive Strength of Test Specimens. Polish Committee for Standardization: Warszawa, Poland, 2019.
31. PN-EN 12390-8; Concrete Research—Part 8: Depth of Water Penetration under Pressure. Polish Committee for Standardization: Warszawa, Poland, 2019.
32. NT BUILD-443; Concrete Hardened: Accelerated Chloride Penetration. Nordtest: Espoo, Finland, 1995; pp. 1–5.
33. ASTM C1556; Standard Test Method for Determining the Apparent Chloride Diffusion Coefficient of Cementitious Mixtures by Bulk Diffusion. ASTM International: West Conshohocken, PA, USA, 2011.
34. ASTM C1202; Electrical Indication of Concrete's Ability to Resist Chloride Ion Penetration. ASTM International: West Conshohocken, PA, USA, 2000.
35. ASTM C1202; ASTM I American Society for Testing and Materials. ASTM International: West Conshohocken, PA, USA, 1997; pp. 1–6.
36. NT Build 492; Concrete, Mortar and Cement-Based Repair Materials: Chloride Migration Coefficient from Non-Steady-State Migration Experi-Ments. NORDTEST: Espoo, Finland, 1999; pp. 1–8.
37. Szweda, Z.; Ponikiewski, T.; Katzer, J. A study on replacement of sand by granulated ISP slag in SCC as a factor formatting its durability against chloride ions. *J. Clean. Prod.* **2017**, *156*, 569–576. [CrossRef]
38. PN-EN 206-1; Beton Część 1: Wymagania, Właściwości i Zgodność. Polish Committee for Standardization: Warszawa, Poland, 2004; pp. 1–70. (In Polish)
39. ASTM-C 867–91; Standard Test Method for Half-Cell Potentials of Uncoated Reinforcing Steel in Concrete C 876–91 (Reapproved 1999). ASTM Int: West Conshohocken, PA, USA, 1999.
40. Raczkiewicz, W. Use of polypropylene fi bres to increase the resistance of reinforcement to chloride corrosion in concretes. *Sci. Eng. Compos. Mater.* **2021**, *22*, 555–567. [CrossRef]

41. Brandt, F.; Bosbach, D. Bassanite ($CaSO_4 \cdot 0.5H_2O$) dissolution and gypsum ($CaSO_4 \cdot 2H_2O$) precipitation in the presence of cellulose ethers. *J. Cryst. Growth* **2001**, *233*, 837–845. [CrossRef]
42. Raczkiewicz, W.; Koteš, P.; Konečný, P. Influence of the Type of Cement and the Addition of an Air-Entraining Agent on the Effectiveness of Concrete Cover in the Protection of Reinforcement against Corrosion. *Materials* **2021**, *14*, 4657. [CrossRef]

Disclaimer/Publisher's Note: The statements, opinions and data contained in all publications are solely those of the individual author(s) and contributor(s) and not of MDPI and/or the editor(s). MDPI and/or the editor(s) disclaim responsibility for any injury to people or property resulting from any ideas, methods, instructions or products referred to in the content.

Article

The Constituent Phases and Micromechanical Properties of Steel Corrosion Layers Generated by Hyperbaric-Oxygen Accelerated Corrosion Test

Baozhen Jiang [1,*], Kotaro Doi [2] and Koichi Tsuchiya [2,*]

1 School of Materials, Sun Yat-sen University, Shenzhen 518107, China
2 Research Center for Structural Materials, National Institute for Materials Science, Tsukuba 305-0047, Japan; doi.kotaro@nims.go.jp
* Correspondence: jiangbzh@mail.sysu.edu.cn (B.J.); tsuchiya.koichi@nims.go.jp (K.T.)

Abstract: Hyperbaric oxygen-accelerated corrosion testing (HOACT) is a newly developed method to study in the labor the corrosion behavior of steel bars in concrete. This work aimed to intensively investigate the mechanical properties and microstructures of HOACT-generated corrosion products by means of nano-indentation tests, Raman micro-spectrometry, and scanning electron microscopy. The local elastic modulus and nanohardness varied over wide ranges of 6.8–75.2 GPa and 0.38–4.44 GPa, respectively. Goethite, lepidocrocite, maghemite, magnetite, and akageneite phases were identified in the corrosion products. Most regions of the rust layer were composed of a complex and heterogeneous mix of different phases, while some regions were composed of maghemite or akageneite only. The relationship between the micromechanical properties and typical microstructural features is finally discussed at the micro-scale level. It was found that the porosity of corrosion products can significantly influence their micromechanical properties.

Keywords: accelerated corrosion; steel reinforced concrete; corrosion products; microstructure; mechanical properties; nano-indentation

1. Introduction

Reinforced concrete structures have been widely used all over the world. Outstanding properties and fair cost can be achieved by the coupling of steel with concrete. During the service life of reinforced concrete structures, the corrosion of steel bars in concrete is unavoidable. Moreover, it has been thought to be the most critical factor leading to the deterioration of reinforced concrete structures [1–3]. A very complex composition, consisting of a mixture of ferric oxyhydroxides and iron oxides, can be found in the resulting corrosion products, which form at the steel/concrete interface. Compared with the steel lost, the specific volumes of different types of corrosion products are 2–7 times larger [4,5]. This type of expansional stress caused by corrosion products can result in crack initiation, propagation, and even spalling of concrete cover.

Corrosion products play roles in the mechanical interactions between rebar and concrete [6]. Therefore, the mechanical properties of corrosion products, especially the elastic modulus, are essential parameters for modeling the cracking process. According to previous studies, which are summarized in Figure 1, the elastic moduli reported cover several orders of magnitude for corrosion products [7–20]. The first reason for such varying values consists of the different methods with which the elastic modulus was investigated. An elastic modulus obtained through an inverse analysis using finite element simulations is of the order of 0.1 GPa [7,8]. Such low input values are required in simulations for the purpose of obtaining realistic predictions of the cracking process of concrete cover. Zhao et al. [11] performed cyclic low-compression testing and oedometer testing of corrosion products, reporting that the elastic modulus was also 0.1 GPa. However, the corrosion products had been ground into small particles before testing, the measured elastic modulus of 0.1 GPa might be not real for bulk corrosion products. In contrast, as indicated by most of the solid circles in Figure 1, the elastic modulus obtained through direct experimental measurements is much higher. Since the corrosion product layer is usually very thin, depth sensing indentation is an effective experimental method for characterizing the elastic modulus of corrosion products [13–20]. The corrosion products investigated by indentation methods in previous studies can be divided into two types: those generated by anodically accelerated corrosion and those generated in nature. The former corrosion products exhibited an elastic modulus of 47 GPa on average in [14], 56 GPa on average in [19], 49.4–67.9 GPa in [13], and 54–110 GPa in [16]. For the latter corrosion products, the elastic modulus was 61–86 GPa in [14], 64–159 GPa in [19], 50–125 GPa in [17], and 67–158 GPa in [20]. The elastic moduli obtained by indentation tests are still quite different. It is believed that the corrosion products generated under different exposed environments and operating conditions should have different constituent phases and morphologies.

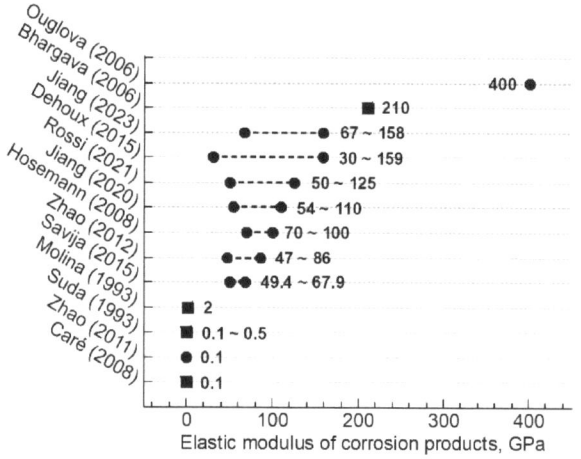

Figure 1. Elastic modulus of corrosion products reported in the literature [7–17,19,20]. Estimated and/or modeled values are indicated by squares, and experimentally measured values are indicated by circles.

To understand the complex steel bar corrosion in reinforced concrete structures, accelerated corrosion test methods are usually performed in the lab. The conventional methods include the impressed current method and the wet–dry cycle method. However, there are some issues related to these two methods. For example, the corrosion products generated by the impressed current method present a higher expansion ratio than the natural corrosion products because this type of artificial corrosion product usually contains a large amount of Cl^-, which is not found in nature [21]. The corrosion products generated by the wet–dry method are similar to those generated in nature. However, it takes a long time to achieve the corrosion of steel bar because it the inside of the concrete cover is slow to dry [22]. Recently, a new method called hyperbaric oxygen-accelerated corrosion testing (HOACT) was developed [23–25]. The oxygen reduction reaction on the steel surface can be promoted by sufficient oxygen. Meanwhile, the anodic reaction can be promoted by the Cl^- added to the concrete. This process relates to the breakdown of passive film and the dissolution of steel. The above two factors contribute to the rapid corrosion of steel in concrete. More importantly, the composition of corrosion products generated by HOACT is almost identical to that generated in the real environment. Therefore, it is believed that HOACT is a better method for studying reinforcing corrosion in concrete.

As a newly developed accelerated corrosion method, HOACT-induced steel corrosion products still lack a systematic and in-depth study. Therefore, the present study aimed to provide an intensive analysis of the micromechanical properties and microstructures of corrosion products generated by HOACT. A series of nano-indentation tests were performed to investigate the elastic modulus and nanohardness of local corrosion products. At the same time, the phase compositions and morphologies of these local corrosion products were investigated by Raman micro-spectrometry, energy dispersive X-ray spectrometry (EDS), and scanning electron microscopy (SEM). These observations help us to comprehensively understand the corrosion products at the micro-scale level and then to clarify the relationships between microstructural features and mechanical properties.

2. Materials and Methods

2.1. Specimen Preparation

A commercial steel bar of Japanese Industrial Standard (JIS) SD345, with a chemical composition of Fe-0.26C-0.19Si-0.88Mn-0.029P-0.018S (wt.%), was used in this study. This steel is equivalent to Grade 40 from American Society for Testing and Materials (ASTM) A615. The diameter of this steel bar is 19 mm. It was embedded in a cylindrical mortar block. This block was 30 mm in diameter, 30 mm in height, and 5.5 mm in cover thickness, and its upper and lower surfaces were sealed with epoxy resin. The weight ratio of 2.06 M NaCl solution cement: fine aggregates was 0.6:1:3 in the mortar. The mortar block was subjected to a curing process for 28 days at room temperature and at 95% relative humidity. Then, it was placed in a high-pressure container. The schematic illustration of the HOACT apparatus was reported by Doi et al. [25]. During HOACT, the internal oxygen pressure was kept at 0.6 MPa, the relative humidity was greater than 95%, and the dissolved oxygen concentration of the mortar block increased. This accelerated corrosion test was performed for 14 days (S1) and 56 days (S2), respectively. After HOACT, the specimen was embedded in cold mounting resin. Grinding (SiC abrasive papers of 280, 400, 600, and 1000 grit) and polishing (polishing suspensions with particle sizes of 9 µm, 6 µm, 3 µm, and 0.06 µm) were performed on an Ecomet 250 Pro Grinder-Polisher machine to produce a mirror surface. The block was cleaned with acetone for 20 min by an ultrasonic cleaner and then dried.

2.2. Nano-Indentation Tests

Nano-indentation tests were conducted to investigate the micromechanical properties of the corrosion product layer. The Hysitron Triboindenter TI950 with a Berkovich tip was used in the present study. Multiple different areas of the corrosion product layer were selected for nano-indentation tests, and the number indentation points in each area varied between 10 and 20. The indentations were separated from each other by a distance of 10 µm. The total number of indentation points made on S1 and S2 were 66 and 279, respectively. Nano-indentation tests were performed in a load control mode. The load increased at a rate of 250 µN/s to 10 mN, holding for 20 s at this force and then decreasing the load at a rate of 250 µN/s. Based on the Oliver and Pharr method [26], the elastic modulus and nanohardness were calculated from the obtained load-depth curves.

2.3. Microstructural Characterization

The phase compositions of corrosion products were identified by micro-Raman analysis (inVia Reflex, Renishaw). Raman measurements were performed at a laser power of 1% of the maximum and at an excitation wavelength of 532 nm. The spectra were recorded using a 50× objective lens with an acquisition time of 5 min. The acquisition and treatment of the spectra were performed with WiRE 5 software. These obtained Raman spectra were analyzed by comparing them with the reference spectra previously reported in the literature [27–29]. SEM observations in backscatter electron (BSE) mode and EDS analysis were conducted at 20 kV with an FEI Quanta FEG 250. The composition of each indentation was quantitatively investigated by EDS spot analysis.

3. Results and Discussion

3.1. Microstructural Characterization of Corrosion Products

Figure 2 shows the BSE micrographs of corrosion products at the steel/mortar interface. The thickness of the corrosion product layer was approximately 55 μm for S1 and 350 μm for S2. There were many cracks with different sizes in the corrosion product layer. Some cracks were parallel to the corrosion product layer, while some cracks were perpendicular to the corrosion product layer. A bright layer with a thickness of around 30 μm could be found inside the corrosion products, as indicated by the black arrows. It was the original mill scale that was not generated during HOACT process but that formed on the steel surface during the manufacturing process, such as hot rolling or forging [20,25]. The mill scale contained some initial defects, including cracks, exfoliation, and so on. It has been reported that crevice corrosion between the steel substrate and the mill scale is induced by Cl^-, which can intrude through these initial defects in the mill scale [25,30]. With the gradual generation of corrosion products, the resulting volume expansion can break down the mill scale. The mill scale is visible near to or far from the steel surface, depending on how much the corrosion has propagated. In addition, as shown in Figure 2c, the volume expansion caused by corrosion products also initiated cracks in the mortar, and some corrosion products penetrated into the mortar along cracks. The corrosion of steel bar will progress more rapidly after cracking in the mortar because the cracks can provide a fast-transportation path for oxygen and Cl^-.

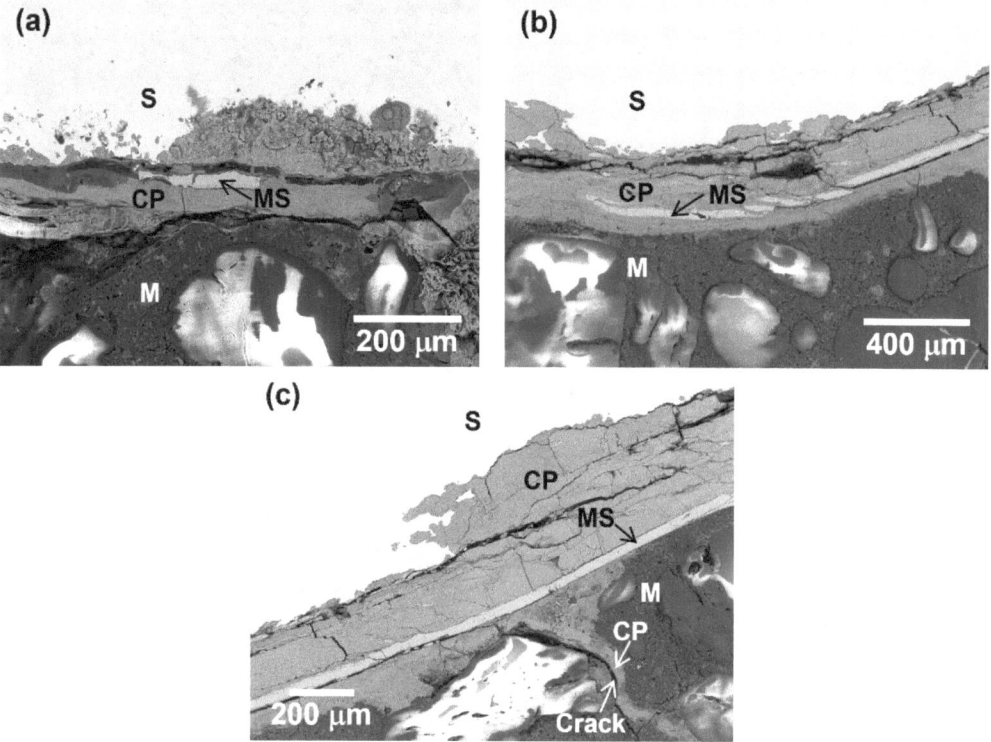

Figure 2. SEM-BSE micrograph of corrosion products generated by HOACT: (**a**) S1; and (**b**) and (**c**) S2. S: steel; M: mortar; CP: corrosion products; MS: mill scale.

Based on the micro-Raman analysis, the phases identified in S1 included lepidocrocite (γ-FeOOH), maghemite (γ-Fe$_2$O$_3$), and magnetite (Fe$_3$O$_4$). The detailed morphologies of corrosion products in S1 and corresponding Raman spectra are shown in Figure 3. These two BSE images show that the corrosion products were loose and porous. For the Raman spectra shown in Figure 3c, the presence of the sharp peak at 702 cm^{-1} can be attributed to maghemite, while the presence of the peak at 687 cm^{-1} is due to a combination of scattering intensities from magnetite and maghemite. Similar results have been previously discussed in the literature [27]. Therefore, the microstructure shown in Figure 3a consists of a mix of maghemite and magnetite. However, it is difficult to distinguish magnetite from maghemite in this BSE image. Figure 3b shows a microstructure consisting of a mix of maghemite and lepidocrocite. The needle-shaped crystals in the corrosion products are lepidocrocite phase, which usually forms in the early period of exposure and is unstable [31–34].

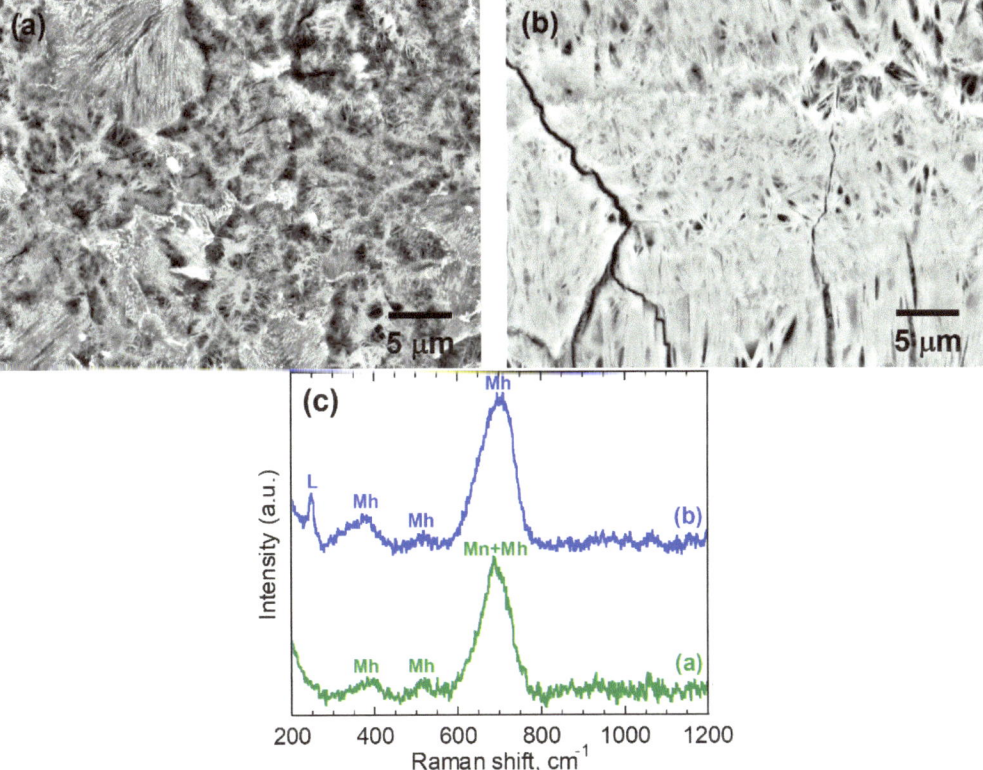

Figure 3. Microstructural characterization of the corrosion products in S1: (**a**,**b**) BSE-SEM micrograph; and (**c**) corresponding Raman spectra. L: lepidocrocite; Mh: maghemite and Mn: magnetite.

The corrosion products in S2 presented more diverse morphologies. Most regions of the corrosion products in S2 contained a mixture of several phases. In addition to the three phases identified in S1, the stable goethite (α-FeOOH) phase was also found in S2. Figure 4 shows three representative BSE morphologies and the corresponding Raman spectra. Maghemite and goethite were identified in all these microstructures. In addition, the microstructure shown in Figure 4a,c also contains magnetite and lepidocrocite, respectively.

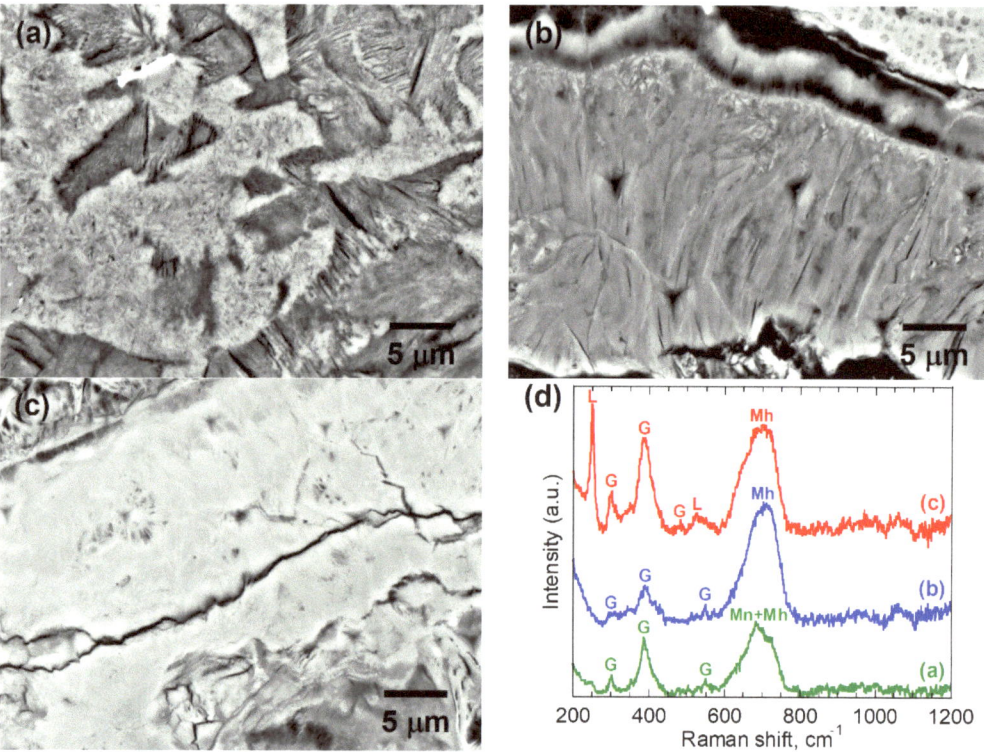

Figure 4. Microstructural characterization of the corrosion products in S2: (**a**–**c**) BSE-SEM micrograph; and (**d**) corresponding Raman spectra. L: lepidocrocite; G: goethite; Mh: maghemite and Mn: magnetite.

Moreover, some local areas of the corrosion products in S2 contained a single phase. Figure 5a exhibits the BSE morphology containing only akageneite (β-FeOOH). Akageneite is an iron oxyhydroxide mineral with a tunnel structure stabilized by inclusion of chloride. The NaCl solution mixed in the concrete contributed to the generation of the akageneite phase. Both BSE morphologies shown in Figure 5b,c contain only maghemite. However, it is clear that these two morphologies are totally different. Figure 5c presents a more compact and denser microstructure compared with Figure 5b.

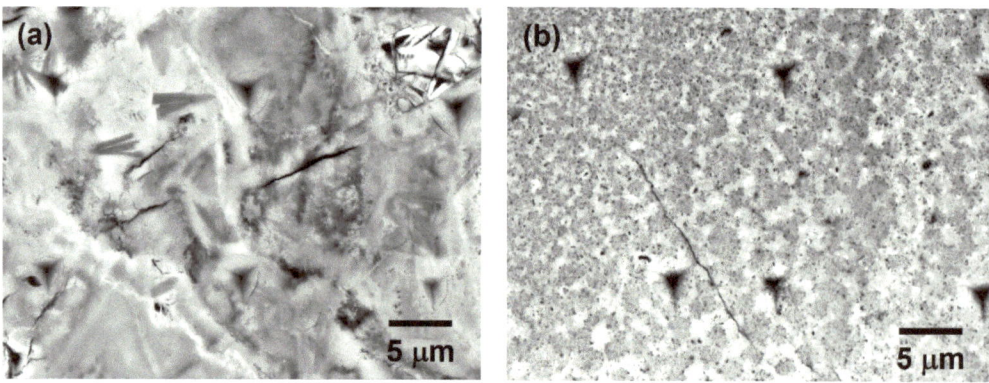

Figure 5. *Cont.*

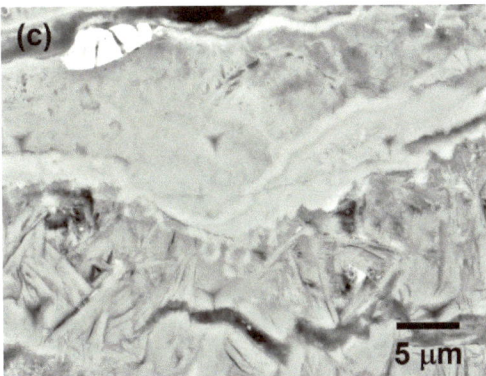
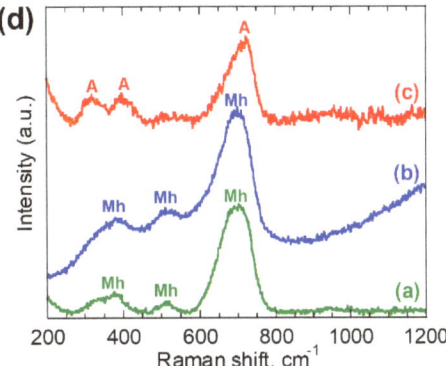

Figure 5. Microstructural characterization of the corrosion products in S2: (**a**–**c**) BSE-SEM micrograph and (**d**) corresponding Raman spectra. A: akageneite and Mh: maghemite.

3.2. Characterization of the Mechanical Properties

Figure 6a shows some typical nano-indentation load-depth curves obtained from different corrosion products. The inserted table lists the elastic modulus E and nanohardness H calculated from corresponding load-depth curve. A steeper curve in the unloading part suggests a higher elastic modulus, and a larger residual indentation depth suggests a lower hardness. The nano-indentation results for all the investigated corrosion products are plotted in Figure 6b. Clearly, a rising trend in the elastic modulus can be seen as the nanohardness increases. The number of indentation points made on S1 was 66 points. The elastic modulus was between 14.5 and 40.5 GPa (25.9 ± 6.4 GPa on average), while nanohardness was between 0.43 and 1.96 GPa (0.98 ± 0.40 GPa on average). The number of indentation points made on S2 was 279 points. Nano-indentation investigations covered more various microstructures, leading to a much wider range in the values of elastic modulus and nanohardness. The elastic modulus varied between 6.8 and 75.2 GPa (31.5 ± 14.7 GPa on average), while nanohardness varied between 0.38 and 4.44 GPa (1.55 ± 0.88 GPa on average).

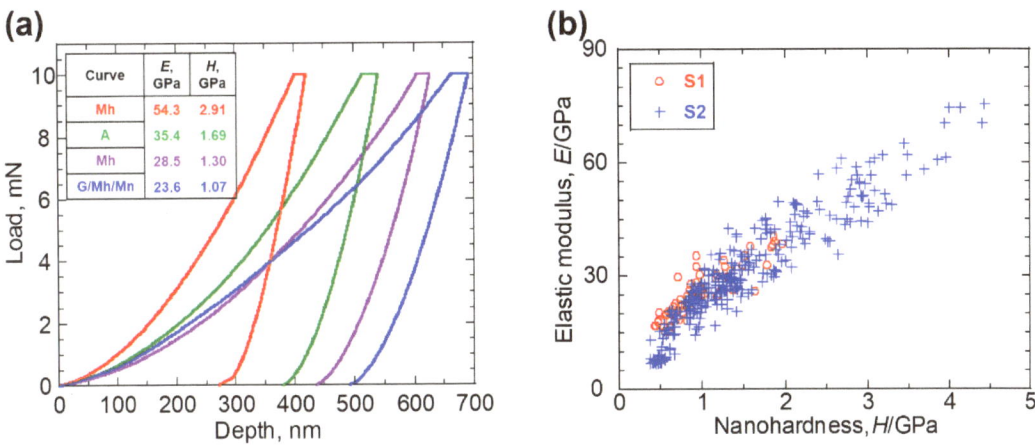

Figure 6. (**a**) Typical nano-indentation load-depth curves obtained from different corrosion products; and (**b**) elastic modulus E and nanohardness H measured on all the indentations.

3.3. Relationship between Local Microstructures and Mechanical Properties

As mentioned above, the HOACT generated corrosion products consisting of several phases, presenting a complex and heterogeneous microstructure and a wide range of mechanical properties. The mechanical properties per phase should be different. Although the related studies have been limited, the elastic modulus and nanohardness of each phase showed an ascending order of goethite < maghemite < magnetite [16,19,20]. For the microstructures mixing with various phases, there is a rough tendency that the elastic modulus and nanohardness become greater by increasing the Fe/O atomic ratio, although the values at certain Fe/O atomic ratios still show a broad range [16]. However, this study showed inconsistent results. Table 1 lists the mechanical properties corresponding to the microstructures shown in Figures 3–5. It is clear that there was no significant, direct correlation between the mechanical properties and the phase compositions (or the Fe/O atomic ratio) for the studied corrosion products.

Table 1. Composition and average values of elastic modulus E and nanohardness H, corresponding to the microstructures shown in Figures 3–5. L: lepidocrocite; G: goethite; A: akaganeite; Mh: maghemite and Mn: magnetite.

Figure		Fe/O Atomic Ratio	Phases	E (GPa)	H (GPa)
3	(a)	0.66 ± 0.06	Mh/Mn	22.6 ± 4.5	0.79 ± 0.25
	(b)	0.60 ± 0.02	L/Mh	31.5 ± 5.2	1.27 ± 0.40
4	(a)	0.65 ± 0.02	G/Mh/Mn	23.5 ± 3.4	1.02 ± 0.27
	(b)	0.62 ± 0.01	G/Mh	40.2 ± 3.0	2.14 ± 0.29
	(c)	0.57 ± 0.02	L/G/Mh	51.8 ± 5.5	3.18 ± 0.41
5	(a)	0.51 ± 0.02	A	35.4 ± 3.3	1.55 ± 0.27
	(b)	0.65 ± 0.01	Mh	28.3 ± 1.8	1.32 ± 0.07
	(c)	0.65 ± 0.02	Mh	54.6 ± 5.2	2.79 ± 0.42

By examining the morphologies of corrosion products, it is believed that the microstructural porosity greatly influences the mechanical properties of corrosion products. Based on these BSE images, the porosity of some corrosion products was measured using the ImageJ software package (v1.8.0.112). For the microstructures composed of several phases, the microstructures shown in Figure 3a,b and Figure 4a have high porosity, which is approximate 17.5%, 14.5%, and 12.7%, respectively. The corresponding elastic modulus and nanohardness listed in Table 1 are quite low. Meanwhile, a more compact and denser microstructure shown in Figure 4b,c has a higher elastic modulus and nanohardness. This kind of phenomenon can also be found in the microstructures composed of single maghemite phase. The more porous microstructure shown in Figure 5b, which has a porosity of about 7.4%, has a lower elastic modulus and nanohardness than the microstructure shown in Figure 5c.

The elastic moduli for maghemite phase reported by previous studies [16,17,19,20], as well as by the current study, are summarized in Figure 7. The same phase investigated in different studies exhibited different elastic moduli, and a much lower elastic modulus is reported in the current study. The corrosion products generated under different environments should have different porosities. Gao et al. [35] reported that a higher corrosion rate resulted in higher porosity. The natural corrosion process is slow, so the corrosion products would densely accumulate at the interface between the steel bar and the cement. The accelerated corrosion test method results in a rapid formation of corrosion products and consequent loose morphologies. This fact can explain why the elastic modulus and nanohardness reported in the literature were higher for natural corrosion products than for artificial corrosion products. Another noteworthy phenomenon is that the porosity of the corrosion product layer likely decreases over time as its thickness increases [35–38]. The much thinner corrosion product layer obtained in the present study is more porous and looser than those reported by previous researchers, exhibiting a much lower elastic modulus and nanohardness.

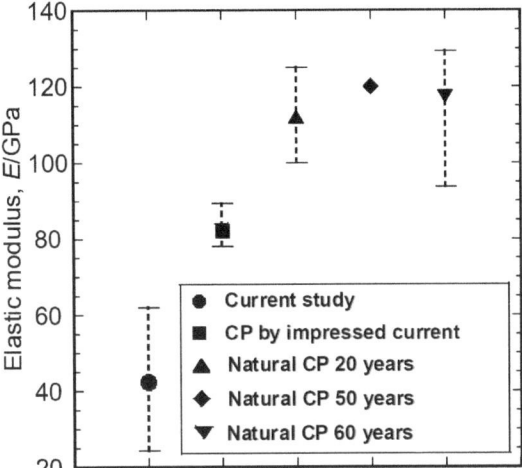

Figure 7. Elastic modulus of maghemite phase reported in previous studies and the present study [16,17,19,20]. The whiskers show the maximum and minimum values, while the solid symbols indicate the average values. CP: corrosion products.

In addition, some researchers have speculated that the mechanical properties might also be affected by some other microstructural parameters, one of which is the crystallinity of corrosion products [19,20]. Unfortunately, this issue cannot be assessed by the present investigation. Based on the previous studies, as well as the present study, the characteristic mechanical properties of corrosion products should not be regarded as constants. To predict accurate values of elastic modulus and nanohardness, many microstructural parameters intrinsic to corrosion products, such as phase composition, porosity, crystallinity, and so on, should be considered.

4. Conclusions

The corrosion of steel bar in concrete was accomplished using a newly developed accelerated corrosion method, HOACT. The nano-indentation method was applied to explore the micromechanical properties of the corrosion products. SEM and micro-Raman analysis were conducted to characterize the microstructures of the corrosion products. The following conclusions were drawn.

1. The phases identified in the corrosion products included goethite, lepidocrocite, maghemite, magnetite, and akageneite, presenting themselves as mixtures in most regions of the corrosion product layer at the microscopic level. Some local regions were composed of a single maghemite or akageneite phrase were also found.
2. The micromechanical properties of HOACT-generated corrosion products covered a wide range. The elastic modulus varied between 6.8 and 75.2 GPa, while the nanohardness varied between 0.38 and 4.44 GPa.
3. The microstructural porosity greatly influenced the micromechanical properties of the corrosion products. A more porous and looser microstructure usually corresponds to a lower elastic modulus and nanohardness.
4. According to the previous studies, as well as the current study of corrosion products, the characteristic mechanical properties per phase should not be considered a constant value. They can change with corrosion conditions.

Author Contributions: Conceptualization, B.J.; methodology, B.J. and K.D.; investigation, B.J.; resources, K.T.; writing—original draft preparation, B.J.; writing—review and editing, K.D. and K.T.; supervision, K.T.; project administration, K.T.; funding acquisition, B.J. and K.T. All authors have read and agreed to the published version of the manuscript.

Funding: This work was supported in part by the Cross-ministerial Strategic Innovation Promotion Program (SIP) "Infrastructure maintenance, renovation and management" of the Council for Science, Technology and Innovation (CSTI) (Funding agency: JST). B.Z. Jiang acknowledges funding support from Sun Yat-Sen University (Funding No. 76180-12230028). The APC was funded by 76180-12230028 from Sun Yat-Sen University.

Institutional Review Board Statement: Not applicable.

Informed Consent Statement: Not applicable.

Data Availability Statement: The raw/processed data required to reproduce these findings cannot be shared online at this time due to technical limitations, but they are available by contacting the corresponding author.

Conflicts of Interest: The authors declare no conflict of interest.

References

1. Bertolini, L.; Elsener, B.; Pedeferri, P.; Redaelli, E.; Polder, R.B. *Corrosion of Steel in Concrete: Prevention, Diagnosis, Repair*, 2nd ed.; John Wiley & Sons: Hoboken, NJ, USA, 2013.
2. Ahmad, S. Reinforcement corrosion in concrete structures, its monitoring and service life prediction—A review. *Cem. Concr. Compos.* **2003**, *25*, 459–471. [CrossRef]
3. Wong, H.S.; Zhao, Y.X.; Karimi, A.R.; Buenfeld, N.R.; Jin, W.L. On the penetration of corrosion products from reinforcing steel into concrete due to chloride-induced corrosion. *Corros. Sci.* **2010**, *52*, 2469–2480. [CrossRef]
4. Marcotte, T.D. Characterization of Chloride-Induced Corrosion Products that Form in Steel-Reinforced Cementitious Materials. Ph.D. Thesis, University of Waterloo, Waterloo, ON, Canada, 2001.
5. Jaffer, S.J.; Hansson, C.M. Chloride-induced corrosion products of steel in cracked-concrete subjected to different loading conditions. *Cem. Concr. Res.* **2009**, *74*, 116–125. [CrossRef]
6. Berrocal, C.G.; Fernandez, I.; Rempling, R. The interplay between corrosion and cracks in reinforced concrete beams with non-uniform reinforcement corrosion. *Mater. Struct.* **2022**, *55*, 120. [CrossRef]
7. Caré, S.; Nguyen, Q.T.; L'Hostis, V.; Berthaud, Y. Mechanical properties of the rust layer induced by impressed current method in reinforced mortar. *Cem. Concr. Res.* **2008**, *38*, 1079–1091. [CrossRef]
8. Suda, K.; Misra, S.; Motohashi, K. Corrosion products of reinforcing bars embedded in concrete. *Corros. Sci.* **1993**, *35*, 1543–1549. [CrossRef]
9. Molina, F.J.; Alonso, C.; Andrade, C. Cover cracking as a function of bar corrosion: Part 2—Numerial model. *Mater. Struct.* **1993**, *26*, 532–548. [CrossRef]
10. Bhargava, K.; Ghosh, A.K.; Mori, Y.; Ramanujam, S. Model for cover cracking due to rebar corrosion in RC structures. *Eng. Struct.* **2006**, *28*, 1093–1109. [CrossRef]
11. Zhao, Y.X.; Dai, H.; Ren, H.Y.; Jin, W.L. Experimental study of the modulus of steel corrosion in a concrete port. *Corros. Sci.* **2012**, *56*, 17–25. [CrossRef]
12. Ouglova, A.; Berthaud, Y.; Francois, M.; Foct, F. Mechanical properties of an iron oxide formed by corrosion in reinforced concrete structures. *Corros. Sci.* **2006**, *48*, 3988–4000. [CrossRef]
13. Savija, B.; Lukovic, M.; Hosseini, S.A.S.; Pacheco, J.; Schlangen, E. Corrosion induced cover cracking studied by X-ray computed tomography, nanoindentation, and energy dispersive X-ray spectrometry (EDS). *Mater. Struct.* **2015**, *48*, 2043–2062. [CrossRef]
14. Zhao, Y.X.; Dai, H.; Jin, W.L. A study of the elastic moduli of corrosion products using nano-indentation techniques. *Corros. Sci.* **2012**, *65*, 163–168. [CrossRef]
15. Hosemann, P.; Swadener, J.G.; Welch, J.; Li, N. Nano-indentation measurement of oxide layers formed in LBE on F/M steels. *J. Nucl. Mater.* **2008**, *377*, 201–205. [CrossRef]
16. Jiang, B.Z.; Doi, K.; Tsuchiya, K.; Kawano, Y.; Kori, A.; Ikushima, K. Micromechanical properties of steel corrosion products in concrete studied by nano-indentation technique. *Corros. Sci.* **2020**, *163*, 108304. [CrossRef]
17. Rossi, E.; Zhang, H.Z.; Garcia, S.J.; Bijleveld, J.; Nijland, T.G.; Copuroglu, O.; Polder, R.B.; Savija, B. Analysis of naturally-generated corrosion products due to chlorides in 20-year old reinforced concrete: An elastic modulus-mineralogy characterization. *Corros. Sci.* **2021**, *184*, 109356. [CrossRef]
18. Dehoux, A.; Bouchelaghem, F.; Berthaud, Y.; Neff, D.; L'Hostis, V. Micromechanical study of corrosion products layer. Part I: Experimental characterization. *Corros. Sci.* **2012**, *54*, 52–59. [CrossRef]
19. Dehoux, A.; Bouchelaghem, F.; Berthaud, Y. Micromechanical and microstructural investigation of steel corrosion layers of variable age developed under impressed current method, atmospheric or saline conditions. *Corros. Sci.* **2015**, *97*, 49–61. [CrossRef]

20. Jiang, B.Z.; Doi, K.; Tsuchiya, K. Characterization of microstructures and micromechanical properties of naturally generated rust layer in 60-year old reinforced concrete. *Constr. Build. Mater.* **2023**, *366*, 130203. [CrossRef]
21. Takaya, S.; Okuno, S.; Honda, M.; Kawakami, K.; Satoh, S.; Hamura, Y.; Yamamoto, Y.; Miyagawa, T. Formation process of steel corrosion products in alkaline environment and corrosion environment of reinforcement in concrete. *Zairyo/J. Soc. Mater. Sci. Japan* **2017**, *66*, 545–552. [CrossRef]
22. Nishikata, A.; Yamashita, Y.; Katayama, H.; Tsuru, T.; Usami, A.; Tanabe, K.; Mabuchi, H. An electrochemical impedance study on atmospheric corrosion of steels in a cyclic wet-dry condition. *Corros. Sci.* **1995**, *37*, 2059–2069. [CrossRef]
23. Doi, K.; Hiromoto, S.; Akiyama, E. Hyperbaric-oxygen accelerated corrosion test for iron in cement paste and mortar. *Mater. Trans.* **2018**, *59*, 927–934. [CrossRef]
24. Doi, K.; Hiromoto, S.; Katayama, H.; Akiyama, E. Effects of oxygen pressure and chloride ion concentration on corrosion of iron in mortar exposed to pressurized humid oxygen gas. *J. Electrochem. Soc.* **2018**, *165*, 582–589. [CrossRef]
25. Doi, K.; Hiromoto, S.; Shinohara, T.; Tsuchiya, K.; Katayama, H.; Akiyama, E. Role of mill scale on corrosion behavior of steel rebars in mortar. *Corros. Sci.* **2020**, *177*, 108995. [CrossRef]
26. Oliver, W.C.; Pharr, G.M. An improved technique for determining hardness and elastic modulus using load displacement sensing indentation experiments. *J. Mater. Res.* **1992**, *7*, 1564–1583. [CrossRef]
27. Maslar, J.E.; Hurst, W.S.; Bowers, W.J.; Hendricks, J.H.; Aquino, M.I. In situ raman spectroscopic investigation of aqueous iron corrosion at elevated temperatures and pressures. *J. Electrochem. Soc.* **2000**, *147*, 2532–2542. [CrossRef]
28. Bellot-Gurlet, L.; Neff, D.; Reguer, S.; Monnier, J.; Saheb, M.; Dillmann, P. Raman studies of corrosion layers formed on archaeological irons in various media. *J. Nano Res.* **2009**, *8*, 147–156. [CrossRef]
29. Morcillo, M.; Chico, B.; Alcantara, J.; Diaz, I.; Wolthuis, R.; Fuente, D. SEM/Micro-Raman characterization of the morphologies of marine atmospheric corrosion products formed on mild steel. *J. Electrochem. Soc.* **2016**, *163*, 426–439. [CrossRef]
30. Ming, J.; Shi, J.J. Distribution of corrosion products at the steel-concrete interface: Influence of mill scale properties, reinforcing steel type and corrosion inducing method. *Constr. Build. Mater.* **2019**, *229*, 116854. [CrossRef]
31. Alcántara, J.; Chico, B.; Simancas, J.; Díaz, I.; Fuente, D.; Morcillo, M. An attempt to classify the morphologies presented by different rust phases formed during the exposure of carbon steel to marine atmospheres. *Mater. Charact.* **2016**, *118*, 65–78. [CrossRef]
32. Singh, J.K.; Singh, D.D.N. The nature of rusts and corrosion characteristics of low alloy and plain carbon steels in three kinds of concrete pore solution with salinity and different pH. *Corros. Sci.* **2012**, *56*, 129–142. [CrossRef]
33. Song, Y.R.; Jiang, G.M.; Chen, Y.; Zhao, P.; Tian, Y.M. Effects of chloride ions on corrosion of ductile iron and carbon steel in soil environments. *Sci. Rep.* **2017**, *7*, 6865. [CrossRef]
34. Zhang, Y.J.; Yuan, R.; Yang, J.H.; Xiao, D.H.; Luo, D.; Zhou, W.H.; Niu, G. Effect of tempering on corrosion behavior and mechanism of low alloy steel in wet atmosphere. *J. Mater. Res. Technol.* **2022**, *20*, 4077–4096. [CrossRef]
35. Gao, M.; Pang, X.; Gao, K. The growth mechanism of CO_2 corrosion products films. *Corros. Sci.* **2011**, *53*, 557–568. [CrossRef]
36. Wang, C.L.; Hua, Y.; Nadimi, S.; Taleb, W.; Barker, R.; Li, Y.X.; Chen, X.H.; Neville, A. Determination of thickness and air-void distribution within the iron carbonate layers using X-ray computed tomography. *Corros. Sci.* **2021**, *179*, 109153. [CrossRef]
37. Wang, Z.H.; Zhang, X.; Cheng, L.; Liu, J.; Wu, K.M. Role of inclusion and microstructure on corrosion initiation and propagation of weathering steels in marine environment. *J. Mater. Res. Technol.* **2021**, *10*, 306–321. [CrossRef]
38. Mishra, P.; Yavas, D.; Bastawros, A.F.; Hebert, K.R. Electrochemical impedance spectroscopy analysis of corrosion product layer formation on pipeline steel. *Electrochim. Acta* **2020**, *346*, 136232. [CrossRef]

Disclaimer/Publisher's Note: The statements, opinions and data contained in all publications are solely those of the individual author(s) and contributor(s) and not of MDPI and/or the editor(s). MDPI and/or the editor(s) disclaim responsibility for any injury to people or property resulting from any ideas, methods, instructions or products referred to in the content.

Article

Influence of the Graphene Oxide on the Pore-Throat Connection of Cement Waste Rock Backfill

Zhangjianing Cheng [1,2], Junying Wang [1], Junxiang Hu [1], Shuaijie Lu [1], Yuan Gao [1,3,4], Jun Zhang [1,*] and Siyao Wang [1,*]

1. School of Transportation and Civil Engineering, Nantong University, Nantong 226019, China; czjn@tongji.edu.cn (Z.C.); 2133110305@stamil.ntu.edu.cn (J.W.); 2233110392@stmail.ntu.edu.cn (J.H.); lushuaijie@stmail.ntu.edu.cn (S.L.); y.gao@ntu.edu.cn (Y.G.)
2. College of Civil Engineering, Tongji University, Shanghai 200092, China
3. Nantong Key Laboratory of Intelligent Civil Engineering and Digital Construction, Nantong University, Nantong 226019, China
4. Nantong Taisheng Blue Lsland Offshore Co., Ltd., Nantong 226200, China
* Correspondence: 13951411616@139.com (J.Z.); wangsiyao@ntu.edu.cn (S.W.)

Abstract: The pore-throat characteristics significantly affect the consolidated properties, such as the mechanical and permeability-related performance of the cementitious composites. By virtue of the nucleation and pore-infilling effects, graphene oxide (GO) has been proven as a great additive in reinforcing cement-based materials. However, the quantitative characterization reports of GO on the pore-throat connection are limited. This study applied advanced metal intrusion and backscattered electron (BSE) microscopy scanning technology to investigate the pore-throat connection characteristics of the cement waste rock backfill (CWRB) specimens before and after GO modification. The results show that the microscopic pore structure of CWRB is significantly improved by the GO nanosheets, manifested by a decrease in the total porosity up to 31.2%. With the assistance of the GO, the transfer among internal pores is from large equivalent pore size distribution to small equivalent pore size distribution. The fitting relationship between strength enhancement and pore reinforcement efficiency under different pore-throat characteristics reveals that the 1.70 μm pore-throat owns the highest correlation in the CWRB specimens, implying apply GO nanosheets to optimizing the pore-throat under this interval is most efficient. Overall, this research broadens our understanding of the pore-throat connection characteristics of CWRB and stimulates the potential application of GO in enhancing the mechanical properties and microstructure of CWRB.

Keywords: cement waste rock backfill; nano-modification; pore-throat connection; metal intrusion; equivalent pore size distribution

1. Introduction

Employing the cemented waste rock backfill (CWRB, which is composed of waste rock, cementitious materials, and water) to fill mining can reduce surface subsidence [1], control rock strata movement [2], reduce water inrush disaster [3] and groundwater resource loss caused by overlying rock failure [4]. Additionally, the CWRB also promotes waste source reduction and resource recycling [5], thereby solving land resource waste caused by waste rocks, additional economic burdens, and potential security threats [3]. The reinforcing effects of the CWRB on the rock movement mainly depend on the mechanical properties of the backfill [6]. If the mechanical parameters of the CWRB cannot meet the expected requirements of the projects, it will cause a series of issues influencing mining safety [7]. Therefore, in recent decades, many scholars devoted themselves to strengthening the mechanical behavior of the CWRB [8–10].

It is generally believed that the key to influencing the mechanical properties of cemented backfill materials is the hydration products constituting its bearing structure [11].

Some researchers studied the direction of optimizing the hydration process or products to reinforce the compressive strength of the CWRB [12]. GO is a derivative of graphene, generally obtained by oxidation of natural graphite powder with various oxidants in acidic media [13], which can be viewed as a layer of graphene grafted with oxygen functional groups [14]. Moreover, the thickness of single-layer GO can be several nanometers, whereas Young's modulus is up to 1 TPa [15], the intrinsic strength is 130 GPa [15], and the specific surface area is about 700–1500 m^2/g [16]. Recently, by virtue of its great mechanical properties, abundant oxygen-containing functional groups, ultrahigh specific surface area, and ultralow-containing cementitious composites, graphene oxide (GO) have become a special CWRB-enhanced additive [17,18]. Gao et al. [3] reported that applying the industrial GO combined with fly ash could replace 20 wt% cement in CWRB, producing high-performance mine backfill materials. Compared with the plain CWRB, the compressive strength and impermeability properties of the GO-fly ash-modified specimens improved by 5.1–16.9% and 32.7–38.9% by only mixing approximately 0.007 wt% GO [3]. In addition, due to the reduction of cement consumption and utilization of solid waste, this special GO-enhanced CWRB is considered to contribute to environmental protection. The industrial GO modification approach also opens a new pathway for more cost-effective backfill materials. By the way, there are also many modification strategies that can effectively optimize the pore structure and improve the durability of cementitious composites. The direct and effective way is to reduce the water-cement ratio. Using more cement means higher strength and a denser matrix microstructure, which also leads to high costs and CO_2 emissions. In addition, adding the mineral particles, such as silica fume [19], lime powder [20], and bentonite [21], to cementitious composite materials can improve their mechanical performance and ion penetration resistance. However, these modifications also have certain disadvantages, such as a complex preparation process, high cost, and fluidity impact. In view of the good prospects of the GO-modified CWRB, it is significant to better understand the reinforcing mechanisms of the GO on CWRB for further optimization of the materials.

GO has such great enhancement effects on cementitious composites mainly due to two benefits the nucleation effects [22,23] and pore infilling effects [24,25]. Due to GO owning the abundant oxygen-containing functional groups, the growth of hydration products in the cement matrix could be significantly promoted in the hydration process, resulting in more hydration products being formed in the hardened matrix and reinforcing the pore structure of the cementitious composites [26,27]. Additionally, in the hardened cement matrix, the GO nanosheets would further act as a "wall-like" role to assist the hydration products in optimizing the cement matrixes' microstructure [28]. Benefited from these two reinforcing mechanisms, GO could reinforce the pore structure of CWRB specimens in two aspects: (I) microstructural refinement of the interfacial transition zone (ITZ) and cement matrix to a denser status and (II) forms barriers at the pore throat junctions to decrease the connectivity among pores [3]. However, in the previous research [29], these mechanisms were mainly analyzed from a qualitative point of view. With the assistance of a new technology called metal intrusion, Gao et al. [29] analyzed the optimization degree of the pore structure of CWRB by GO, and established the relationship between the optimization of pore structure and the improvement of mechanical properties. However, there are few studies on the quantitative characterization of CWRB pore-throat connection and its relationship with mechanical properties.

The degree of cement pore-throat size correlation is an important property affecting the cementitious composites' mechanical and permeability-related properties [30]. Wardlaw et al. [31] revealed that the pore-throat structure consisted of pores of different sizes arranged randomly in a network and connected by throats, all of which were smaller than the smallest pore. Therefore, pore throat size is correlated, and characterizing the pore-throat of the CWRB is very important but challenging. Hence, in this study, we investigated the pore-throat connection characteristics of the CWRB specimens before and after GO modification with the help of metal intrusion and backscatter electron (BSE) microscope scanning technologies based on the author's previous research. Firstly, the plain CWRB specimens and the GO-fly ash-modified CWRB specimens were prepared. After that, the 28-day compressive strength and the pore structure of the corresponding age of the CWRB specimens were measured. Then, the effects of the GO nanosheets on the different levels of pore throats in the cement matrix of the CWRB were discussed in detail. Finally, the relationship between different pore-throat modifications and mechanical properties enhancement was further analyzed.

2. Methods

2.1. Sample Preparation

The preparation processes of the CWRB specimens and the related specimen preparation diagram can be found in the authors' previous study [29]. Firstly, the industrial GO mixed with the polycarboxylate superplasticizer powders were dissolved in distilled water and dispersed by ultrasonication treatment. The ultrasonication treatment was performed using a VCX-500 W ultrasonicator with a 13 cm probe. The whole ultrasonication process was treated in an ice-water environment to ensure the GO suspension would not be overheated and affect the properties of the GO. The ultrasonic mode was selected as the 3 s pulse and 3 s pause mode with a 150 W ultrasonication power and 10 min ultrasonication time [32]. The mass ratio of the industrial GO and polycarboxylate superplasticizer powders were chosen as 0.08 wt% and 0.64 wt% of the suspension. A previous study [33] suggested that this mixing ratio would trade off the reinforcing effects and the economic benefits of the GO nanosheets. A reference group (Ref-group), with the same mixed proportion of the PC without GO, was used to better investigate the effects of GO on the pore-throat connection and the corresponding properties of the CWRB.

After the industrial GO dispersion, for the GO-reinforced group, the ordinary Portland cement (OPC, type P.O. 42.5) and fly ash powders were mixed together under the mass ratio of 4:1 with a 5 min dry stirring to avoid the uneven distribution in the hardened CWRB specimens. After that, the prepared GO suspension was poured into the powders' mixture with a 0.6 water-to-cement (W/C) ratio to generate the cementitious slurry. The Ref-group suspension was also mixed with the plain OPC powder under the same W/C ratio. The detailed mix design of the GO- and Ref-group is shown in Table 1. Afterward, the fresh cementitious slurry was poured into the coal gangue. The coal gangue, with a density of 2.25 g/cm^3, was employed as the rock backfill aggregates in this work. Then, the fresh CWRB mixtures were poured into the cylindrical mold with the size of 50 × 100 mm^2 for the preparation of the compressive mechanical properties tests.

Table 1. Mix proportions of the Ref and GO-FA hybrid modified CWRB specimens [29].

Specimen	W/C	G/s (wt%)	P/s (wt%)	F/SL (wt%)	C/SL (wt%)	C/B (wt%)	G/B (wt%)
Ref-group	0.6	--	0.64	--	62.5	16.2	--
GO-group	0.6	0.08	0.64	12.5	50.0	13.0	0.008

(Note: G/s and P/s represent GO and PC to suspension weight percentages; F/SL and C/SL represent FA and cement powder to mixing cement-based slurry weight percentages; C/B and G/B represent cement and GO to backfill specimens weight percentages).

According to the ASTM C192/C192M-20a [34], the diameter of the test specimens must be at least three times higher than the biggest aggregate particle to avoid the side effects affecting the mechanical behavior. Therefore, in this study, the coal gangues were sieved into seven levels followed by the previous report [35] as 0–0.5 mm, 0.5–1.0 mm, 1.0–1.5 mm, 1.5–2.5 mm, 2.5–5 mm, 5–8 mm, and 8–10 mm, respectively. The Talbot gradation theory [36] was then applied to partition the gradation of the backfill aggregates. Four Talbot indices, 0.2, 0.4, 0.6, and 0.8, were adopted in CWRB specimen preparation.

After the corresponding weight of the coal gangue aggregates was selected, the gangues were then mixed with prepared fresh cementitious slurry with a concentration ratio of 1:0.35 using a concrete mixer and stirred for 5 min, and poured into a 50 × 100 mm^2 under another 3 min vibration to ensure the compactness of the CWRB. The specimens, after pouring them, were placed in a concrete curing box for 28 days of curing with a temperature and humidity of 20 ± 1 °C and 95%, respectively.

2.2. Measurement

To investigate the influence of the graphene oxide on the pore-throat connection of cement waste rock backfill, the CWRB specimens, after 28 days of curing, were soaked in an ethanol solution to avoid the CWRB continuing hydration. Afterward, the cylindrical specimens were cut into small cubes with the size of 5 × 5 × 5 mm^3. After that, the cut CWRB cubes were placed in a ventilated oven for 24 h at 105 ± 2 °C for drying. Then, the metal intrusion method was applied to fill the CWRB specimens with Field's metal. The purpose of the metal intrusion is to let the pore structure in CWRB could be better displayed under a scanning electron microscope (SEM). The specific operation steps and principles of metal intrusion follow the literature [37]. After metal intrusion, the cubes were soaked in the epoxy impregnation for the preparation of the SEM characterization. Before scanning electron microscope tests, the specimens were first polished. The polishing was divided into two main steps by using six grades of coarse sandpaper (178 μm, 61 μm, 38 μm, 23 μm, 13 μm, and 5.5 μm) and four fine grades of diamond grit (2.5 μm, 1 μm, 0.5 μm, and 0.25 μm), respectively [36].

The polished CWRB specimens were then characterized using a high-resolution field emission SEM with a backscattered electron (BSE) detector. The dwell time and scanning electron for the BSE characterization were selected as ten μs and five keV, respectively. To alleviate the influence of electrical charges during the imaging process, one two-nanometer-thickness gold film was deposited on the surface of polished specimens. The magnification of each BSE image is set to 320 times, with 1280 × 960 pixels square. More details concerning the metal intrusion and BSE image characterization have followed the literature [37]. By the way, three specimens for each group (four Ref-group and four GO-group) were tested in BSE image processing, and 8 BSE images were taken for each specimen. After that, the cylindrical CWRB specimens at 28-day age were used for the compressive strength tests according to the authors' previous reports [29]. The mechanical test results are listed in the literature [29].

3. Results and Discussion

3.1. Pore-Throat Characterization

Considering the pore-throat connection of the CWRB significantly influence the related mechanical performance, permeability-related properties, and durability of the hardened specimens. The modified mechanisms of the graphene oxide nanosheets on the micro/nanopore structure of the CWRB were first investigated. Figure 1a shows one typical BSE image of the CWRB specimens after metal intrusion treatment. A previous study suggested that the cementitious composites belong to a porous polyphase structure, the chemical composition of each phase is different, the BSE electron image can be used to distinguish different phases, the cement-based materials phase qualitative and quantitative analysis [38]. The image magnification is 320 times with 1280 × 960 pixels. In order to better distinguish the different CWRB terms in the BSE image, especially the substrate materials and the pores, the image was first treated with binarization. The binarization of the image is conducive to the further processing of the image, making the CWRB images simple. The amount of data was reduced, which could highlight the outline of the objects in the cementitious matrix. The IsoData thresholding method was used to calculate the optimal threshold, referring to the literature [39]. The image after binarization treatment is exhibited in Figure 1b, where black represents the cement matrix and white represents the pores in CWRB specimens. Afterward, the pore structure, pore size distribution, and pore-throat connection of the CWRB specimens could be well investigated.

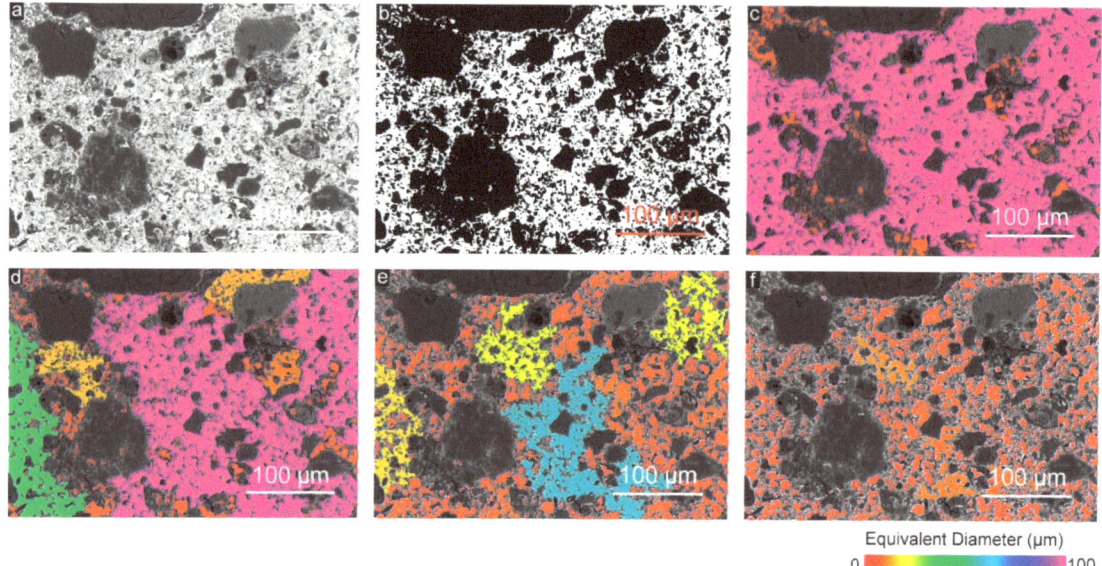

Figure 1. (**a**) One BSE image of the cement matrix area of the CWRB specimens; (**b**) The BSE image after binarization treatment. Equivalent pore size distribution of the imaged samples with different pore-throat characterization; (**c**) 0.34 μm; (**d**) 1.02 μm; (**e**) 1.7 μm; and (**f**) 2.38 μm indicated using a color bar.

In order to more intuitively display the pore-structure characteristics such as pore size, distribution, and pore-throat connection in the BSE image, the pores of different sizes in the CWRB specimens are further distinguished by color mapping, and the equivalent diameter (representing the corresponding diameter of the circle which has the same surface area of pores [40]) was used to characterize the pore size. As shown in Figure 1c, the original white pores are filled with purple and red, where the purple part represents pores with an equivalent diameter close to or much larger than 100 μm, while the red part represents pores with an equivalent diameter of 0–14.3 μm. In addition, due to the existence of pore-throats, the original small pores are interconnected by throats and become larger pores, which greatly changes the pore structure of specimens.

Four kinds of pore throats with equivalent diameters (0.34, 1.02, 1.70, and 2.38 μm) were defined to investigate the effect of pore-throat size on the pore structure of the specimen and the effect of GO on the pore-throat and the corresponding pore structure. Taking a pore-throat with the size of 1 pixel (0.34 μm) as an example, we consider that the pores are independent of each other when the distance between pores (pore-throat size) is less than 0.34 μm, and the pores are connected when the distance between pores (pore-throat size) is larger than 0.34 μm. The equivalent diameter of the smallest pore-throat is 0.34 μm (the actual size corresponding to a single pixel), which is calculated from the pixel size (1280 × 960 pixels square) and the actual size (435.2 × 326.4 microns square) of the BSE image.

The equivalent pore size distribution of CWRB specimens under different pore throat characteristics was presented in Figure 1c–f. As the equivalent diameter of pore-throat increases, the equivalent diameter of pores in specimens decreases significantly, and a small part of the original 100 μm pores (marked in purple) are firstly divided into 15 μm pores (marked in orange) and 30 μm pores (marked in green). Subsequently, most of the 100 μm pores (marked in purple) are divided into 10 μm pores (marked in red), 20 μm pores (marked in yellow), and 60 μm pores (marked in light blue). At the same time, 30 μm pores (marked in green) turned into pores marked in yellow and red, and 15 μm pores (marked in orange) almost all turned red. When the equivalent diameter of the pore-throat is 2.38 μm, the BSE image of specimens is entirely covered by pores with an equivalent pore diameter of 0–20 μm. When the equivalent diameter of the pore-throat is 2.38 μm, the BSE image of specimens is entirely covered by pores (marked in red and orange) with an equivalent pore diameter of 0–15 μm.

3.2. Effect of GO on Pore-Throat of CWRB Specimens

To further reveal microstructural alteration of CWRB specimens caused by GO, the changes in pore size, equivalent pore size distribution, and porosity of samples with different Talbot gradation indexes before and after adding GO were compared. As shown in Figure 2a, the BSE image of the Ref-group specimen presents a single interpenetrating pore accompanied by partially separated pores with low equivalent diameters under the pore-throat equaling 0.34 μm. An apparent pore-splitting phenomenon appears after the addition of GO, especially at the Talbot gradation index $n = 0.2$ (i.e., Group-1), the through pores are divided into many pores of different sizes, and the ability of GO to segment pores gradually weakens with the increasing Talbot gradation index. Figure 2b shows the equivalent pore size distribution of Ref- and GO-enhanced CWRB specimens under 0.34 μm pore-throat characterization. The pores in the Ref-1 specimen are mainly concentrated in the range of 100–250 μm, the pores in Ref-2 and Ref-4 specimens are mainly concentrated in the range of 200–250 μm while those of Ref-3 specimen are in the range of 150–200 μm. With the addition of GO, 100–250 μm pores in the Ref-1 specimen were transformed into a large number of 0–100 μm pores, and some of the 200–250 μm pores in the Ref-2 specimen were transformed into 150–200 μm pores. Compared with GO-1 and GO-2 specimens, the pore distribution in GO-3 and GO-4 did not change significantly, which confirms that GO has a limited effect on optimizing the microstructure of specimens with high Talbot gradation index.

The relationship between cumulative porosity and equivalent diameters of pores in the matrix is shown in Figure 2c. The first half of most curves are flat, indicating that there are few pores with equivalent diameters corresponding to those ranges, while the latter half of the curve exhibits an apparent plunge, indicating that most of the pores are large pores, and most of them are distributed in the range of 200–250 μm, which is consistent with the conclusion that the pores in CWRB specimens are mostly large pores with a single penetration, as concluded in Figure 2a,b. Additionally, the cumulative percentage of the porosity in the CWRB matrix with different aggregate Talbot indices under 0.34 μm pore-throat characterization is shown in Figure 4d. It can be found that with a certain amount of GO incorporation, the porosity of the CWRB specimen with aggregate Talbot index $n = 0.6$ (i.e., Group-3) remains unchanged. For the other aggregate Talbot indices, the porosity decreases to some extent, ranging from 6.0% to 21.1% (reaching the maximum at Group-2). The above phenomena indicate that GO effectively optimizes the microstructure of CWRB specimens, and as previous studies have concluded, GO significantly changes the pore structure of CWRB specimens through the filling and nucleation effect. On the one hand, due to the tiny physical size of GO, it can effectively fill the micro-pores in cement consolidation, thereby optimizing the microstructure of specimens (filling effect). On the other hand, GO has an ultra-high specific surface area and abundant oxygen-containing functional groups, which can provide additional nucleation sites for the early hydration of cement, significantly promote the hydration reaction of cement, and make the hydration products grow along the surface of GO, and act together with the produced hydration products in the cement matrix (nucleation effect). The dense hydration products that grow on the surface of GO will gradually develop into a unique wall-like structure, which promotes the division of large pores into smaller ones in the cement matrix.

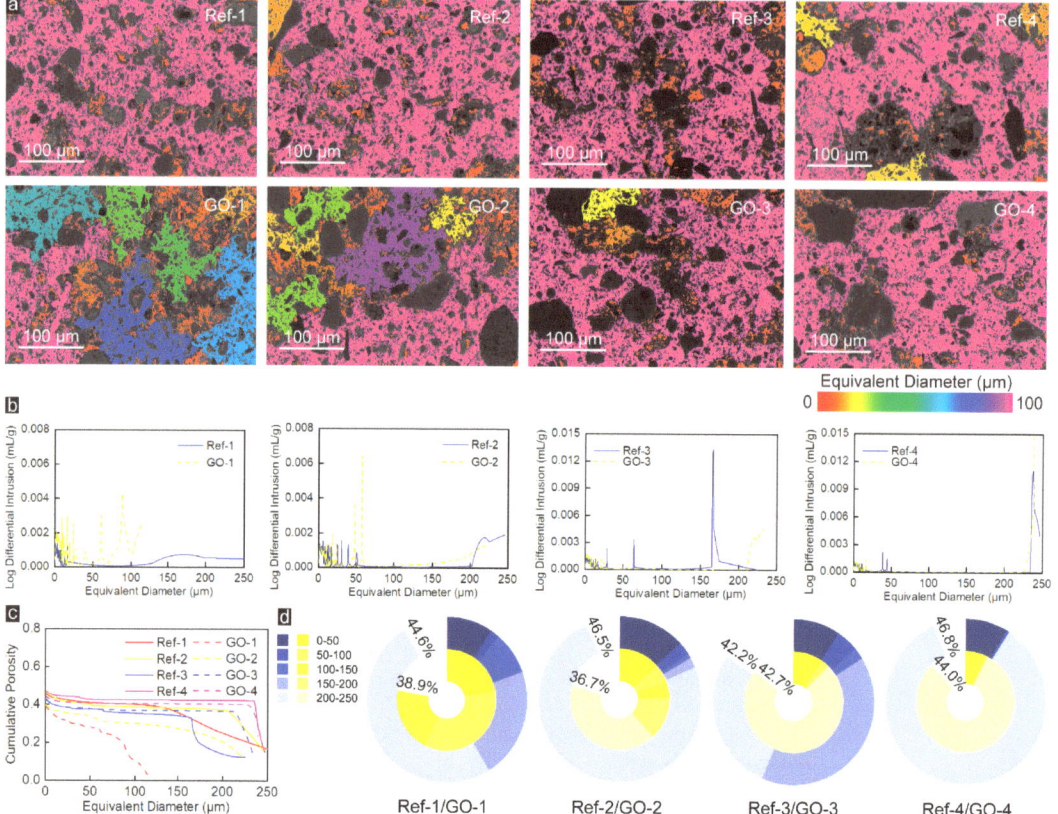

Figure 2. (**a**) The pore sizes are indicated by the colormaps under the pore-throat equaling 0.34 μm. (**b**) The corresponding equivalent pore size distribution of the Ref- and GO-reinforced CWRB specimens with different aggregate Talbot indices under 0.34 μm pore-throat characterization. (**c**) The cumulative porosity versus the characterized equivalent pore diameter in the tested cement matrix. (**d**) The cumulative percentage of the porosity in the CWRB matrix with different aggregate Talbot indices under 0.34 μm pore-throat characterization.

When the pore-throat size is 3 pixels (which is equal to 1.02 μm), as shown in Figure 3a,b, different from the single connected large pores in Figure 2a, under the lower Talbot gradation index (n = 0.2), many separation pores with different sizes appear in the BSE images of the Ref-group specimens, and the equivalent diameter of these pores is almost less than 100 μm. With the increase of aggregate Talbot index, on the one hand, the BSE image is gradually covered by pores with an equivalent diameter more prominent than 100 μm and a small number of pores with an equivalent pore diameter smaller than 20 μm. On the other hand, the number of pores with large equivalent diameters increases significantly. The pore structure of the CWRB specimens changed significantly after adding GO, where the pore distribution characteristics shifted from the sizeable equivalent pore size distribution to the small equivalent pore size distribution. The appearance of smaller equivalent diameter pores in the BSE images corroborates this perspective. Moreover, as represented in Figure 3c,d, GO has the best optimization effect on the porosity of the Group-1 specimen, with an optimization range of 21.1%, followed by Group-4, and there is no noticeable optimization effect in Group-3, while for Group-4 has a deteriorating effect.

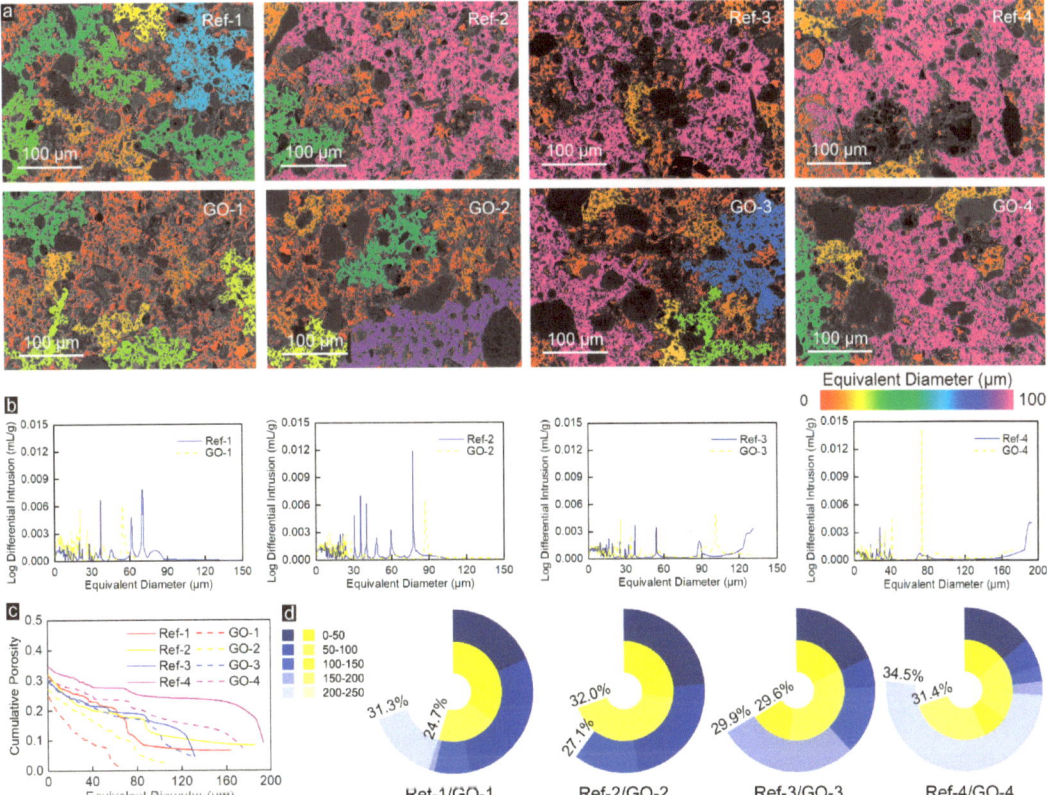

Figure 3. (**a**) The pore sizes are indicated by the colormaps under the pore-throat equaling 1.02 μm. (**b**) The corresponding equivalent pore size distribution of the Ref- and GO-reinforced CWRB specimens with different aggregate Talbot indices under 1.02 μm pore-throat characterization. (**c**) The cumulative porosity versus the characterized equivalent pore diameter in the tested cement matrix. (**d**) The cumulative percentage of the porosity in the CWRB matrix with different aggregate Talbot indices under 1.02 μm pore-throat characterization.

When the pore-throat size is 5 pixels (which is equal to 1.70 μm), as shown in Figure 4a, only a small number of large pores (equivalent diameter is 60–80 μm) are present in BSE images of Ref-4 specimen, and the BSE images in the other Ref-Groups are covered by a red and orange pore (equivalent diameter are 0–20 μm) and a few yellow and green pores (equivalent diameter are 2–40 μm). Similar to Ref-Group, the BSE images of GO-Group specimens (excluding GO-4) are almost entirely covered by red and orange pores. In contrast, a small number of blue pores (equivalent diameter ~60 μm) are present in GO-4 specimens. On the other hand, unlike Figures 2a and 3a, when the pore-throat size is 5 pixels (which is equal to 1.70 μm), in the BSE image of CWRB specimens, there are a large number of tiny pores (equivalent diameter is 20–40 μm) rather than connected large pores (equivalent diameter is 60–1000 μm). This significant change is due to the presence of a large number of pore throats (greater than 1.02 μm and less than 1.70 μm) in CWRB specimens. As the size of pore-throat increases to 1.70 μm, the large number of 1.02–1.70 μm pore-throats that act as connecting pores are no longer recognized as pore channels, resulting in a large number of large pores being split into tiny pores.

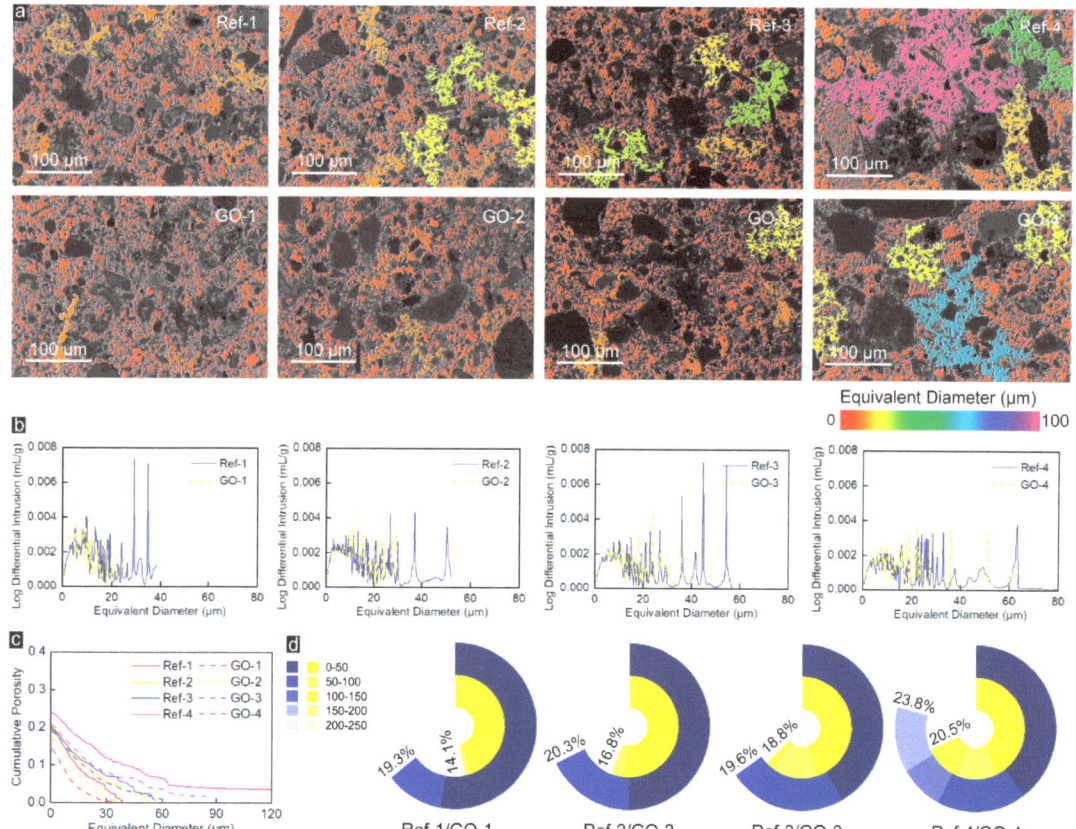

Figure 4. (**a**) The pore sizes are indicated by the colormaps under the pore-throat equaling 1.70 μm. (**b**) The corresponding equivalent pore size distribution of the Ref- and GO-reinforced CWRB specimens with different aggregate Talbot indices under 1.70 μm pore-throat characterization. (**c**) The cumulative porosity versus the characterized equivalent pore diameter in the tested cement matrix. (**d**) The cumulative percentage of the porosity in the CWRB matrix with different aggregate Talbot indices under 1.7 μm pore-throat characterization.

Moreover, comparing the BSE images of Ref-Group and GO-Group, it is easy to find that the incorporation of GO leads to a decrease in the number of pores with large equivalent diameters and an increase in the number of pores with small equivalent diameters. The results in Figure 4b correspond to the BSE image, which can be clearly observed that the addition of GO leads to an abnormally significant decrease in the mercury intrusion into pores with large equivalent diameters, and the pore distribution characteristics are significantly transferred from large equivalent pore size distribution to small equivalent pore size distribution. Furthermore, the results of porosity and pore distribution in Figure 4c,d also indicate that, no matter which Talbot index is, GO is always effective in optimizing the microscopic pore structure of CWRB specimens, and the enhancement efficiency is 4~26.9%. Therefore, based on the above reasons, it is considered that GO has the gradely enhancement effect on the microscopic pore-structure of CWRB specimens when the pore-throat is equal to 1.70 μm.

When the pore-throat size is 7 pixels (which is equal to 2.38 μm), most areas in the BSE image of the Ref-Group specimens are covered by red pores with an equivalent diameter of less than 10 μm and small areas were covered by orange pores (equivalent diameter are 0–20 μm). However, the BSE image of the Ref-4 sample presents partial yellow and green pores with an equivalent diameter of 20–40 μm, as shown in Figure 5a. With the incorporation of GO, only red and orange pores remained in the BSE image of all CWRB specimens. Unlike Figure 4b, the equivalent pore size distribution characteristics of CWRB specimens (except for Group-4) with different Talbot indices in Figure 5b are similar and do not alter due to GO blending. In addition, comparing the porosity changes of CWRB specimens before and after GO modification in Figures 2c and 5c and Figures 2d and 5d, it can be found that GO has the best enhancement effect on the microstructure of the CWRB specimens under pore-throat equaling 2.38 μm with reinforcing ratio varying from 7.4% to 31.2% and better enhancement effect on large pore-throat (equaling 1.70 and 2.38 μm) than small pore-throat (equaling 1.70 and 2.38 μm).

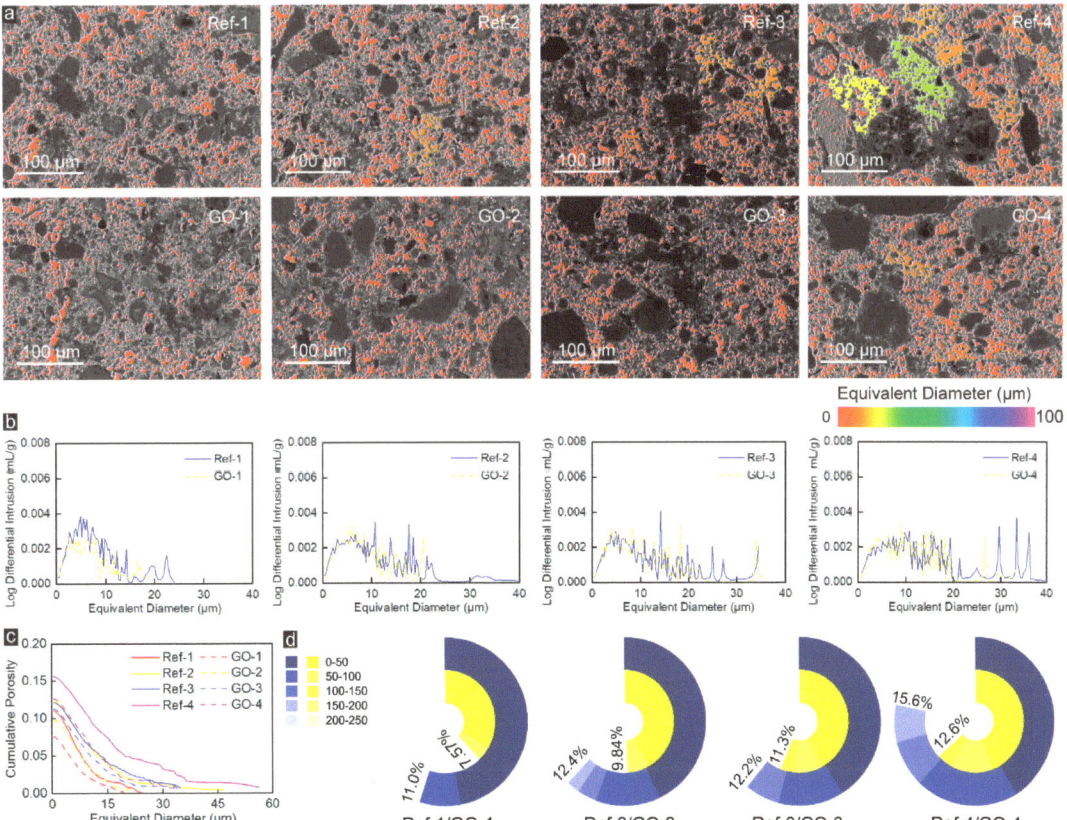

Figure 5. (**a**) The pore sizes are indicated by the colormaps under the pore-throat equaling 2.38 μm. (**b**) The corresponding equivalent pore size distribution of the Ref- and GO-reinforced CWRB specimens with different aggregate Talbot indices under 2.38 μm pore-throat characterization. (**c**) The cumulative porosity versus the characterized equivalent pore diameter in the tested cement matrix. (**d**) The cumulative percentage of the porosity in the CWRB matrix with different aggregate Talbot indices under 2.38 μm pore-throat characterization.

3.3. Effect of GO on the Number of Pores with Different Equivalent Pore Sizes of CWRB Specimens

To further investigate the influence of GO on the number of pores with different equivalent pore sizes under different pore-throat characterization, the change ratio of pores with different equivalent pore sizes was calculated and counted as shown in Figure 6, where a change ratio greater than 0 indicates an increase and less than 0 indicates a decrease in pore number. Figure 6a shows the change in pores number with different equivalent pore sizes when pore-throat size is equal to 0.34 μm. It is clear that GO has the best improvement in CWRB specimens for Group 1. Its 150–200 μm and 200–250 μm pores completely disappear (the change ratio is −100%) and resulted in a significant increase in the number of pores in the range of 0–150 μm, especially in the range of 50–100 μm by a staggering 2130%. At the same time, that also means that the pores in CWRB specimens were segmented into smaller pores caused by GO. Compared with Group 1, the effect of GO on pore structure improvement for Group 2 was weakened. The pore number of 200–250 μm pores decreased by 44.7%, while the pore number of 50–100, 100–150, and 150–200 μm pores increased by 130.8%, 474.3%, and 563.3%, respectively.

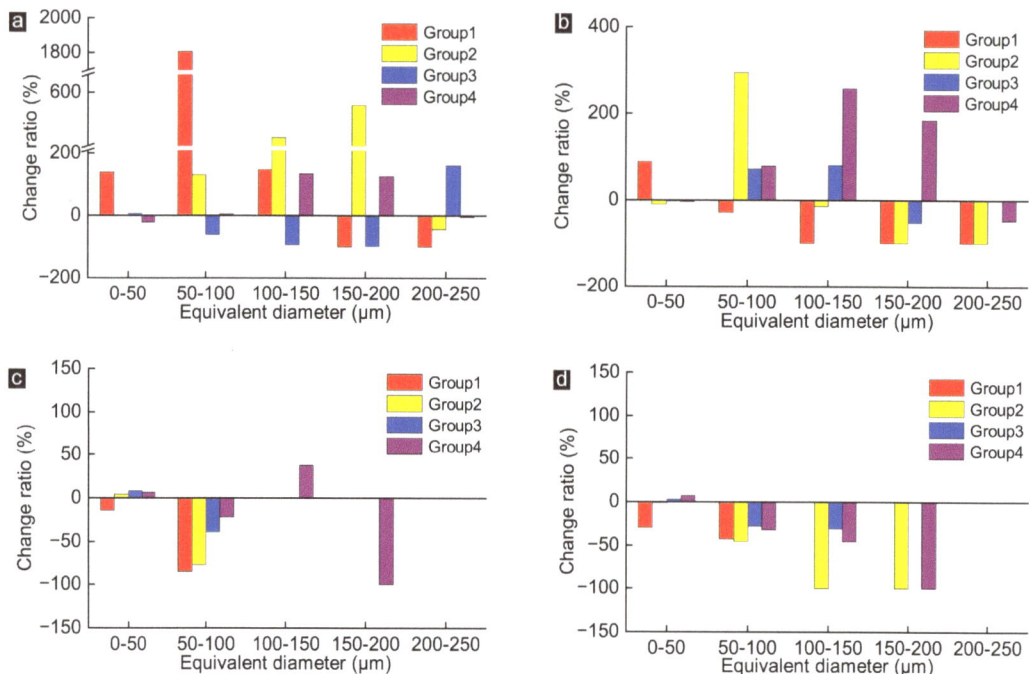

Figure 6. Change ratio of pores with different equivalent pore sizes of CWRB specimens under the pore-throat equaling (**a**) 0.34 μm. (**b**) 1.02 μm. (**c**) 1.70 μm. and (**d**) 2.38 μm.

Similar to Group 2, the number of 200–250 μm pores in Group 4 decreased slightly, and the number of 0–50 μm pores increased slightly, different from 0 to 50 μm pores in Group 2, which remained constant. However, GO caused the deterioration of the microscopic pore structure of Group 4, manifested by the decrease of 50–200 μm pores and the increase of 200–250 μm pores. When the pore throat is equal to 1.02 μm (result as represented in Figure 5b), GO can enhance the microstructure of all CWRB specimens to a certain extent, as evidenced by the significant reduction in the number of large pores and the significant increase in the number of pores with relatively small equivalent diameters. As the pore throat increases to 1.70 μm (as shown in Figure 5c), except for Group 4, the 100–200 μm pores in other CWRB samples nearly disappear, which is the reason for the change rate of 0. Notably, the number of pores with all equivalent diameters in the Group 1 sample decreases, which means that the total porosity decreases. The conclusions of the other groups are consistent with that pore-throat is equal to 1.02 μm. In addition, when the pore-throat is equal to 2.38 μm, the pore-structure of CWRB specimens still exhibits a decrease of large pores and an increase of tiny pores, which is different from the pore-throat equal to 0.34 μm, 1.02 μm, and 1.70 μm, with a decrease in all pores more significant than 50 μm and an increase in pores smaller than 50 μm (Group 1 shows the decrease in 0–50 μm pores).

3.4. Relationship between Pore-Throat Modifications and Mechanical Properties of CWRB Specimens

In order to clarify the interaction law of compressive strength and porosity of CWRB specimens before and after GO modification under different aggregate Talbot indices, the mechanical test results of CWRB specimens in the previous study [22] (as shown in Figure 7a) were used to establish the fitting relationship between compressive strength enhancement efficiency and the Porosity enhancement efficiency under different pore-throat characterization (as shown in Figure 7b). The fitting coefficient is the highest at 0.987 under pore-throat equaling 1.70 μm. Therefore, we consider that 1.70 μm pore-throat characterization can accurately describe the microscopic pore-structure of the CWRB specimens, and recommend this parameter to be adopted in future studies.

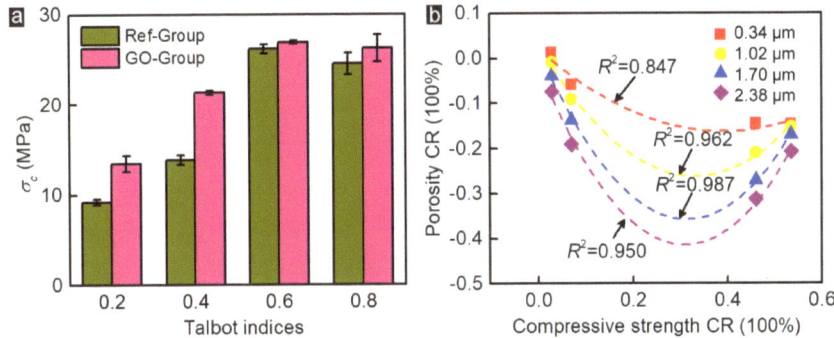

Figure 7. (**a**) The compressive strength of Ref- and GO-reinforced CWRB specimens with different aggregate Talbot indices. (**b**) Fitting relationship between compressive strength CR and porosity CR of CWRB specimens.

4. Conclusions

In this study, advanced metal intrusion and backscattered electron (BSE) microscopy scanning techniques were applied to investigate the pore-throat connectivity of CWRB specimens before and after GO modification with different pore-throat characteristics and the fitting relationship between the intensity enhancement efficiency and pore enhancement efficiency was established, with following conclusions:

(1) GO is able to optimize the microstructure of CWRB specimens under different pore-throat characteristics, and optimization amplitude hits up to approximately 32.1%, which is due to the nucleation effect that promotes the growth of hydration products and the pore-infilling effect that fills and divides the pores and pore-throat, both of which contribute to the optimization of microstructure;

(2) With the increase of pore-throat size, the porosity of CWRB specimens showed a decreasing trend, and the pores shifted from large equivalent pore size distribution to small equivalent pore size distribution, which was due to the fact that the original small pore throat was no longer used as a channel to connect the pores, but split the original large pores;

(3) A large number of pore-throat with equivalent diameters of 1.02–1.70 μm existed in the CWRB specimens, and GO significantly reduced the number of pore-throats in this size, effectively partitioning the pores and improving the microscopic pore-structure of CWRB;

(4) The 1.70 μm pore-throat characteristic can reflect the microscopic pore-structure of CWRB specimens more accurately than others, which is recommended for future studies.

Author Contributions: Z.C., Y.G., J.Z. and S.W. conceived and designed the simulations; Z.C., J.W., J.H., S.L. and S.W. performed the experiments; Z.C., J.W., J.H., Y.G. and S.W. analyzed the data; Z.C., J.W., J.H., J.Z. and S.W. wrote the paper; and all seven authors revised the paper. All authors have read and agreed to the published version of the manuscript.

Funding: This study was supported by the National College Student Innovation Training Program (No. 202310304064Z), the Natural Science Foundation of the Jiangsu Higher Education Institutions of China (No. 22KJB560010), and the Nantong Basic Science Research Program of China (No. JC12022098).

Institutional Review Board Statement: Not applicable.

Informed Consent Statement: Not applicable.

Data Availability Statement: The data presented in this study are available on request from the corresponding author.

Conflicts of Interest: The authors declare no conflict of interest.

References

1. Wu, J.; Jing, H.; Yin, Q.; Yu, L.; Meng, B.; Li, S. Strength prediction model considering material, ultrasonic and stress of cemented waste rock backfill for recycling gangue. *J. Clean. Prod.* **2020**, *276*, 123189. [CrossRef]
2. Li, Y.; Qiu, B. Investigation into Key Strata Movement Impact to Overburden Movement in Cemented Backfill Mining Method. *Procedia Eng.* **2012**, *31*, 727–733. [CrossRef]
3. Gao, Y.; Sui, H.; Yu, Z.; Wu, J.; Chen, W.; Jing, H.; Ding, M.; Liu, Y. Industrial graphene oxide-fly ash hybrid for high-performance cemented waste rock backfill. *Constr. Build. Mater.* **2022**, *359*, 129484. [CrossRef]
4. Akbar, A.; Kodur, V.K.R.; Liew, K.M. Microstructural changes and mechanical performance of cement composites reinforced with recycled carbon fibers. *Cem. Concr. Compos.* **2021**, *121*, 104069. [CrossRef]
5. Puyol, D.; Flores-Alsina, X.; Segura, Y.; Molina, R.; Padrino, B.; Fierro, J.; Gernaey, K.; Melero, J.; Martinez, F. Exploring the effects of ZVI addition on resource recovery in the anaerobic digestion process. *Chem. Eng. J.* **2018**, *335*, 703–711. [CrossRef]
6. Gao, Y.; Yu, Z.; Cheng, Z.; Chen, W.; Zhang, T.; Wu, J. Influence of industrial graphene oxide on tensile behavior of cemented waste rock backfill. *Constr. Build. Mater.* **2023**, *371*, 130787. [CrossRef]
7. Gren, G.; Wang, H.; Chen, C.; Wang, J. An energy conservation and environmental improvement solution-ultra-capacitor/battery hybrid power source for vehicular applications. *Sustain. Energy Technol. Assess.* **2021**, *44*, 100998.
8. Ezquerro, C.S.; Laspalas, M.; Aznar, J.M.G.; Miñana, C.C. Monitoring interactions through molecular dynamics simulations: Effect of calcium carbonate on the mechanical properties of cellulose composites. *Cellulose* **2023**, *30*, 705–726. [CrossRef]
9. Yu, Z.; Jing, H.; Gao, Y.; Xu, X.; Zhu, G.; Sun, S.; Wu, J.; Liu, Y. Effect of CNT-FA hybrid on the mechanical, permeability and microstructure properties of gangue cemented rockfill. *Constr. Build. Mater.* **2023**, *392*, 131978. [CrossRef]
10. Yan, R.-F.; Liu, J.-M.; Yin, S.-H.; Zou, L.; Kou, Y.-Y.; Zhang, P.-Q. Effect of polypropylene fiber and coarse aggregate on the ductility and fluidity of cemented tailings backfill. *J. Central South Univ.* **2022**, *29*, 515–527. [CrossRef]
11. Wang, Y.; Wu, J.; Ma, D.; Yang, S.; Yin, Q.; Feng, Y. Effect of aggregate size distribution and confining pressure on mechanical property and microstructure of cemented gangue backfill materials. *Adv. Powder Technol.* **2022**, *33*, 103686. [CrossRef]
12. Wu, J.; Yin, Q.; Gao, Y.; Meng, B.; Jing, H. Particle size distribution of aggregates effects on mesoscopic structural evolution of cemented waste rock backfill. *Environ. Sci. Pollut. Res.* **2021**, *28*, 16589–16601. [CrossRef] [PubMed]
13. Singh, R.K.; Kumar, R.; Singh, D.P. Graphene oxide: Strategies for synthesis, reduction and frontier applications. *RSC Adv.* **2016**, *6*, 64993–65011. [CrossRef]
14. Zhao, L.; Guo, X.; Song, L.; Song, Y.; Dai, G.; Liu, J. An intensive review on the role of graphene oxide in cement-based materials. *Constr. Build. Mater.* **2020**, *241*, 117939. [CrossRef]
15. Lee, C.; Wei, X.; Kysar, J.W.; Hone, J. Measurement of the elastic properties and intrinsic strength of monolayer graphene. *Science* **2008**, *321*, 385–388. [CrossRef] [PubMed]
16. Zhu, Y.; Murali, S.; Cai, W.; Li, X.; Suk, J.W.; Potts, J.R.; Ruoff, R.S. Graphene and graphene oxide: Synthesis, properties, and applications. *Adv. Mater.* **2010**, *22*, 3906–3924. [CrossRef]
17. Wu, Y.-Y.; Que, L.; Cui, Z.; Lambert, P. Physical Properties of Concrete Containing Graphene Oxide Nanosheets. *Materials* **2019**, *12*, 1707. [CrossRef]
18. Wang, S.; Zhang, S.; Wang, Y.; Sun, X.; Sun, K. Reduced graphene oxide/carbon nanotubes reinforced calcium phosphate cement. *Ceram. Int.* **2017**, *43*, 13083–13088. [CrossRef]
19. Lü, Q.; Qiu, Q.; Zheng, J.; Wang, J.; Zeng, Q. Fractal dimension of concrete incorporating silica fume and its correlations to pore structure, strength and permeability. *Constr. Build. Mater.* **2019**, *228*, 116986. [CrossRef]
20. Wang, D.H.; Shi, C.J.; Farzadnia, N.; Shi, Z.G.; Jia, H.F.; Ou, Z.H. A review on use of limestone powder in cement-based materials: Mechanism, hydration and microstructures. *Constr. Build. Mater.* **2018**, *181*, 659–672. [CrossRef]

21. Yang, H.; Long, D.; Zhenyu, L.; Yuanjin, H.; Tao, Y.; Xin, H.; Jie, W.; Zhongyuan, L.; Shuzhen, L. Effects of bentonite on pore structure and permeability of cement mortar. *Constr. Build. Mater.* **2019**, *224*, 276–283. [CrossRef]
22. Chintalapudi, K.; Pannem, R.M.R. Enhanced Strength, Microstructure, and Thermal properties of Portland Pozzolana Fly ash-based cement composites by reinforcing Graphene Oxide nanosheets. *J. Build. Eng.* **2021**, *42*, 102521. [CrossRef]
23. Mohammed, A.; Al-Saadi NT, K.; Sanjayan, J. Inclusion of graphene oxide in cementitious composites: State-of-the-art review. *Aust. J. Civ. Eng.* **2018**, *16*, 81–95. [CrossRef]
24. Danish, A.; Mosaberpanah, M.A.; Salim, M.U.; Fediuk, R.; Rashid, M.F.; Waqas, R.M. Reusing marble and granite dust as cement replacement in cementitious composites: A review on sustainability benefits and critical challenges. *J. Build. Eng.* **2021**, *44*, 102600. [CrossRef]
25. Yang, H.; Cui, H.; Tang, W.; Li, Z.; Han, N.; Xing, F. A critical review on research progress of graphene/cement based composites. *Compos. Part A Appl. Sci. Manuf.* **2017**, *102*, 273–296. [CrossRef]
26. Lee, S.-J.; Jeong, S.-H.; Kim, D.-U.; Won, J.-P. Effects of graphene oxide on pore structure and mechanical properties of cementitious composites. *Compos. Struct.* **2020**, *234*, 111709. [CrossRef]
27. Lee, S.-J.; Jeong, S.-H.; Kim, D.-U.; Won, J.-P. Graphene oxide as an additive to enhance the strength of cementitious composites. *Compos. Struct.* **2020**, *242*, 112154. [CrossRef]
28. Gao, Y.; Jing, H.; Fu, G.; Zhao, Z.; Shi, X. Studies on combined effects of graphene oxide-fly ash hybrid on the workability, mechanical performance and pore structures of cementitious grouting under high W/C ratio. *Constr. Build. Mater.* **2021**, *281*, 122578. [CrossRef]
29. Gao, Y.; Sui, H.; Yu, Z.; Wu, J.; Chen, W.; Liu, Y. Cemented waste rock backfill enhancement via fly ash-graphene oxide hybrid under different particle size distribution. *Constr. Build. Mater.* **2023**, *394*, 132162. [CrossRef]
30. Li, X.; Xu, Y. Microstructure-Based Modeling for Water Permeability of Hydrating Cement Paste. *J. Adv. Concr. Technol.* **2019**, *17*, 405–418. [CrossRef]
31. Wardlaw, N.C. The effects of pore structure on displacement efficiency in reservoir rocks and in glass micromodels. In Proceedings of the SPE/DOE Enhanced Oil Recovery Symposium, Tulsa, OK, USA, 20–23 April 1980.
32. YGao, Y.; Jing, H.W.; Chen, S.J.; Du, M.R.; Chen, W.Q.; Duan, W.H. Influence of ultrasonication on the dispersion and enhancing effect of graphene oxide–carbon nanotube hybrid nanoreinforcement in cementitious composite. *Compos. Part B Eng.* **2019**, *164*, 45–53.
33. Gao, Y.; Jing, H.; Yu, Z.; Li, L.; Wu, J.; Chen, W. Particle size distribution of aggregate effects on the reinforcing roles of carbon nanotubes in enhancing concrete ITZ. *Constr. Build. Mater.* **2022**, *327*, 126964. [CrossRef]
34. Assadollahi, A.; Moore, C. Effects of Crumb Rubber on Concrete Properties When Used as an Aggregate in Concrete Mix Design. In *Structures Congress 2017*, American Society of Civil Engineers: Reston, VA, USA, 2017; pp. 308–314
35. Wu, J.; Ma, D.; Pu, H.; Yang, S.; Wang, Y.; Yin, Q.; Jing, H. Effect of fractal gangue on macroscopic and mesoscopic mechanical properties of cemented waste rock backfill. In *Managing Mining and Minerals Processing Wastes*; Elsevier: Amsterdam, The Netherlands, 2023; pp. 19–43.
36. Zhu, S.; Feng, Y.M.; Feng, S.R.; Chen, W.Y. Particles Gradation Optimization of Blasting Rockfill Based on Fractal Theory. *Adv. Mater. Res.* **2012**, *366*, 469–473. [CrossRef]
37. Chen, S.J.; Tian, Y.; Li, C.Y.; Duan, W.H. A new scheme for analysis of pore characteristics using centrifuge driven non-toxic metal intrusion. *Geomech. Geophys. Geo-Energy Geo-Resour.* **2016**, *2*, 173–182. [CrossRef]
38. Georget, F.; Wilson, W.; Scrivener, K.L. edxia: Microstructure characterisation from quantified SEM-EDS hypermaps. *Cem. Concr. Res.* **2021**, *141*, 106327. [CrossRef]
39. Ridler, T.W.; Calvard, S. Picture thresholding using an iterative selection method. *IEEE Trans. Syst. Man Cybern* **1978**, *8*, 630–632.
40. Bryk, M. Computer-aided image analysis as a tool for examination of soil structure. *Acta Agrophysica* **2020**, *2001*, 41–45.

Disclaimer/Publisher's Note: The statements, opinions and data contained in all publications are solely those of the individual author(s) and contributor(s) and not of MDPI and/or the editor(s). MDPI and/or the editor(s) disclaim responsibility for any injury to people or property resulting from any ideas, methods, instructions or products referred to in the content.

Article

Influence of Nano-SiO$_2$ Content on Cement Paste and the Interfacial Transition Zone

Shaofeng Zhang [1,2,3], **Ronggui Liu** [2], **Chunhua Lu** [2], **Junqing Hong** [1,3], **Chunhong Chen** [2,4] and **Jiajing Xu** [1,*]

[1] School of Transportation and Civil Engineering, Nantong University, Nantong 226019, China; zsfgh@ntu.edu.cn (S.Z.); hongjq@ntu.edu.cn (J.H.)
[2] Faculty of Civil Engineering and Mechanics, Jiangsu University, Zhenjiang 212013, China; liurg@ujs.edu.cn (R.L.); lch79@ujs.edu.cn (C.L.); chench@cczu.edu.cn (C.C.)
[3] Nantong Key Laboratory of Intelligent Civil Engineering and Digital Construction, Nantong University, Nantong 226019, China
[4] School of Urban Construction, Changzhou University, Changzhou 213164, China
* Correspondence: xujiajing@ntu.edu.cn

Abstract: Nano-SiO$_2$ (NS) is widely used in cement-based materials due to its excellent physical properties. To study the influence of NS content on a cement paste and the interfacial transition zone (ITZ), cement paste samples containing nano content ranging from 0 to 2% (by weight of cement) were prepared, and digital image correlation (DIC) technology was applied to test the mechanical properties. Finally, the optimal NS content was obtained with statistical analysis. The mini-slump cone test showed that, with the help of superplasticizer and ultrasonic treatment, the flowability decreased continuously, as the NS content increased. The DIC experimental results showed that NS could effectively improve the mechanical properties of the cement paste and the ITZ. Specifically, at the content level of 1%, the elastic modulus of cement paste and ITZ was 20.95 GPa and 3.20 GPa, respectively. When compared to that without nanomaterials, the increased amplitude was 73.50% and 90.50%, respectively. However, with the further increase in NS content, the mechanical properties decreased, which was mainly caused by the agglomeration of nanomaterials. Additionally, the NS content did not exhibit a significant effect on the thickness of the ITZ, and its value was maintained at 76.91–91.38 μm. SEM confirmed that NS would enhance the microstructure of both cement paste and ITZ.

Keywords: nano-SiO$_2$ content; interfacial transition zone; digital image correlation; elastic modulus; ITZ thickness

1. Introduction

Cement as an important cementitious material has been widely used in engineering–construction industry for a long time. This is mainly attributed to its advantages of low cost, easy availability, good mechanical properties, etc. [1,2]. However, the properties of concrete structures, especially the bearing capacity and durability in various corrosive environments, have become a major concern [3]. For example, the rapid invasion of aggressive ions in the marine environment can permeate into concrete, causing premature corrosion of steel bars in a concrete structure and reducing its service life [4,5].

Therefore, researchers have attempted to add different kinds of nanomaterials to cement-based materials to improve their mechanical properties and durability [6,7]. Wang et al. [8] proved that nanomaterials can improve various properties of cement and concrete. Alireza et al. [9] found that the rheological characteristics reduced as increase in nanoparticles and micropores in ITZ were reduced with the addition of nano-Al$_2$O$_3$. Chen et al. [10] presented that nano-TiO$_2$ was able to provide sufficient nucleation sites to promote hydration and fill the porous structure, despite an absence of the pozzolanic effect. Gao et al. [11] revealed that carbon nanotubes not only generated higher hydration reaction, but also created net-form distributions in the hardened concrete to reinforce the ITZ.

Zhu et al. [12] reported that carbon nanofiber significantly improved the elastic modulus of the ITZ, thus enhancing the Young's modulus of concrete.

When compared to the above nanomaterials, NS has the greatest advantage in the potential pozzolanic reaction with cement hydration products [3,13]. Quercia et al. [14] demonstrated that NS efficiently used in self-compacting concrete can improve mechanical properties and durability. Zhang et al. [15] reported that when 3% of NS was added in concrete, the compressive and flexural strength increased by 15.5% and 27.3%, respectively. Aleem et al. [16] concluded that 5% NS could lead to the formation of homogeneous, dense, and compact microstructure and improve the 28 d compressive strength by 60%. Additionally, Gong et al. [17] found that when the NS content was 4%, the 28 days compressive strength of foam concrete was increased by 12%, and when NS content was 15%, the compressive strength was only amplified by 9.1%. Wu et al. [18] concluded that excess addition of nano-SiO_2 and nano-$CaCO_3$ could result in a reduction in mechanical properties, due to difficulties in ensuring sufficient dispersion. Above all, it is very effective to use NS to improve the performance of cement-based materials, but it should be mentioned that the degree of dispersion is also very important [19].

Numerical simulation has been widely used in concrete. When conducting numerical simulation of concrete, it is usually considered to contain three phases: cement paste, ITZ, and aggregates [20]. Several kinds of numerical models have been established to study the properties of cementitious materials, such as elastic modulus [12], compressive strength [21], tensile strength [22], plastic-damage response [23], and chloride ion transport [24]. Properties of each phase will affect the acquisition of accurate conclusions. Especially the ITZ will have a deep impact on the performance of concrete [25,26]. Furthermore, Wang et al. [27] revealed that NS enhances the mechanical properties of ITZ and mortar, through the numerical simulation, and the concrete added with 2% NS has the better resistance to external force than that without NS.

As for the experimental method related to the cement paste with nanomaterials, compressive, flexural, and tensile properties are often used to verify the macroscopic mechanical properties [28–30]. Concrete water absorption [11,31], chloride penetration, and freezing–thawing resistance [15,32] are used to test the durability. Moreover, nanoindentation is mainly applied to test the micromechanical properties [12,33,34].

In this study, the DIC technology was applied to carry out the micromechanical properties of the cement paste with nanomaterials. Different proportions were added to study the effect of NS on the cement paste. By comparing the effects of different NS content, the optimal dosage can be obtained. Finally, the microstructure of the cement paste was also observed using SEM.

2. Experimental

2.1. Materials

Ordinary Portland cement (OPC) P·I 42.5 (Fushun Cement Co., Ltd. (Fushun, China)) and NS (Shanghai Yuanjiang Chemical Co., Ltd. (Shanghai, China)) were used as raw materials, their chemical compositions, and their physical properties are listed in Tables 1–3, respectively. Granite slabs were used as aggregate, and their surface maintained the natural roughness after cutting. Aggregate surface roughness has a significant effect on the ITZ [35]. Therefore, the granite used for the specimen was kept the same, so that the influence caused by roughness could be minimized. Surface roughness is determined by the parameter R_a, which represents the arithmetical mean deviation of surface profile [36]. The R_a value of granite was measured using a precision roughness meter, as shown in Figure 1. The test result was 2.931 ± 0.359 μm (mean ± SD).

Table 1. Chemical composition of OPC (%).

Oxides						Other Components	Ignition Loss
SiO_2	Al_2O_3	Fe_2O_3	CaO	MgO	SO_3	1.436	2.14
20.96	4.13	3.03	62.32	2.90	2.38		

Table 2. Physical properties of OPC.

Specific Surface Area (m^2/kg)	Density (kg/m^3)	Standard Consistency (%)	Setting Time (min)		Flexural Strength (MPa)	Compressive Strength (MPa)
			Initial	Final	3 days	3 days
358	3110	26.00	107	167	5.9	28.0

Table 3. Physical properties of NS.

Diameter (nm)	Surface-to-Volume Ratio (m^2/g)	Density (g/cm^3)	Purity (%)
50	210	2.41	99.9

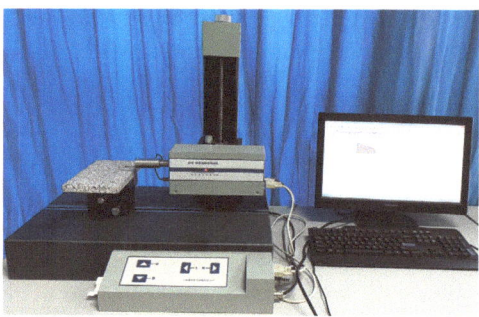

Figure 1. Precision roughness meter.

The mixed proportion of the specimens is shown in Table 4. For all groups, the water/binder (w/b) ratio was maintained at 0.5, and the superplasticizer (Sobute New Materials Co., Ltd. (Nanjing, China)) dosage was maintained at 0.5% by weight of cement. The proportion of nanomaterials was considered as the only variable among each group.

Table 4. Mix proportion of the specimens (kg/m^3).

No.	w/b	Mix Proportion of the Specimens			
		Water (g)	Cement (g)	NS (g)	Superplasticizer (% by Weight of Cement)
NS0	0.5	225	450	-	0.5
NS0.5	0.5	225	447.75	2.25	0.5
NS1	0.5	225	445.5	4.5	0.5
NS1.5	0.5	225	443.25	6.75	0.5
NS2	0.5	225	441	9	0.5

It should be noted that, in the DIC test, cement paste was used instead of concrete as the base material, which can reduce the impact of other factors. Nevertheless, in the SEM test, to better observe the microstructure of the ITZ, the mortar sample was prepared in accordance with GB/T 17671-2021 [37]. Standard sand was used for the sample preparation, and the mix proportion of other components remained unchanged.

2.2. Specimen Preparation

The specimen preparation in this paper was consistent with that of reference [38]. Broadly speaking, the cement paste with different mix proportions was poured onto the granite slab. After curing for 28 d, samples were cut into small specimens for the DIC test. The main difference lay in the treatment of NS. First of all, NS and superplasticizer were added to water, and then sonicated for 15 min. The addition of superplasticizer not only contributed to the dispersion of flocculation structure, but also mainly reduced the adsorption of water molecules of cement particle surface [39]. Before casting, they were added to the cement for further mixing with an electric mixer.

After standard curing for 28 d, specimens were cut to a size suitable for the axial compression test. Ultimately, six specimens per group were required for the loading test, half of them were used for the cement paste properties test and half for the ITZ test. The method of making artificial speckles was the same as in previous research.

2.3. Experimental Procedure

2.3.1. Flowability

The mini-slump cone test, according to GB/T8077-2012 [40], was carried out to assess the flowability of the cement paste. The prepared cement paste was loaded into the standard truncated cone mold, then the mold was vertically lifted, and the maximum diameters of two orthogonal directions of the flow spread were measured 30 s later. The average value was taken as the flowability of the cement paste.

2.3.2. Axial Compression Test

Based on the reference [38], the axial compression test was carried out by a universal testing machine (20 kN), and the loading rate was 0.05 mm/min. According to the bearing capacity test, the final load was also 10,000 N. During the compression test, the CCD camera was applied to collect digital images of the corresponding observation area, as illustrated in Figure 2.

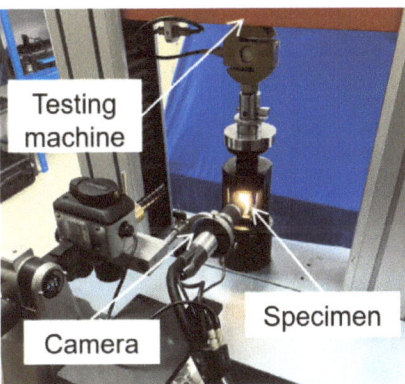

Figure 2. The experimental setup.

For specimens used to test the cement properties, the observation area was placed in the central area of the cement paste, as shown in Figure 3. Similarly, for the ITZ, the observation area was at the center of the interface. Each specimen was loaded only once.

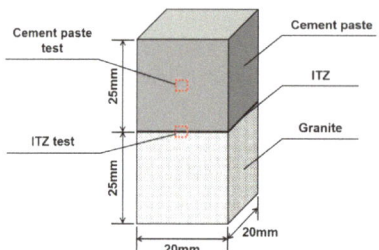

Figure 3. Geometry and observation areas of the specimen.

2.3.3. Basics of DIC

By calculating the correlation of digital images observed before and after deformation, the corresponding displacement and strain fields are obtained [41]. Specifically, a subset was determined in the reference digital image before deformation. Then, the mathematical function, called the cross-correlation coefficient, was used to find the most similar image subset on the deformed image. In this paper, the zero-mean normalized cross-correlation function can be expressed as follows:

$$C = \frac{\sum\sum \left[f(x,y) - \overline{f}\right] \cdot \left[g(x',y') - \overline{g}\right]}{\sqrt{\sum\sum \left[f(x,y) - \overline{f}\right]^2 \cdot \sum\sum \left[g(x',y') - \overline{g}\right]^2}} \quad (1)$$

where $f(x, y)$ is the gray value at coordinate (x, y) for the reference image; $g(x', y')$ has an equivalent definition but for the deformed image; and $\overline{f}$ and $\overline{g}$ are the average gray values of the reference and deformed images, respectively.

After the target subset was matched, the displacement function of the subset is most commonly expressed as follows:

$$\begin{cases} x' = x + u + \frac{\partial u}{\partial x}\Delta x + \frac{\partial u}{\partial y}\Delta y \\ y' = y + v + \frac{\partial v}{\partial x}\Delta x + \frac{\partial v}{\partial y}\Delta y \end{cases} \quad (2)$$

where u and v are the x- and y-directional displacement components of the reference sub-region center towards the deformed position. Also, Δx and Δy are the distances from the point (x', y') to the center of the image sub-region (x, y). Meanwhile, $\partial u/\partial x$, $\partial u/\partial y$, $\partial v/\partial x$, and $\partial v/\partial y$ are the first derivatives of the displacements.

It is noteworthy that by smoothing the computed displacement field first and then calculating the strain, the accuracy of strain estimation will be improved [42]. The local least-squares fitting technology for strain estimation was used to compute the strain. With the application of the Vic-2D v6 software (Correlated Solutions, Inc., Irmo, SC, USA), the displacement and strain field can be calculated.

2.3.4. SEM

Field emission scanning electron microscopy (Zeiss Gemini SEM 300) (Carl Zeiss Ltd., Cambridge, UK) was used to characterize the microstructure of hardened cement paste, which would further confirm the conclusions of mechanical experiments.

3. Results and Discussion

3.1. Influence of NS Content on the Flowability

Water is the fundamental factor affecting the flowability of the cement paste. Water first filled the packing voids of cement particles and then formed the water coating film. As the amount of water increased, the excess water would increase the paste flowability [39]. Figure 4 presented the change of flowability of the cement paste with NS. With the increase in NS content, the flowability decreased continuously. Specifically, when the content is 0.5% and 1.5%, respectively, the flowability of cement paste was significantly reduced compared with the previous state. Comparing the flowability of the NS0 and NS2, the value decreased from 312.2 mm to 112.8 mm. As mentioned in references [43–45], the nanoparticles formed loose flocculated and coated layers around cement particles, which absorbed some free water that originally contributed to flowability. It is worth noting that the incorporation of superplasticizer may affect the flowability of the paste. A classical explanation can be that the added superplasticizer was adsorbed on the surface of cement and nanomaterial particles, yielding a negative surface charge. The electrostatic repulsion hindered aggregation and decreased the flocculation effect, and the water inside the agglomeration was released [39,46].

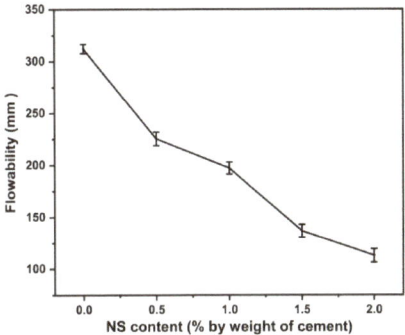

Figure 4. Influence of NS content on flowability.

3.2. DIC Analysis

The vertical displacement (v) and strain (ε_{yy}) values in the observation area were obtained with DIC technology. Typical vertical displacement (v) and strain (ε_{yy}) field of NS0.5 were shown in Figure 5. In the vertical displacement field, because of the difference in mechanical properties of each phase, their displacement changes were different as well. Same as in reference [38], the displacement data were extracted and fitted by the piecewise function. As shown in Figure 6, the jumping area between two horizontal lines represents the ITZ. The y-direction between the two inflection points is the thickness of the ITZ. In the strain field, strain values under different compressive loads were extracted from the DIC analysis results. By fitting the curve during the elastic stage, the corresponding elastic modulus can be calculated. This method was also applicable to cement paste and the ITZ.

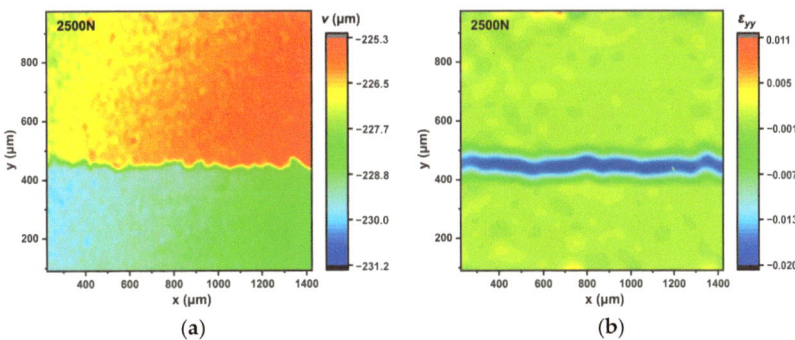

Figure 5. DIC results for NS0.5 under the load of 2500 N. (**a**) Displacement field (v) in the y-direction. (**b**) Strain field (ε_{yy}) in the y-direction.

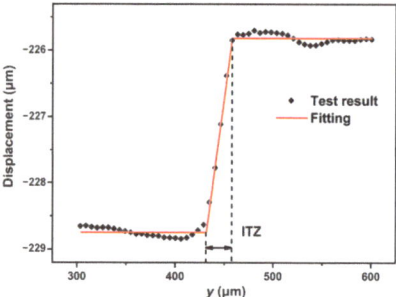

Figure 6. Piecewise fitting of displacement curve.

It should be pointed out that the DIC experimental results showed a high dispersion. Therefore, it was necessary to use statistical methods for further analysis, and make sure that the sample size of each group was not less than 30. The high dispersion may be caused by the following reasons: (1) cement paste and granite are heterogeneous materials and (2) the experiment was carried out on the mesoscopic level; hence, small changes could also lead to a high dispersion.

3.2.1. Elastic Modulus of the Cement Paste with NS

As can be seen in Figure 7 and Table 5, the elastic modulus of the cement paste increased with the gradual increase in NS content. According to the Shapiro–Wilk test, the data sample did not follow the normal distribution, so the median was used as the result of each group. Groups NS0.5, NS1, NS1.5, and NS2 showed an increase of 26.42%, 73.50%, 65.05%, and 63.86%, respectively, when compared to NS0. The maximum value among them was 20.95 GPa from Group NS1. The numerical range of the elastic modulus obtained using the DIC experiment was consistent with references [47,48]. Statistical analysis implied that there were significant differences between Group NS0 and other groups ($p < 0.05$), except for NS0.5 ($p = 0.916 > 0.05$). This result confirmed that after the addition of NS, the enhanced effect was gradually produced, and the best effect was achieved at 1%. It was attributed to the filling effect and the high pozzolanic reactivity of NS to improve the strength [49]. When the NS content exceeded 1%, the elastic modulus of the cement paste slightly decreased. This may be related to the agglomeration that was difficult to disperse, even with the addition of superplasticizer and ultrasonic treatment. It should be pointed out that superplasticizer will have an impact on the mixing of cement slurry, the dispersion of nanomaterials, and even the mechanical properties of cement paste. Therefore, we took the content of NS as the only variable and kept the content of superplasticizer as a constant, so as to reduce the additional effect of superplasticizer.

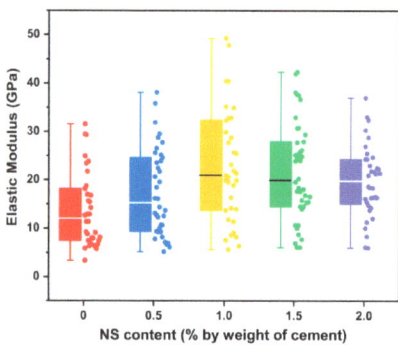

Figure 7. Influence of NS content on the elastic modulus of the cement paste.

Table 5. Statistical results for the elastic modulus of the cement paste.

NS Content (% by Weight of Cement)	0	0.5	1.0	1.5	2.0
Mean	13.82	17.45	22.79	21.80	19.76
Standard deviation	7.70	9.01	11.70	10.11	7.40
Minimum	3.38	5.18	5.67	6.09	5.99
Median	12.08	15.27	20.95	19.93	19.79
Maximum	31.61	38.13	49.33	42.32	36.99
Interquartile range (Q3–Q1)	10.82	15.33	18.64	13.82	9.34
Range (Maximum–Minimum)	28.23	32.95	43.66	36.23	30.99

3.2.2. Elastic Modulus of the ITZ with NS

Figure 8 and Table 6 summarized the statistical analysis results of NS content on the elastic modulus of the ITZ. As the NS content increased, the elastic modulus of the ITZ increased first, and then showed a decreasing trend when the NS content exceeded 1%. It was obvious that NS has a significant effect on the mechanical properties of the ITZ, since the p-values between Group NS0 and the other four groups were less than 0.05. Moreover, there was a significant difference between Group NS1 and NS2 ($p < 0.05$), which implied that excess nanomaterials would have significant adverse effects. To be specific, Group NS0.5, NS1, NS1.5, and NS2 increased by 60.72%, 90.50%, 78.95%, and 51.22%, respectively, when compared to NS0. The maximum elastic modulus was 3.20 GPa, which was derived from group NS1. The reasons for this phenomenon were as follows: (1) The smaller particle size of the NS filled the pores between the cement particles. (2) NS can consume excess CH, promote the hydration of cement, and then improve the mechanical properties of the paste [26]. (3) When the NS content is larger than 1%, the agglomeration of nanomaterials could become significant and have a negative effect. Thus, for the ITZ, the optimal addition range is about 1%.

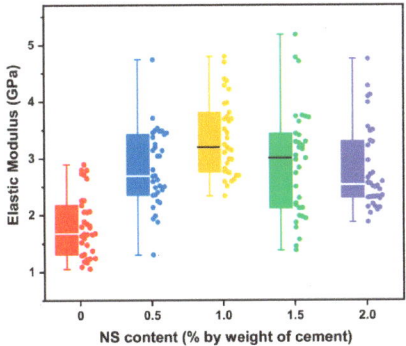

Figure 8. Influence of NS content on the elastic modulus of the ITZ.

Table 6. Statistical results for the elastic modulus of the ITZ.

NS Content (% by Weight of Cement)	0	0.5	1.0	1.5	2.0
Mean	1.82	2.83	3.36	2.90	2.81
Standard deviation	0.55	0.67	0.67	0.95	0.73
Minimum	1.06	1.30	2.34	1.38	1.88
Median	1.68	2.70	3.20	3.01	2.54
Maximum	2.90	4.74	4.80	5.18	4.75
Interquartile range (Q3–Q1)	0.86	1.07	1.05	1.32	1.00
Range (Maximum–Minimum)	1.84	3.44	2.46	3.80	2.88

Compared with the previous section, the increased amplitude of the elastic modulus of the ITZ was larger than that of the cement paste, except for Group NS2. This is because there are many more pores and defects in the ITZ, so NS can provide more remarkable improvements. Similar trends were also observed by Deependra et al. [2].

3.2.3. Thickness of the ITZ with NS

The influence of NS on the thickness of ITZ is presented in Figure 9 and Table 7. Statistical analysis showed that differences between each group were not significant. As explained in the previous sections, NS would certainly promote the mechanical properties of the ITZ. Even so, the elastic modulus of the ITZ remained smaller than that of the cement paste. Therefore, the boundary of each phase can still be distinguished by the change of displacement. Xu et al. [50] also confirmed the conclusion that the influence of NS on the interface width was not significant, and its impact on nanomechanical properties of the interface was marked.

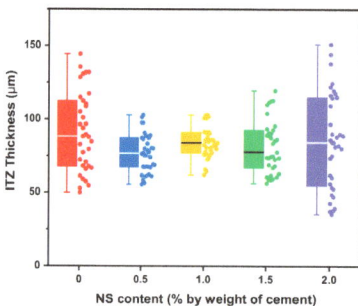

Figure 9. Influence of NS content on the thickness of the ITZ.

Table 7. Statistical results for the thickness of the ITZ.

NS Content (% by Weight of Cement)	0	0.5	1.0	1.5	2.0
Mean	91.38	76.91	84.25	81.80	84.71
Standard deviation	27.07	13.13	10.44	18.09	32.65
Minimum	49.98	55.55	62.11	56.39	35.45
Median	88.36	76.83	84.21	77.84	84.57
Maximum	144.34	102.98	103.14	119.63	151.16
Interquartile range (Q3–Q1)	44.63	19.94	13.91	25.92	60.93
Range (Maximum–Minimum)	94.36	47.43	41.03	63.24	115.71

As determined with the Shapiro–Wilk test, the results of each group followed a normal distribution; thus, the mean value was considered as the ITZ thickness. And the thickness value was in the range of 76.91–91.38 µm.

3.3. SEM Analysis

SEM analysis is beneficial to study the cause of the mechanical characteristics' change of the cement paste and the ITZ. As illustrated in Figure 10a–e, when NS content changed from 0–1.0%, the microstructure of the paste became denser. It is well known that NS particles can not only be used as void fillers to improve the microstructure, but also to promote the pozzolanic reaction [2,9,51]. However, when the NS dosage was larger than 1%, some pores filled with needle-hydrates can be observed. This phenomenon can be explained by the agglomeration of excess nanomaterials, limiting the formation of uniform hydrate microstructure and leading to low strength [52]. The decrease in the elastic modulus obtained in the previous sections has a good relationship with the SEM study.

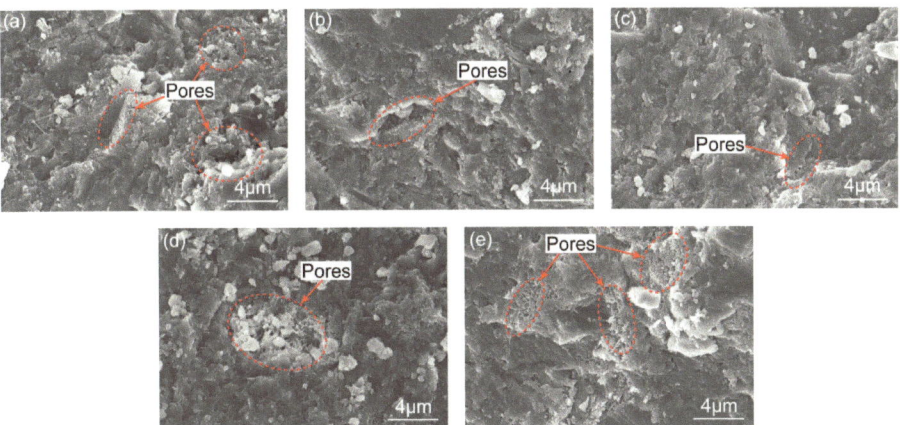

Figure 10. SEM images of the cement paste: (**a**) NS0; (**b**) NS0.5; (**c**) NS1; (**d**) NS1.5; and (**e**) NS2.

To study the microstructure of ITZ, the broken pieces of the mortar sample were selected for SEM. Since the diameter of sand particles is much larger than that of cement and nanomaterial particles, ITZ can also be generated on the surface of sand particles. As revealed in Figure 11, we mainly focused on the edge of the small pit formed after the sand particles were removed. From this perspective, both the porous structure of the ITZ and the microstructure of the interface between the cement paste and sand particle can be observed.

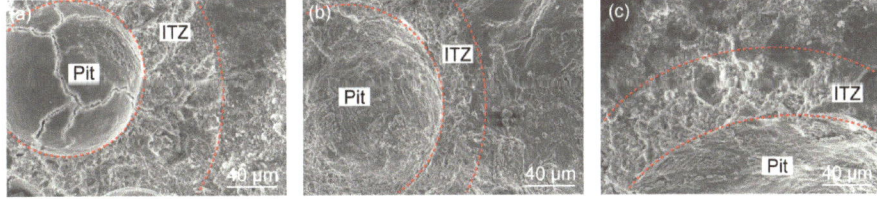

Figure 11. SEM images of the ITZ: (**a**) NS0; (**b**) NS1; and (**c**) NS2.

Compared with the cement paste, the porous structure around the sand grains was more obvious, as can be seen from Figure 11. The cement paste gradually became dense with an increase in the distance from the sand particle's surface [53,54]. The interior of the sand particle pit was dense and smooth on the whole. This is mainly due to the "wall effect", and the local water is sufficient for the hydration reaction so that the sand grains can be well wrapped by cement paste [55]. Yet there were some cracks, pores, and unhydrated cement particles, which is inevitable. Moreover, some wide cracks in the sand particle pit might be caused by the self-shrinking of the cement paste. Figure 11a–c displayed the typical ITZ microstructure at the NS content of 0, 1%, and 2%, respectively. The ITZ became denser, after the NS was added in. However, because of the complex porous structure, it is difficult to quantify the thickness of ITZ using SEM.

4. Conclusions

In this paper, the influence of NS content on the cement paste and ITZ was investigated with the DIC technology. The main conclusions can be summarized below:

1. The flowability of the cement paste decreased continuously, as the NS content increased, under the condition of superplasticizer and ultrasonic treatment. Comparing the flowability of the NS0 and NS2, the value decreased from 312.2 mm to 112.8 mm. In the content range of 0.5–1%, the flowability showed good working performance.

2. NS can effectively improve the mechanical properties of cement paste. When the NS content was 1%, the elastic modulus increased the most, about 73.50%, compared with Group NS0. The maximum value among them was 20.95 GPa from group NS1. When the content exceeded 1%, the elastic modulus slightly decreased.
3. For the ITZ, NS can greatly promote its mechanical properties. When 1.0% NS was incorporated, its elastic modulus increased by 90.50%. The maximum value is 3.20 GPa, also from NS1. However, NS content has no significant effect on the thickness of the ITZ, and the thickness value was in the range of 76.91–91.38 μm.
4. SEM confirmed that NS densified the microstructure of the cement paste and the ITZ. Compared with the cement paste, the porous structure around the sand grains was more obvious. The cement paste gradually became dense with the increase in the distance from the sand particle's surface.

To sum up, when applying NS to enhance the mechanical properties of the cement paste, 1% can be used as the optimal content. Of course, treatments conducive to the dispersion of NS are also necessary. In this paper, the results about the cement paste and ITZ can provide reference for relevant numerical simulations. Likewise, based on the DIC technology, effects of various nanomaterials on cementitious materials can be deeply studied. This work will provide the basis for the application of nanomaterials in the field of engineering.

Author Contributions: S.Z.: Conceptualization, Investigation, Validation, and Writing—original draft. R.L.: Supervision, Writing—review, and Resources. C.L.: Investigation and Writing—review and editing. J.H.: Investigation, Validation, and Resources. C.C.: Data curation and Software. J.X.: Software, Methodology, and Writing—original draft. All authors have read and agreed to the published version of the manuscript.

Funding: This study was funded by the National Natural Science Foundation of China (Nos. 51778272, 51878319, 52108190) and the Science and Technology Planning Project of Nantong (No. JC22022102). We thank Nantong University Analysis & Testing Center for the technical support.

Institutional Review Board Statement: Not applicable.

Informed Consent Statement: Not applicable.

Data Availability Statement: Not applicable.

Acknowledgments: This study was supported by the National Natural Science Foundation of China (Nos. 51778272, 51878319, 52108190) and the Science and Technology Planning Project of Nantong (No. JC22022102). We thank Nantong University Analysis & Testing Center for the technical support.

Conflicts of Interest: The authors declare that they have no known competing financial interest or personal relationships that could have appeared to influence the work reported in this paper.

References

1. Sunantha, B.; Patel, J.; Poojalakshmi, E.S.; Sudhakumar, J.; Ramaswamy, K.; Khan, R.A.; Nair, P.S.; Thomas, B. A comprehensive review on the properties of engineered cementitious composite with a self-healing material. *Mater. Today Proc.* **2023**, *in press*. [CrossRef]
2. Bhatta, D.P.; Singla, S.; Garg, R. Experimental investigation on the effect of Nano-silica on the silica fume-based cement composites. *Mater. Today Proc.* **2022**, *57*, 2338–2343. [CrossRef]
3. Du, H.J.; Du, S.H.; Liu, X.M. Durability performances of concrete with nano-silica. *Constr. Build. Mater.* **2014**, *73*, 705–712. [CrossRef]
4. Song, J.; Li, Y.; Xu, W.; Liu, H.; Lu, Y. Inexpensive and non-fluorinated superhydrophobic concrete coating for anti-icing and anti-corrosion. *J. Colloid Interface Sci.* **2019**, *541*, 86–92. [CrossRef] [PubMed]
5. Yin, B.; Xu, T.; Hou, D.; Zhao, E.; Hua, X.; Han, K.; Zhang, Y.; Zhang, J. Superhydrophobic anticorrosive coating for concrete through in-situ bionic induction and gradient mineralization. *Constr. Build. Mater.* **2020**, *257*, 119510. [CrossRef]
6. Reches, Y. Nanoparticles as concrete additives: Review and perspectives. *Constr. Build. Mater.* **2018**, *175*, 483–495. [CrossRef]
7. Bai, S.; Guan, X.C.; Li, H.; Ou, J.P. Effect of the specific surface area of nano-silica particle on the properties of cement paste. *Powder Technol.* **2021**, *392*, 680–689. [CrossRef]
8. Wang, X.; Dong, S.; Ashour, A.; Zhang, W.; Han, B. Effect and mechanisms of nanomaterials on interface between aggregates and cement mortars. *Constr. Build. Mater.* **2020**, *240*, 117942. [CrossRef]

9. Joshaghani, A.; Balapour, M.; Mashhadian, M.; Ozbakkaloglu, T. Effects of nano-TiO_2, nano-Al_2O_3, and nano-Fe_2O_3 on rheology, mechanical and durability properties of self-consolidating concrete (SCC): An experimental study. *Constr. Build. Mater.* **2020**, *245*, 118444. [CrossRef]
10. Chen, J.; Kou, S.; Poon, C. Hydration and properties of nano-TiO_2 blended cement composites. *Cem. Concr. Compos.* **2012**, *34*, 642–649. [CrossRef]
11. Gao, Y.; Jing, H.W.; Zhao, Z.L.; Shi, X.S.; Li, L. Influence of ultrasonication energy on reinforcing-roles of CNTs to strengthen ITZ and corresponding anti-permeability properties of concrete. *Constr. Build. Mater.* **2021**, *303*, 124451. [CrossRef]
12. Zhu, X.Y.; Gao, Y.; Dai, Z.W.; Corr, D.J.; Shah, S.P. Effect of interfacial transition zone on the Young's modulus of carbon nanofiber reinforced cement concrete. *Cem. Concr. Res.* **2018**, *107*, 49–63. [CrossRef]
13. Sharkawi, A.M.; Abd-Elaty, M.A.; Khalifa, O.H. Synergistic influence of micro-nano silica mixture on durability performance of cementitious materials. *Constr. Build. Mater.* **2018**, *164*, 579–588. [CrossRef]
14. Quercia, G.; Spiesz, P.; Hüsken, G.; Brouwers, H. Scc modification by use of amorphous nano-silica. *Cem. Concr. Compos.* **2014**, *45*, 69–81. [CrossRef]
15. Zhang, P.; Sha, D.; Li, Q.; Zhao, S.; Ling, Y. Effect of nano silica particles on impact resistance and durability of concrete containing coal fly ash. *Nanomaterials* **2021**, *11*, 1296. [CrossRef] [PubMed]
16. Aleem, S.A.E.; Heikal, M.; Morsi, W.M. Hydration characteristic, thermal expansion and microstructure of cement containing nano-silica. *Constr. Build. Mater.* **2014**, *59*, 151–160. [CrossRef]
17. Gong, J.; Zhu, L.; Li, J.; Shi, D. Silica Fume and Nanosilica Effects on Mechanical and Shrinkage Properties of Foam Concrete for Structural Application. *Adv. Mater. Sci. Eng.* **2020**, *2020*, 3963089. [CrossRef]
18. Wu, Z.; Shi, C.; Khayat, K.H.; Wan, S. Effects of different nanomaterials on hardening and performance of ultra-high strength concrete (UHSC). *Cem. Concr. Compos.* **2016**, *70*, 24–34. [CrossRef]
19. Gdoutos, E.E.; Konsta-Gdoutos, M.S.; Danoglidis, P.A. Portland cement mortar nanocomposites at low carbon nanotube and carbon nanofiber content: A fracture mechanics experimental study. *Cem. Concr. Compos.* **2016**, *70*, 110–118. [CrossRef]
20. Du, X.L.; Jin, L.; Ma, G.W. A meso-scale numerical method for the simulation of chloride diffusivity in concrete. *Finite Elem. Anal. Des.* **2014**, *85*, 87–100. [CrossRef]
21. Wang, J.M.; Jivkov, A.P.; Li, Q.M.; Engelberg, D.L. Experimental and numerical investigation of mortar and ITZ parameters in meso-scale models of concrete. *Theor. Appl. Fract. Mec.* **2020**, *109*, 102722. [CrossRef]
22. Maleki, M.; Rasoolan, I.; Khajehdezfuly, A.; Jivkov, A.P. On the effect of ITZ thickness in meso-scale models of concrete. *Constr. Build. Mater.* **2020**, *258*, 119639. [CrossRef]
23. SKim, M.; Al-Rub, R.K.A. Meso-scale computational modeling of the plastic-damage response of cementitious composites. *Cem. Concr. Res.* **2011**, *41*, 339–358. [CrossRef]
24. Liu, Q.F.; Easterbrook, D.; Yang, J.; Li, L.Y. A three-phase, multi-component ionic transport model for simulation of chloride penetration in concrete. *Eng. Struct.* **2015**, *86*, 122–133. [CrossRef]
25. Xiao, J.Z.; Li, W.G.; Corr, D.J.; Shah, S.P. Effects of interfacial transition zones on the stress–strain behavior of modeled recycled aggregate concrete. *Cem. Concr. Res.* **2013**, *52*, 82–99. [CrossRef]
26. Scrivener, K.L.; Crumbie, A.K.; Laugesen, P. The interfacial transition zone (ITZ) between cement paste and aggregate in concrete. *Interface Sci.* **2004**, *12*, 411–421. [CrossRef]
27. Wang, C.; Zhang, M.Y.; Wang, Q.C.; Dai, J.P.; Luo, T.; Pei, W.S.; Melnikov, A.; Zhang, Z. Research on the influencing mechanism of nano-silica on concrete performances based on multi-scale experiments and micro-scale numerical simulation. *Constr. Build. Mater.* **2022**, *318*, 125873. [CrossRef]
28. Reches, Y.; Thomson, K.; Helbing, M.; Kosson, D.S.; Sanchez, F. Agglomeration and reactivity of nanoparticles of SiO_2, TiO_2, Al_2O_3, Fe_2O_3 and clays in cement pastes and effects on compressive strength at ambient and elevated temperatures. *Constr. Build. Mater.* **2018**, *167*, 860–873. [CrossRef]
29. Li, W.G.; Long, C.; Tam, V.W.; Poon, C.S.; Duan, W.H. Effects of nano-particles on failure process and microstructural properties of recycled aggregate concrete. *Constr. Build. Mater.* **2017**, *142*, 42–50. [CrossRef]
30. Rupasinghe, M.; Mendis, P.; Ngo, T.; Nguyen, T.N.; Sofi, M. Compressive strength prediction of nano-silica incorporated cement systems based on a multiscale approach. *Mater. Des.* **2017**, *115*, 379–392. [CrossRef]
31. Grzeszczyk, S.; Jurowski, K.; Bosowska, K.; Grzymek, M. The role of nanoparticles in decreased washout of underwater concrete. *Constr. Build. Mater.* **2019**, *203*, 670–678. [CrossRef]
32. Zhang, M.H.; Li, H. Pore structure and chloride permeability of concrete containing nano-particles for pavement. *Constr. Build. Mater.* **2011**, *25*, 608–616. [CrossRef]
33. Lahayne, O.; Zelaya-Lainez, L.; Buchner, T.; Eberhardsteiner, J.; Füssl, J. Influence of nanoadditives on the Young's modulus of cement. *Mater. Today Proc.* **2022**, *62*, 2488–2494. [CrossRef]
34. Xu, J.; Wang, B.B.; Zuo, J.Q. Modification effects of nanosilica on the interfacial transition zone in concrete: A multiscale approach. *Cem. Concr. Compos.* **2017**, *81*, 1–10. [CrossRef]
35. Qudoos, A.; Rehman, A.; Kim, H.G.; Ryou, J.S. Influence of the surface roughness of crushed natural aggregates on the microhardness of the interfacial transition zone of concrete with mineral admixtures and polymer latex. *Constr. Build. Mater.* **2018**, *168*, 946–957. [CrossRef]

36. Gu, X.L.; Hong, L.; Wang, Z.L.; Lin, F. Experimental study and application of mechanical properties for the interface between cobblestone aggregate and mortar in concrete. *Constr. Build. Mater.* **2013**, *46*, 156–166. [CrossRef]
37. *GB/T 17671-2021*; Test Method of Cement Mortar Strength. National Standards of People's Republic of China: Beijing, China, 2021.
38. Zhang, S.F.; Liu, R.G.; Lu, C.H.; Gao, Y.; Xu, J.J.; Yao, L.; Chen, Y. Application of digital image correlation to study the influence of the water/cement ratio on the interfacial transition zone in cement-based materials. *Constr. Build. Mater.* **2023**, *367*, 130167. [CrossRef]
39. Zhao, M.; Zhang, X.; Zhang, Y.J. Effect of free water on the flowability of cement paste with chemical or mineral admixtures. *Constr. Build. Mater.* **2016**, *111*, 571–579. [CrossRef]
40. *GB/T 8077-2012*; Methods for Testing Uniformity of Concrete Admixture. National Standards of People's Republic of China: Beijing, China, 2012.
41. Pan, B.; Xie, H.M. Full-field strain measurement based on least-square fitting of local displacement for digital image correlation method. *Acta Opt. Sin.* **2007**, *27*, 1980–1986. (In Chinese)
42. Pan, B.; Qian, K.M.; Xie, H.M.; Asundi, A. Two-dimensional digital image correlation for in-plane displacement and strain measurement: A review. *Meas. Sci. Technol.* **2009**, *20*, 062001. [CrossRef]
43. Kong, D.Y.; Corr, D.J.; Hou, P.K.; Yang, Y.; Shah, S.P. Influence of colloidal silica sol on fresh properties of cement paste as compared to nano-silica powder with agglomerates in micron-scale. *Cem. Concr. Compos.* **2015**, *63*, 30–41. [CrossRef]
44. Kong, D.Y.; Su, Y.; Du, X.F.; Yang, Y.; Wei, S.; Shah, S.P. Influence of nano-silica agglomeration on fresh properties of cement pastes. *Constr. Build. Mater.* **2013**, *43*, 557–562. [CrossRef]
45. Pourjavadi, A.; Fakoorpoor, S.M.; Hosseini, P.; Khaloo, A. Interactions between superabsorbent polymers and cement-based composites incorporating colloidal silica nanoparticles. *Cem. Concr. Compos.* **2013**, *37*, 196–204. [CrossRef]
46. Li, Q.C.; Fan, Y.F. Rheological evaluation of nano-metakaolin cement pastes based on the water film thickness. *Constr. Build. Mater.* **2022**, *324*, 126517. [CrossRef]
47. Shen, P.; Lu, L.; He, Y.; Wang, F.; Hu, S. The effect of curing regimes on the mechanical properties, nano-mechanical properties and microstructure of ultrahigh performance concrete. *Cem. Concr. Res.* **2019**, *118*, 1–13. [CrossRef]
48. Foley, E.M.; Kim, J.J.; Reda Taha, M.M. Synthesis and nano-mechanical characterization of calcium-silicate-hydrate (C-S-H) made with 1.5 CaO/SiO_2 mixture. *Cem. Concr. Res.* **2012**, *42*, 1225–1232. [CrossRef]
49. Wang, X.L.; Gong, C.C.; Lei, J.G.; Dai, J.; Lu, L.C.; Cheng, X. Effect of silica fume and nano-silica on hydration behavior and mechanism of high sulfate resistance Portland cement. *Constr. Build. Mater.* **2021**, *279*, 122481. [CrossRef]
50. Xu, J.; Corr, D.J.; Shah, S.P. Nanomechanical investigation of the effects of $nanoSiO_2$ on C-S-H gel/cement grain interfaces. *Cem. Concr. Compos.* **2015**, *61*, 7–17. [CrossRef]
51. Gao, Y.; Yu, Z.; Chen, W.; Yin, Q.; Wu, J.; Wang, W. Recognition of rock materials after high-temperature deterioration based on SEM images via deep learning. *J. Mater. Res. Technol.* **2023**, *25*, 273–284. [CrossRef]
52. Lothenbach, B.; Scrivener, K.; Hooton, R.D. Supplementary cementitious materials. *Cem. Concr. Res.* **2011**, *41*, 1244–1256. [CrossRef]
53. Aghaee, K.; Khayat, K.H. Effect of internal curing and shrinkage-mitigating materials on microstructural characteristics of fiber-reinforced mortar. *Constr. Build. Mater.* **2023**, *386*, 131527. [CrossRef]
54. Nandhini, K.; Ponmalar, V. Investigation on nano-silica blended cementitious systems on the workability and durability performance of self-compacting concrete. *Mater. Express* **2020**, *10*, 10–20. [CrossRef]
55. Sun, Z.H.; Garboczi, E.J.; Shah, S.P. Modeling the elastic properties of concrete composites: Experiment, differential effective medium theory and numerical simulation. *Cem. Concr. Compos.* **2007**, *29*, 22–38. [CrossRef]

Disclaimer/Publisher's Note: The statements, opinions and data contained in all publications are solely those of the individual author(s) and contributor(s) and not of MDPI and/or the editor(s). MDPI and/or the editor(s) disclaim responsibility for any injury to people or property resulting from any ideas, methods, instructions or products referred to in the content.

9. Joshaghani, A.; Balapour, M.; Mashhadian, M.; Ozbakkaloglu, T. Effects of nano-TiO_2, nano-Al_2O_3, and nano-Fe_2O_3 on rheology, mechanical and durability properties of self-consolidating concrete (SCC): An experimental study. *Constr. Build. Mater.* **2020**, *245*, 118444. [CrossRef]
10. Chen, J.; Kou, S.; Poon, C. Hydration and properties of nano-TiO_2 blended cement composites. *Cem. Concr. Compos.* **2012**, *34*, 642–649. [CrossRef]
11. Gao, Y.; Jing, H.W.; Zhao, Z.L.; Shi, X.S.; Li, L. Influence of ultrasonication energy on reinforcing-roles of CNTs to strengthen ITZ and corresponding anti-permeability properties of concrete. *Constr. Build. Mater.* **2021**, *303*, 124451. [CrossRef]
12. Zhu, X.Y.; Gao, Y.; Dai, Z.W.; Corr, D.J.; Shah, S.P. Effect of interfacial transition zone on the Young's modulus of carbon nanofiber reinforced cement concrete. *Cem. Concr. Res.* **2018**, *107*, 49–63. [CrossRef]
13. Sharkawi, A.M.; Abd-Elaty, M.A.; Khalifa, O.H. Synergistic influence of micro-nano silica mixture on durability performance of cementitious materials. *Constr. Build. Mater.* **2018**, *164*, 579–588. [CrossRef]
14. Quercia, G.; Spiesz, P.; Hüsken, G.; Brouwers, H. Scc modification by use of amorphous nano-silica. *Cem. Concr. Compos.* **2014**, *45*, 69–81. [CrossRef]
15. Zhang, P.; Sha, D.; Li, Q.; Zhao, S.; Ling, Y. Effect of nano silica particles on impact resistance and durability of concrete containing coal fly ash. *Nanomaterials* **2021**, *11*, 1296. [CrossRef] [PubMed]
16. Aleem, S.A.E.; Heikal, M.; Morsi, W.M. Hydration characteristic, thermal expansion and microstructure of cement containing nano-silica. *Constr. Build. Mater.* **2014**, *59*, 151–160. [CrossRef]
17. Gong, J.; Zhu, L.; Li, J.; Shi, D. Silica Fume and Nanosilica Effects on Mechanical and Shrinkage Properties of Foam Concrete for Structural Application. *Adv. Mater. Sci. Eng.* **2020**, *2020*, 3963089. [CrossRef]
18. Wu, Z.; Shi, C.; Khayat, K.H.; Wan, S. Effects of different nanomaterials on hardening and performance of ultra-high strength concrete (UHSC). *Cem. Concr. Compos.* **2016**, *70*, 24–34. [CrossRef]
19. Gdoutos, E.E.; Konsta-Gdoutos, M.S.; Danoglidis, P.A. Portland cement mortar nanocomposites at low carbon nanotube and carbon nanofiber content: A fracture mechanics experimental study. *Cem. Concr. Compos.* **2016**, *70*, 110–118. [CrossRef]
20. Du, X.L.; Jin, L.; Ma, G.W. A meso-scale numerical method for the simulation of chloride diffusivity in concrete. *Finite Elem. Anal. Des.* **2014**, *85*, 87–100. [CrossRef]
21. Wang, J.M.; Jivkov, A.P.; Li, Q.M.; Engelberg, D.L. Experimental and numerical investigation of mortar and ITZ parameters in meso-scale models of concrete. *Theor. Appl. Fract. Mec.* **2020**, *109*, 102722. [CrossRef]
22. Maleki, M.; Rasoolan, I.; Khajehdezfuly, A.; Jivkov, A.P. On the effect of ITZ thickness in meso-scale models of concrete. *Constr. Build. Mater.* **2020**, *258*, 119639. [CrossRef]
23. SKim, M.; Al-Rub, R.K.A. Meso-scale computational modeling of the plastic-damage response of cementitious composites. *Cem. Concr. Res.* **2011**, *41*, 339–358. [CrossRef]
24. Liu, Q.F.; Easterbrook, D.; Yang, J.; Li, L.Y. A three-phase, multi-component ionic transport model for simulation of chloride penetration in concrete. *Eng. Struct.* **2015**, *86*, 122–133. [CrossRef]
25. Xiao, J.Z.; Li, W.G.; Corr, D.J.; Shah, S.P. Effects of interfacial transition zones on the stress–strain behavior of modeled recycled aggregate concrete. *Cem. Concr. Res.* **2013**, *52*, 82–99. [CrossRef]
26. Scrivener, K.L.; Crumbie, A.K.; Laugesen, P. The interfacial transition zone (ITZ) between cement paste and aggregate in concrete. *Interface Sci.* **2004**, *12*, 411–421. [CrossRef]
27. Wang, C.; Zhang, M.Y.; Wang, Q.C.; Dai, J.P.; Luo, T.; Pei, W.S.; Melnikov, A.; Zhang, Z. Research on the influencing mechanism of nano-silica on concrete performances based on multi-scale experiments and micro-scale numerical simulation. *Constr. Build. Mater.* **2022**, *318*, 125873. [CrossRef]
28. Reches, Y.; Thomson, K.; Helbing, M.; Kosson, D.S.; Sanchez, F. Agglomeration and reactivity of nanoparticles of SiO_2, TiO_2, Al_2O_3, Fe_2O_3 and clays in cement pastes and effects on compressive strength at ambient and elevated temperatures. *Constr. Build. Mater.* **2018**, *167*, 860–873. [CrossRef]
29. Li, W.G.; Long, C.; Tam, V.W.; Poon, C.S.; Duan, W.H. Effects of nano-particles on failure process and microstructural properties of recycled aggregate concrete. *Constr. Build. Mater.* **2017**, *142*, 42–50. [CrossRef]
30. Rupasinghe, M.; Mendis, P.; Ngo, T.; Nguyen, T.N.; Sofi, M. Compressive strength prediction of nano-silica incorporated cement systems based on a multiscale approach. *Mater. Des.* **2017**, *115*, 379–392. [CrossRef]
31. Grzeszczyk, S.; Jurowski, K.; Bosowska, K.; Grzymek, M. The role of nanoparticles in decreased washout of underwater concrete. *Constr. Build. Mater.* **2019**, *203*, 670–678. [CrossRef]
32. Zhang, M.H.; Li, H. Pore structure and chloride permeability of concrete containing nano-particles for pavement. *Constr. Build. Mater.* **2011**, *25*, 608–616. [CrossRef]
33. Lahayne, O.; Zelaya-Lainez, L.; Buchner, T.; Eberhardsteiner, J.; Füssl, J. Influence of nanoadditives on the Young's modulus of cement. *Mater. Today Proc.* **2022**, *62*, 2488–2494. [CrossRef]
34. Xu, J.; Wang, B.B.; Zuo, J.Q. Modification effects of nanosilica on the interfacial transition zone in concrete: A multiscale approach. *Cem. Concr. Compos.* **2017**, *81*, 1–10. [CrossRef]
35. Qudoos, A.; Rehman, A.; Kim, H.G.; Ryou, J.S. Influence of the surface roughness of crushed natural aggregates on the microhardness of the interfacial transition zone of concrete with mineral admixtures and polymer latex. *Constr. Build. Mater.* **2018**, *168*, 946–957. [CrossRef]

36. Gu, X.L.; Hong, L.; Wang, Z.L.; Lin, F. Experimental study and application of mechanical properties for the interface between cobblestone aggregate and mortar in concrete. *Constr. Build. Mater.* **2013**, *46*, 156–166. [CrossRef]
37. *GB/T 17671-2021*; Test Method of Cement Mortar Strength. National Standards of People's Republic of China: Beijing, China, 2021.
38. Zhang, S.F.; Liu, R.G.; Lu, C.H.; Gao, Y.; Xu, J.J.; Yao, L.; Chen, Y. Application of digital image correlation to study the influence of the water/cement ratio on the interfacial transition zone in cement-based materials. *Constr. Build. Mater.* **2023**, *367*, 130167. [CrossRef]
39. Zhao, M.; Zhang, X.; Zhang, Y.J. Effect of free water on the flowability of cement paste with chemical or mineral admixtures. *Constr. Build. Mater.* **2016**, *111*, 571–579. [CrossRef]
40. *GB/T 8077-2012*; Methods for Testing Uniformity of Concrete Admixture. National Standards of People's Republic of China: Beijing, China, 2012.
41. Pan, B.; Xie, H.M. Full-field strain measurement based on least-square fitting of local displacement for digital image correlation method. *Acta Opt. Sin.* **2007**, *27*, 1980–1986. (In Chinese)
42. Pan, B.; Qian, K.M.; Xie, H.M.; Asundi, A. Two-dimensional digital image correlation for in-plane displacement and strain measurement: A review. *Meas. Sci. Technol.* **2009**, *20*, 062001. [CrossRef]
43. Kong, D.Y.; Corr, D.J.; Hou, P.K.; Yang, Y.; Shah, S.P. Influence of colloidal silica sol on fresh properties of cement paste as compared to nano-silica powder with agglomerates in micron-scale. *Cem. Concr. Compos.* **2015**, *63*, 30–41. [CrossRef]
44. Kong, D.Y.; Su, Y.; Du, X.F.; Yang, Y.; Wei, S.; Shah, S.P. Influence of nano-silica agglomeration on fresh properties of cement pastes. *Constr. Build. Mater.* **2013**, *43*, 557–562. [CrossRef]
45. Pourjavadi, A.; Fakoorpoor, S.M.; Hosseini, P.; Khaloo, A. Interactions between superabsorbent polymers and cement-based composites incorporating colloidal silica nanoparticles. *Cem. Concr. Compos.* **2013**, *37*, 196–204. [CrossRef]
46. Li, Q.C.; Fan, Y.F. Rheological evaluation of nano-metakaolin cement pastes based on the water film thickness. *Constr. Build. Mater.* **2022**, *324*, 126517. [CrossRef]
47. Shen, P.; Lu, L.; He, Y.; Wang, F.; Hu, S. The effect of curing regimes on the mechanical properties, nano-mechanical properties and microstructure of ultrahigh performance concrete. *Cem. Concr. Res.* **2019**, *118*, 1–13. [CrossRef]
48. Foley, E.M.; Kim, J.J.; Reda Taha, M.M. Synthesis and nano-mechanical characterization of calcium-silicate-hydrate (C-S-H) made with 1.5 CaO/SiO_2 mixture. *Cem. Concr. Res.* **2012**, *42*, 1225–1232. [CrossRef]
49. Wang, X.L.; Gong, C.C.; Lei, J.G.; Dai, J.; Lu, L.C.; Cheng, X. Effect of silica fume and nano-silica on hydration behavior and mechanism of high sulfate resistance Portland cement. *Constr. Build. Mater.* **2021**, *279*, 122481. [CrossRef]
50. Xu, J.; Corr, D.J.; Shah, S.P. Nanomechanical investigation of the effects of $nanoSiO_2$ on C-S-H gel/cement grain interfaces. *Cem. Concr. Compos.* **2015**, *61*, 7–17. [CrossRef]
51. Gao, Y.; Yu, Z.; Chen, W.; Yin, Q.; Wu, J.; Wang, W. Recognition of rock materials after high-temperature deterioration based on SEM images via deep learning. *J. Mater. Res. Technol.* **2023**, *25*, 273–284. [CrossRef]
52. Lothenbach, B.; Scrivener, K.; Hooton, R.D. Supplementary cementitious materials. *Cem. Concr. Res.* **2011**, *41*, 1244–1256. [CrossRef]
53. Aghaee, K.; Khayat, K.H. Effect of internal curing and shrinkage-mitigating materials on microstructural characteristics of fiber-reinforced mortar. *Constr. Build. Mater.* **2023**, *386*, 131527. [CrossRef]
54. Nandhini, K.; Ponmalar, V. Investigation on nano-silica blended cementitious systems on the workability and durability performance of self-compacting concrete. *Mater. Express* **2020**, *10*, 10–20. [CrossRef]
55. Sun, Z.H.; Garboczi, E.J.; Shah, S.P. Modeling the elastic properties of concrete composites: Experiment, differential effective medium theory and numerical simulation. *Cem. Concr. Compos.* **2007**, *29*, 22–38. [CrossRef]

Disclaimer/Publisher's Note: The statements, opinions and data contained in all publications are solely those of the individual author(s) and contributor(s) and not of MDPI and/or the editor(s). MDPI and/or the editor(s) disclaim responsibility for any injury to people or property resulting from any ideas, methods, instructions or products referred to in the content.

Article

Efficiency and Mechanism of Surface Reinforcement for Recycled Coarse Aggregates via Magnesium Phosphate Cement

Siyao Wang [1], Jingtao Hu [1], Zhiyuan Sun [1], Yuan Gao [1], Xiao Yan [2,*] and Xiang Xue [3,*]

[1] School of Transportation and Civil Engineering, Nantong University, Nantong 226019, China; wangsiyao@ntu.edu.cn (S.W.); jingtaohu2003@163.com (J.H.); 15937747640@163.com (Z.S.); y.gao@ntu.edu.cn (Y.G.)

[2] Department of Geotechnical Engineering, College of Civil Engineering, Tongji University, Shanghai 200092, China

[3] School of Civil Engineering, Chongqing University, Chongqing 400044, China

* Correspondence: xiao_yan@tongji.edu.cn (X.Y.); wdmzjxx@live.com (X.X.)

Abstract: Recycled aggregate concrete (RAC) exhibits inferior mechanical and durability properties owing to the deterioration of the recycled coarse aggregate (RCA) surface quality. To improve the surface properties of RCA, the reinforcement efficiency of RAC, and the maneuverability of the surface treatment method, this study used magnesium phosphate cement (MPC), a clinker-free low-carbon cement with excellent bonding properties, to precoat RCA under three-day pre-conditioning. Moreover, variable amounts of fly ash (FA) or granulated blast furnace slag (GBFS) were utilized to partly substitute MPC to enhance the compressive strength and chloride ion penetration resistance. Subsequently, FA–MPC and GBFS–MPC hybrid slurries with the best comprehensive performance were selected to coat the RCA for optimal reinforcement. The crushing value and water absorption of RCA, as well as the mechanical strengths and durability of RAC, were investigated, and microstructures around interfaces were studied via BSE-EDS and microhardness analysis to reveal the strengthening mechanism. The results indicated that the comprehensive property of strengthening paste was enhanced significantly through substituting MPC with 10% FA or GBFS. Surface coating resulted in a maximum reduction of 8.15% in the crushing value, while the water absorption barely changed. In addition, modified RAC outperformed untreated RAC regarding compressive strength, splitting tensile strength, and chloride ion penetration resistance with maximum optimization efficiencies of 31.58%, 49.75%, and 43.11%, respectively. It was also evidenced that the improved MPC paste properties enhanced the performance of modified RAC. Microanalysis revealed that MPC pastes exhibited an excellent bond with RCA or new mortar, and the newly formed interfacial transition zone between MPC and the fresh mortar exhibited a dense microstructure and outstanding micro-mechanical properties supported with an increase in the average microhardness value of 30.2–33.4%. Therefore, MPC pastes incorporating an appropriate mineral admixture have enormous potential to be utilized as effective RCA surface treatment materials and improve the operability of RCA application in practice.

Keywords: recycled aggregate concrete; magnesium phosphate cement; mechanical properties; durability; microstructure

1. Introduction

Making recycled aggregate concrete (RAC) [1–3] is a very efficient approach to using construction and demolition waste [4,5] for resource-saving and environmental protection disposal, in which the natural coarse aggregate (NCA) is partially or wholly replaced with recycled coarse aggregate (RCA) [6–8]. Researchers have shown that RAC's mechanical properties and durability decreased with increased replacement by RCA [9–11], which hinders RAC's sustainable development and application [12–14]. The primary reason for this phenomenon is the high porosity and water absorption of the old mortar attached to the RCA surface [15], causing the new interfacial transition zone (ITZ) to be more porous and the interfacial bond strength between the RCA and new mortar to be weaker.

Accordingly, to expand the application of RAC, many RCA surface treatments have been proposed to enhance the performance of RAC through improving the new ITZ [16]. These include soaking or surface precoating with polyvinyl alcohol (PVA) [17], silane polymers [18,19], sodium sulfate solutions [20], sodium silicate solutions [21], volcanic ash materials [22–24], cement, and other cementitious materials [25–28], as well as accelerated carbonation [29–31] and biological carbonate deposition [29,32,33]. However, these techniques might be hampered by a lack of durability from polymer compounds, a weak bond between the surface-coating paste and the RCA, and uncertainty about the efficacy of the carbonation treatment. To this end, Chen et al. [34] suggested a novel surface treatment technique in which magnesium phosphate cement (MPC) was used as a "bridge" between fresh concrete mortar and RCA. As a low-carbon green cementing material, MPC was generally acknowledged to have a high bonding strength with existing concrete, ranging from 77% to 120% higher than that of ordinary Portland cement [35], as well as high volume stability [36], and excellent durability for application in a diversity of complex environments [37,38]. Therefore, MPC can overcome the poor bonding strength of surface-coating pastes to RCA and effectively improve the mechanical properties of RAC [39–41]. Nevertheless, the modified RCA needed a long curing period for use and thus may delay the duration of construction of RAC application in practice [34]. Moreover, the effect of MPC on the long-term performance [42] of RAC lacks proof, which is likewise a primary concern in engineering applications. Thus, to enhance the operability of this treatment for engineering applications, the physical properties of MPC-modified RCA under short curing ages and the corresponding RAC's mechanical performance and durability need to be further investigated. In addition, interfacial adhesion enhancement between MPC paste and RCA or new mortar, as well as improvement of the new ITZ, have not been thoroughly studied, which is crucial to reveal the enhancement mechanism of surface reinforcement by MPC paste.

Meanwhile, the performance of both RCA and RAC has been confirmed to exhibit a strong correlation with the fundamental property of strengthening pastes. For instance, the water absorption and crushing value of RCA were affected by the strengthening paste's hardened strength and anti-permeability., The strength of RAC was likewise related to the mechanical strength and compactness of the strengthening paste [43–45]. According to studies, mineral admixtures have been frequently employed in MPC systems in appropriate dosages as cost-effective, ecologically friendly components that enhance MPC qualities. For example, utilizing the "ball effect", micro-aggregation effect [46,47], and hydration-induced effect, fly ash (FA) can improve the later mechanical characteristics of FA–MPC [48–50]. Moreover, granulated blast furnace slag (GBFS) could improve the mechanical properties and durability of MPC due to the physical filling and the chemical reactions contributed by the presence of calcium components [51,52]. Accordingly, to acquire the optimum efficacy of surface coating, the mineral admixtures FA or GBFS could be incorporated into MPC materials, and the appropriate dosing amounts need to be investigated.

In this study, a series of experiments were carried out with the aim of revealing the reinforcement efficiency of RCA and RAC via surface treatment with MPC paste under a short pre-conditioning time, as well as the influence of MPC slurry properties on the reinforcement efficiency for RCA and RAC. Firstly, different surface-strengthening pastes were prepared with MPC supplemented by 0%, 5%, 10%, and 15% of FA or GBFS. Subsequently, the properties of surface-strengthening pastes, including compressive strength, chloride ion penetration resistance, and the synergistic mechanism of MPC with FA or GBFS, were examined. Based on the results obtained, the optimal FA or GBFS dosage can be selected, and the corresponding MPC blend pastes were utilized for the surface enhancement of RCA. Afterward, the enhancement effects of the surface coating on RCA and RAC were verified through testing the water absorption and crush value of RCA, as well as the mechanical strength and chlorine ion penetration resistance of RAC. Finally, the microstructural properties and elemental distributions of the bond interface and new ITZ were characterized through conducting backscattered electron and energy dispersive spectroscopy (BSE-EDS) measurements as well as microhardness tests to reveal the microscopic strengthening mechanism of various MPC pastes on RAC macroscopic properties. This study assists in improving the operability of RCA application in practice, promoting the production of high-quality RAC, and thus contributing to fostering sustainable development of the construction industry and yielding environmental benefits.

2. Materials and Methods

2.1. Major Raw Materials

In this study, dead-burned magnesia (MgO), ammonium dihydrogen phosphate ($NH_4H_2PO_4$, abbreviated as ADP), borax ($Na_2B_4O_7 \cdot 10H_2O$, abbreviated as B) as a retardant, and water were combined in precise ratios to create a pure MPC paste, referred to as S-0. MgO and ADP were purchased from Liaoning Yangyang High Tech Materials Co., Ltd. in Yingkou City, and B was purchased from Zhiyuan Chemical Reagent Co., Ltd. in Tianjin. Additionally, 5%, 10%, and 15% of the mass of MgO were replaced with FA or GBFS to create mineral admixture–MPC pastes. FA and GBFS were purchased from Ningdong Thermal Power Co., Ltd. in Yinchuan City and Rongchangsheng Environmental Protection Materials Co., Ltd. in Zhengzhou City, respectively. The chemical compositions of FA, GBFS, and MgO used in this study were determined via X-ray fluorescence (XRF) oxide analysis, and the results are presented in Table 1. Purities of the industrial-grade ADP and B were above 98% and 99.5%, respectively. The particle size distributions of FA, GBFS, and MgO were examined using a laser particle size analyzer, and the average particle sizes were approximately 9 μm, 7 μm, and 12 μm for FA, GBFS, and MgO, respectively.

Table 1. Chemical composition of raw materials (by wt/%).

Raw Materials	Mass Fraction of the Sample (%)									
	SiO_2	Al_2O_3	CaO	Fe_2O_3	K_2O	TiO_2	Na_2O	SO_3	MgO	P_2O_5
FA	49.80	30.69	5.30	5.08	2.23	2.02	1.54	1.25	1.11	0.47
GBFS	35.51	13.11	39.82	0.37	0.29	2.63	0.37	2.26	4.88	0.02
MgO	2.35	1.30	1.31	1.27	0.02	0.03	0.04	-	92.12	0.12

Using an experimental jaw crusher, the untreated RCA used in this study was produced from original concrete with a compressive strength of approximately 35 MPa and labeled as RCA0. Figure 1 displays the gradation information of RCA0 obtained from the sieving method. According to Chinese Standard GB/T 14685-2022 [53], the accuracy of the sieving method could be guaranteed based on the sampling process. The sampling process was specified as follows: first, the sample was formed through randomly selecting aggregates of approximately equal mass from different portions of the aggregate heap; then, the sample was placed on a flat plate, mixed well under natural conditions, and piled up into a heap; afterward, the heap was divided into four equal portions along two diameters perpendicular to each other, and the two diagonal portions of the heap were re-mixed and piled up into a heap; the process was repeated until the amount of sample was reduced to that required for the test. The experiment also utilized natural river sand with a fineness modulus of 2.63 and water absorption of roughly 1.9%, as well as Portland cement (P.O. 42.5), with mechanical and physical parameters shown in Table 2. A water reducer was also included to improve the workability of the concrete.

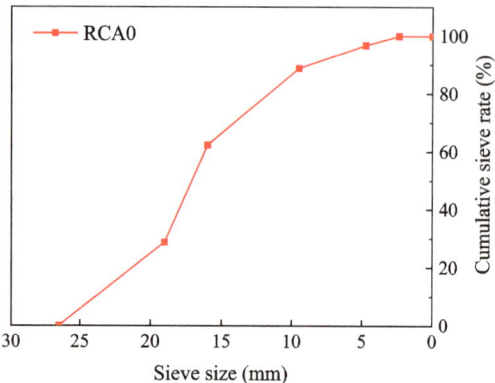

Figure 1. Particle size distribution of RCA0.

Table 2. Properties of Portland cement.

Cement Type	Density (kg/m^3)	Specific Surface Area (m^2/kg)	Setting Time (min)		Compressive Strength (MPa)		Flexural Strength (MPa)	
			Initial Setting	Final Setting	3 d	28 d	3 d	28 d
OPC	3090	398	220	310	25.1	47.3	4.5	7.9

2.2. Preparation Methods

2.2.1. Surface-Strengthening Pastes

Different amounts of FA and GBFS were used to replace MgO to prepare blended MPC pastes. Table 3 shows the mixing ratio of each MPC paste required to modify 1000 kg RCA0. Precast MPC paste was prepared through blending the non-water components of the mixture materials based on the prescribed ratio first, then adding the corresponding amount of water and mixing for 60 s. After that, each type of precast MPC paste was cast in six 40 mm^3 cubic molds and six cylindrical molds with a size of Φ100 × 50 mm^3. Then, all specimens were demolded after 3 days and maintained at 20 ± 2 °C and 64 ± 2% RH for 28 days. The cylindrical specimens were used for the rapid chloride permeability test (RCPT). In addition, the cubic specimens were used for the compressive strength test, and approximately 10 mm^3 pieces were cut from the fractured hardened blocks for BSE-EDS analysis. In detail, for microscopic test sample preparation, the slices were first soaked in ethanol for 24 h to halt cement hydration, then dried and embedded in epoxy resin with a cylindrical rubber mold measuring 20 mm in height and 25 mm in diameter, and finally, the samples were polished to create a smooth surface, dried, and stored in a vacuum chamber before testing.

Table 3. Materials ratios of MPC pastes for modifying 1000 kg RCA0.

Paste Type	ADP (kg)	MgO (kg)	FA (kg)	GBFS (kg)	Water (kg)	B (kg)
S-0	93.2	186.4	0	0	50.9	8.4
S-FA5	93.2	177.08	9.32	0	50.9	8.4
S-FA10	93.2	167.76	18.64	0	50.9	8.4
S-FA15	93.2	158.44	27.96	0	50.9	8.4
S-GBFS5	93.2	177.08	0	9.32	50.9	8.4
S-GBFS10	93.2	167.76	0	18.64	50.9	8.4
S-GBFS15	93.2	158.44	0	27.96	50.9	8.4

Note: S-0 denotes strengthening slurry without mineral admixture; "S-FA" and "S-GBFS" denote strengthening slurries with FA and GBFS, respectively; the numbers "5", "10", and "15" denote the percentage of MgO replaced with mineral admixture.

2.2.2. Surface-Reinforced RCA

The basic properties of strengthening pastes containing different amounts of FA or GBFS were tested regarding compressive strength and chloride ion penetration resistance. Based on the results, the most optimal FA or GBFS dosages leading to higher strength and lower chloride penetration were determined, and the corresponding blended MPC pastes were selected to prepare surface-reinforced RCA. Furthermore, S-0 was chosen as a comparison. It was expected that the physical properties of RCA, including water absorption and crushing value, could be improved through surface strengthening [44,45,54].

After preparation, the strengthening paste was immediately mixed and stirred with RCA0 for 5 min to precoat RCA0. Subsequently, the RCA was removed from the tank, and any excess MPC paste stuck to it was sieved away. Afterward, these treated RCA were exposed to the air with $64 \pm 2\%$ RH and $20 \pm 2\ °C$ for 3 days. A portion of the RCA was used for characterization tests, including water absorption and crushing value; another portion of the RCA was used to prepare RAC.

2.2.3. Concrete

The original RCA0 and treated RCA obtained from Section 2.2.2 were utilized as coarse aggregate to produce concrete to obtain the best modification effect for high-quality RAC and the influence of the strengthening paste properties on the modification efficiency. Since the coated paste amount was negligible compared to the weight of RCA (approximately 2% to 3%) [43], based on a concrete strength grade of C30, the mix proportion for each concrete type was cement:water:sand:coarse aggregate:superplasticizer = 431:247:767:989:2. In order to get superior modification outcomes, this work adopted the double mixing method [55,56] to prepare modified RAC utilizing surface-treated RCA, and it has been demonstrated to reduce the water–to–cement ratio of the new ITZ thereby improving the interface zone, compressive strength, and chloride ion penetration resistance of concrete [55,57,58]. The specific mixing procedure, as shown in Figure 2, is as follows: first, a portion of the water (Water(1)) was added to the aggregates of each group and stirred for 60 s to obtain moist aggregates; then, cement was added and stirred for 120 s to coat the aggregate surfaces with a layer of low water–to–cement ratio cement slurry; finally, the remaining water (Water(2)) was added along with the superplasticizer used, and the fresh concrete was obtained through mixing for 120 s.

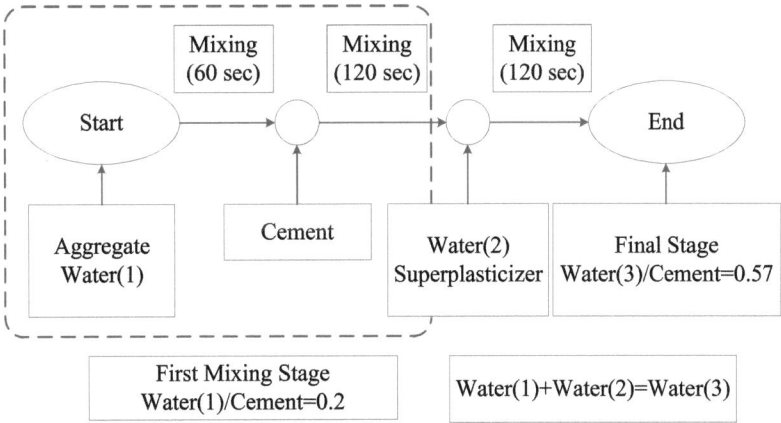

Figure 2. Double mixing method.

The mechanical strength and chloride ion penetration resistance of concrete were compared to understand the difference in the enhancement of RCA with various strengthening pastes. Each type of target concrete specimen consisted of six cubic specimens with dimensions of 100 mm³ and six cylindrical specimens with diameters of 100 ± 1 mm and heights of 50 ± 2 mm. All specimens were cured in a laboratory environment (25 ± 2 °C, 95 ± 2% RH) for 28 days. In order to prepare samples for microstructure analysis, slices were cut from the fractured hardened blocks obtained after the mechanical strength testing, whose surface included the desired testing areas containing the interface. Detailed procedures for sample preparation can be obtained from Section 2.2.1.

2.3. Test Methods

2.3.1. Performance Testing of Strengthening Pastes

Compressive strength testing was conducted on the 2000 kN servo-hydraulic compressional testing machine according to GB/T 17671-2021 [59]. Moreover, RCPT was used to determine the resistance to chloride penetration of each group of strengthening paste. The procedure from specimen preparation to testing is detailed in the ASTM C1202-19 standard [60]. Furthermore, each group's chloride ion permeability of the strengthening paste was qualitatively graded via the mean electrical flux.

Microscopic examination of different types of hardened paste was conducted using scanning electron microscopy (SEM, TESCAN MIRA LMS, Czech Republic) equipped with EDS (Oxford Xplore). BSE-EDS pictures were captured and utilized to investigate the strengthening mechanisms of FA and GBFS on the microstructure of hardened MPC pastes, thus providing more information on the effect of the paste's properties on RAC performance. The imaging machine operated at a 15 mm working distance with a 15 kV voltage.

2.3.2. Characterization Testing of RCA

(1) Water absorption

The water absorption of RCA was derived via the following equation:

$$Water\ absorption\ ratio = \frac{wet\ weight - dry\ weight}{dry\ weight} * 100\% \tag{1}$$

The wet weight and dry weight of aggregates could be measured based on Chinese Standard GB/T 14685-2022 [53].

(2) Crushing value

The crushing value tests of RCAs were carried out based on Chinese Standard GB/T 14685-2022 [53], and the crushing value could be calculated following the formula below:

$$Curshing\ value = \frac{G_2}{G_1} * 100\% \tag{2}$$

G_1 and G_2 were the total weight of aggregates and the weight of crushed aggregates finer than 2.36 mm, respectively.

2.3.3. Macroscopic Properties Testing of Concrete

Each concrete group's compressive and splitting tensile strengths were tested using three cubes, following the guidelines specified in GB/T 50081-2019 [61]. The RCPT was conducted to determine each concrete group's chloride ion penetration resistance after 28 days of curing, and the evaluation was conducted following ASTM C1202-19 [60].

2.3.4. Microscopic Characterization Testing of Interfaces

(1) BSE-EDS testing

To examine the microstructure and precise elemental distribution, BSE-EDS imaging was carried out on the bond interfaces between strengthening pastes and RCA0, as well as the new ITZ regions. This made it possible to disclose the bonding and strengthening mechanisms of the strengthening paste on RCA0 and the new ITZ. The preparation method of the BSE-EDS testing samples has been described in Section 2.2.1.

(2) New ITZ microhardness testing

In addition to the microstructural composition, the microscopic mechanical properties of materials are also critical microstructural characteristics. Microhardness (Vickers hardness) has been used to understand the microscopic mechanical characteristics of RAC [62–64]. Therefore, to further validate the new ITZ performance improvement due to MPC modification, microhardness tests were performed on the regions containing the new ITZ in all RAC specimens, using a digital Vickers microhardness tester equipped with 40 measurement objectives and 10 magnification objectives (HV-1000BZ, Shanghai, China). As shown in Figure 3, the test region size was 240 μm × 250 μm, and a 9 × 6 indent points matrix was applied within the region. The samples employed for the microhardness test are detailed in Section 2.2.3. At least three areas were chosen randomly from two samples of each target concrete for testing. The two-dimensional microhardness distribution maps for each indent region were generated using the Surfer 13's Contour map feature. The new ITZ's boundaries were identified based on the color variations observed in the microhardness distribution maps. The average microhardness values of the new ITZ were determined using statistical analysis according to the microhardness values of each indent point within the boundaries.

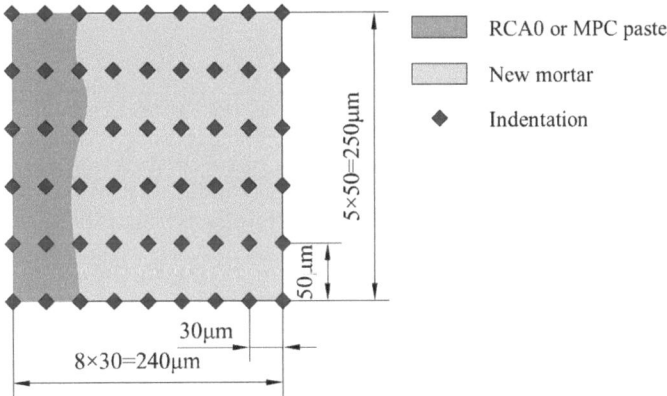

Figure 3. Indent area and corresponding indent matrix for new ITZ's microhardness testing.

3. Results and Discussion

3.1. Performance Characteristics of Strengthening Pastes

3.1.1. Macroscopic Performance of Strengthening Pastes

Based on previous research, the modified RCA and RAC's properties correlate firmly with the surface-strengthening paste's performance, which can be strengthened through adding appropriate amounts of mineral admixtures. Therefore, this section compared the compressive strength and chloride ion penetration resistance of hardened MPC. On this basis, it was expected to select the best-performing FA-doped or GBFS-doped MPC paste for coating RCA.

The compressive strengths of prefabricated MPC pastes are shown in Figure 4a. It can be indicated that the compressive strengths of the blended MPC pastes were higher than that of S-0 without mineral admixture on the condition that the admixture of either FA or GBFS was 5%, 10%, and 15%. Moreover, it can be further seen that the compressive strengths of MPC pastes tended to increase and then decrease with increasing dosages of both mineral admixtures. That is, the optimal dosing for both FA and GBFS is 10%, and in that condition, the compressive strengths were 60.73 MPa and 64.82 MPa, and an increase of 7.49 MPa and 11.58 MPa in comparison with S-0 was exhibited, respectively. S-GBFS10 had the most significant gain among them.

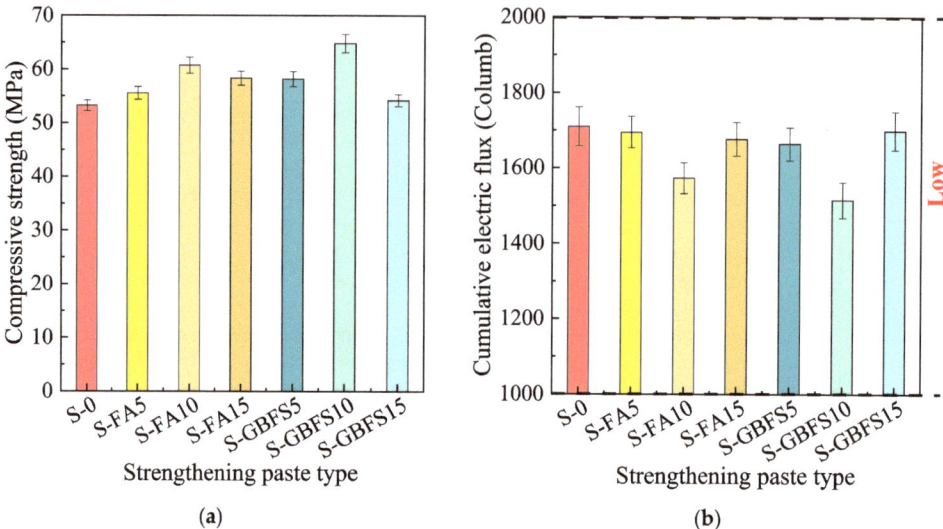

Figure 4. Different MPC pastes' (**a**) compressive strengths and (**b**) cumulative electric fluxes.

Figure 4b displays the cumulative electrical fluxes that passed through all MPC pastes in six hours. It can be seen that the cumulative electrical flux for each type of MPC paste was relatively minimal, categorizing them as "low". Their excellent resistance to chloride ion penetration can be attributed to the low water–to–cementitious material ratio and drying shrinkage of the MPC. Moreover, the graph demonstrates that the electrical fluxes of the MPC pastes exhibited the most significant diminution through adding 10% mineral admixtures. Compared to the S-0 paste at 1710C, S-FA10 and S-GBFS10 exhibited reductions of 137C and 195C, respectively. This indicates that adding additive FA or GBFS in the proper quantity dramatically improved the MPC paste's density and impermeability. Combining these findings with those from the compressive strength test, it was quickly found that a better comprehensive performance could be obtained on the condition that FA or GBFS doping was 10%. The chief reason for this could be inferred as the mineral admixture in the right amount may play the role of physical filling and facilitate the secondary reaction, thus enhancing the microstructure of MPC paste, yet the excessive substitution of MgO with mineral admixture resulted in a decrease in the number of hydration products and thus led to the poor densification of MPC microstructure [65]. Hence, S-FA10 and S-GBFS10 were chosen to reinforce RCA for their excellent comprehensive performance. The strengthening mechanism of FA or GBFS for the mineral admixture–MPC system's microstructure will be interpreted in detail in the following section.

3.1.2. Enhancement Mechanisms for MPC via FA or GBFS

To explore the strengthening mechanism of both mineral admixtures on the macroscopic properties of MPC pastes, Figure 5a,b displays the BSE-EDS images of hardened S-FA10 and S-GBFS10 pastes at high magnification, respectively. It can be recognized that both hardened pastes exhibited the creation of the struvite phase, and unreacted MgO grains were detected throughout the matrix and appear to be the nucleation sites for struvite formation. It was evident from the reaction equation between MgO and ADP that there would be some solid volume expansion from MgO to struvite, resulting in a denser microstructure of the MPC paste. Due to dehydration under vacuum for examination, the embedded struvite particles in the polished parts seemed severely cracked. The spherical particles of various sizes in Figure 5a were FA particles. It can be seen that several medium-sized (10–20 μm) spherical particles were identified as surface depressions, indicating that the particles underwent partial reactions [66], whereas smaller particles with similar erosion depths on their surfaces may have undergone complete reactions or dissolution. In the EDS images of Figure 5a, the region where the elements P and Ca appeared to overlap significantly, marked with yellow wireframes, as well as the region where the elements Mg, P, Si, and Al seemed to coincide, marked with blue coils, also provided evidence of chemical reactions between the aluminosilicate FA particles and the other constituents in the MPC paste. The reaction products were speculated to be calcium phosphate, enstatite, and berlinite [41] based on recent evaluations [51,67]. In Figure 5b, the angular particles of various sizes rich in Ca, Al, and Si elements corresponded to the unreacted calcium aluminosilicate glassy portion in GBFS particles. The regions highlighted with the yellow coils in the elemental maps of Figure 5b indicated that the active calcium oxide in GBFS reacted with phosphate in the matrix, leading to the formation of calcium phosphate gel, which was consistent with the mechanism of GBFS being used as an adsorbent for phosphate removal in wastewater systems [68]. Therefore, it may be inferred that FA or GBFS will form a strong link with the surrounding hydration products and act as aggregates within the matrix due to their dissolution and subsequent reaction. Consequently, the MPC matrix's integrity was improved, and its strength and permeability resistance were significantly boosted.

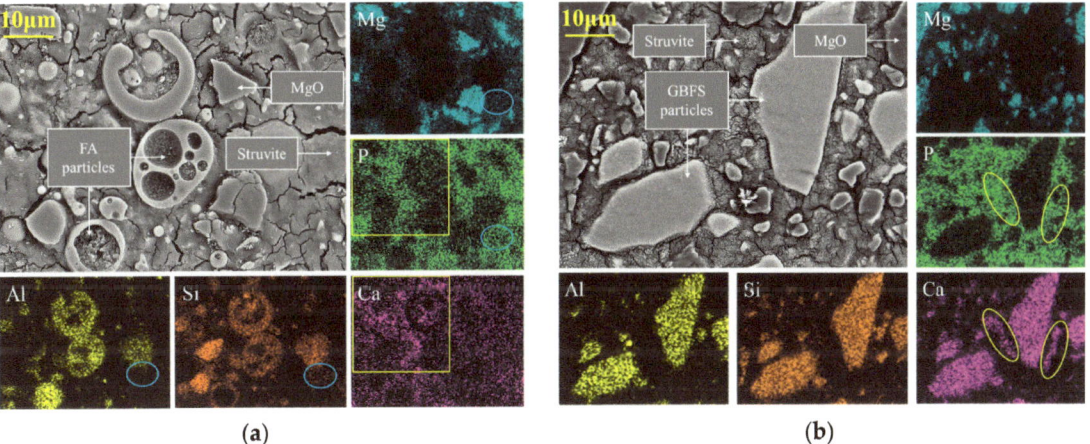

Figure 5. Typical BSE images and elemental maps of hardened (**a**) S-FA10 paste and (**b**) S-GBFS10 paste.

Furthermore, based on the percentages of calcium oxide and aluminosilicate components in the mineral admixtures, as well as the test results of compressive strength and chloride ion penetration resistance, we prefer to believe that the calcium oxide-related reaction dominated the synergistic effect between MPC and FA or GBFS. This finding aligned with the conclusion drawn in the reference [52].

3.2. Performance Characteristics of RCA

The modified RCA obtained from S-0, S-FA10, and S-GBFS10 pastes were labeled R-1, R-2, and R-3, respectively. Figure 6 displays photographs of the modified RCA and untreated RCA0, with a coin diameter of approximately 25 mm. The water absorption and crushing values of RCA before and after the surface reinforcement are shown in Figure 7. The means of the surface-reinforced RCA were observed to be reduced compared to RCA0 in terms of water adsorption ratios and crushing values. Moreover, Figure 7 shows that R-2 and R-3 exhibited a more significant reduction than R-1 in consistency with the analysis result of strengthening paste properties. However, the improvement in water absorption was not salient, with a maximum decrease of 1.8% in comparison to RCA0. Moreover, the chief reasons for this involved the inability of coated pastes to prevent water from infiltrating and suffusing RCA0 due to the almost negligible amount of the paste compared to the weight of RCA [43]. The result obtained was in agreement with the previous study [28].

Figure 6. The modified RCA and RCA0.

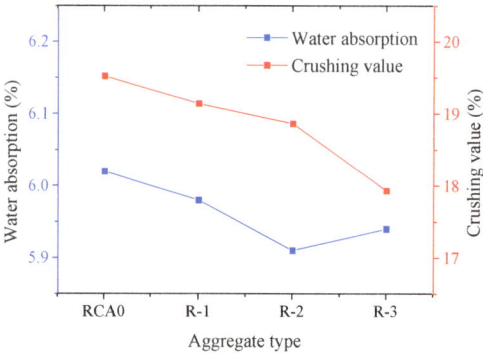

Figure 7. The effects of surface reinforcement on water absorption and crushing value of RCA.

To verify the validity of surface reinforcement, the RAC was prepared from RCA0, R-1, R-2, and R-3, and labeled as C-0, C-1, C-2, and C-3, respectively. As the properties of the concrete exhibited a strong correlation with the moisture state of the aggregate, RCA0, R-1, R-2, and R-3 were dried at above 40 °C for 3 days prior to preparing the concrete. The moisture content of RCA0, R-1, R-2, and R-3 were tested to be 1.2%, 1.2%, 1.3%, and 1.1%, respectively. Moreover, the macroscopic properties of RACs, as well as the microstructural properties of ITZs between the MPC and RCA0 or fresh mortar, were investigated.

3.3. Macroscopic Properties of Concrete

3.3.1. Mechanical Properties

The 28-day compressive and splitting tensile strengths of all RAC are shown in Figure 8a,b, respectively. It can be seen that the RAC samples obtained after the enhancement treatment with different MPC pastes exhibited increased compressive and splitting tensile strengths compared to C-0. The increase in strengths may be attributed to the high bonding performance of the MPC paste to RCA0 or the new mortar, the filling of micro-defects in RCA0, and the strengthening of the new ITZ. This inference will be substantiated in the following sections. On the condition that 10% FA or GBFS was added, the compressive strength of RAC increased compared to C-1, with increments of 11.32% and 24.13%, respectively. Moreover, C-3 exhibited the highest improvement degree. The trend in splitting tensile strength aligned with compressive strength, and C-1, C-2, and C-3 showed progressive increasing values, which rose by 14.43%, 37.31%, and 49.75%, respectively, compared to C-0. The modified RAC's mechanical strength variations aligned with the performance of the MPC paste.

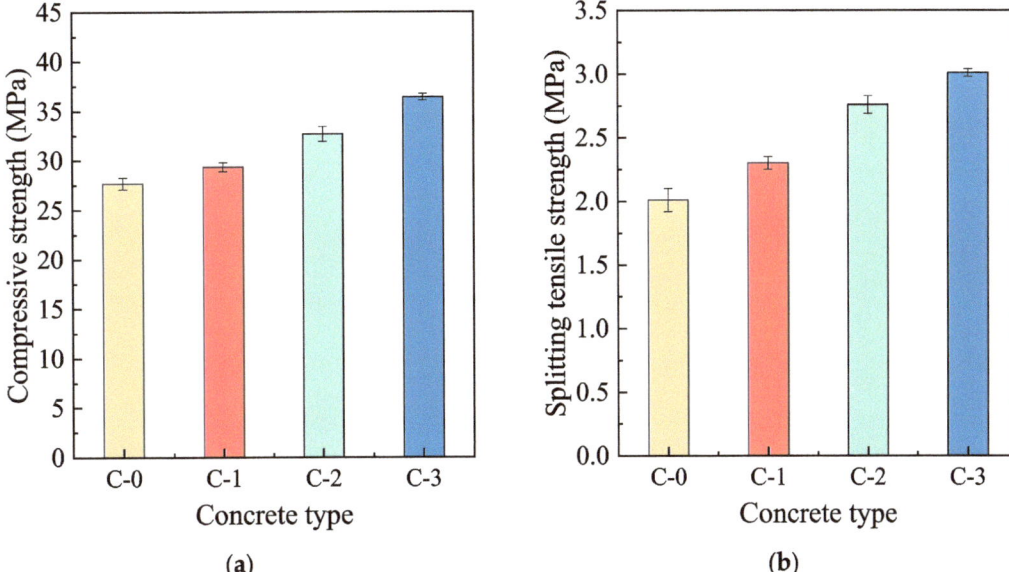

Figure 8. (a) Compressive strengths and (b) splitting tensile strengths of various concrete.

3.3.2. Chloride Ion Penetration Resistance

Figure 9 displays the cumulative electrical flux for each target concrete. As observed in the figure, the precoating for RCA0 with various MPC pastes improved RAC's chloride ion penetration resistance to varying degrees, possibly due to the enhanced bonding of RCA0 to the new mortar as well as the better chloride ion penetration resistance of the MPC pastes. Detailed evidence for this hypothesis will be elaborated in the following sections. Additionally, double mixing had a favorable effect on the performance of the new ITZ in the modified RAC, thereby improving its chloride ion penetration resistance to some extent. The cumulative electric fluxes passed through C-1, C-2, and C-3 were 2015 C, 1804 C, and 1568 C, respectively, showing a decreasing trend. Compared to C-0, these values represented a reduction of approximately 26.9%, 34.5%, and 43.1%, respectively. The observed variation in RAC's chloride ion penetration resistance likewise aligned with the performance trend of MPC paste. According to ASTM C1202, untreated C-0 can be classified as "moderate", whereas the RAC enhanced with S-FA10 or S-GBFS10 was classified as "low". The reduced chloride ion permeability indicated that modifying RAC with MPC enhanced its anticipated durability.

3.4. Interfacial Bond Behavior and Microscopic Characteristics

To investigate the bond efficiency of the strengthening paste as a "bridge" and the mechanisms of filling in RCA0 and strengthening the new ITZ, BSE-EDS images were captured at a magnification of 500 times to analyze the microstructure at the interface between various MPC pastes and the new or old mortar. Figure 10a–c presents the typical BSE-EDS images of the bond interfaces between hardened S-0, S-FA10, or S-GBFS10 pastes and RCA0. The images show that all strengthening pastes exhibited excellent bonding with RCA0, forming relatively dense, robust, and uniform interface regions. The regions highlighted with the yellow coils in Figure 10 demonstrated that the P element was prominently incorporated into the old mortar zones, overlapping with the Ca element. This observation indicated excellent mechanical and chemical interlocking ascribed to MPC pastes' infiltration and filling in RCA0, as well as the reaction between soluble acidic phosphates from the infiltrated MPC pastes and $Ca(OH)_2$ in the old mortar, resulting in a favorable bond between the MPC paste and the RCA0. The characteristics mentioned above likewise aided in improving the pore structure of the RCA0 surface, resulting in surface reinforcement. Moreover, mineral admixtures in MPC pastes can operate as fillers through penetrating the pores of the old mortar and interface, as shown in Figure 10b. They might also have a pozzolanic effect that formed new hydrated products and improved the homogeneity and density of the old mortar and interface [69,70].

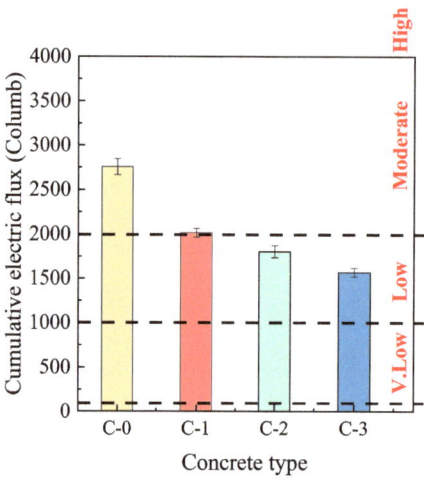

Figure 9. Chloride ion penetration resistances of various concrete.

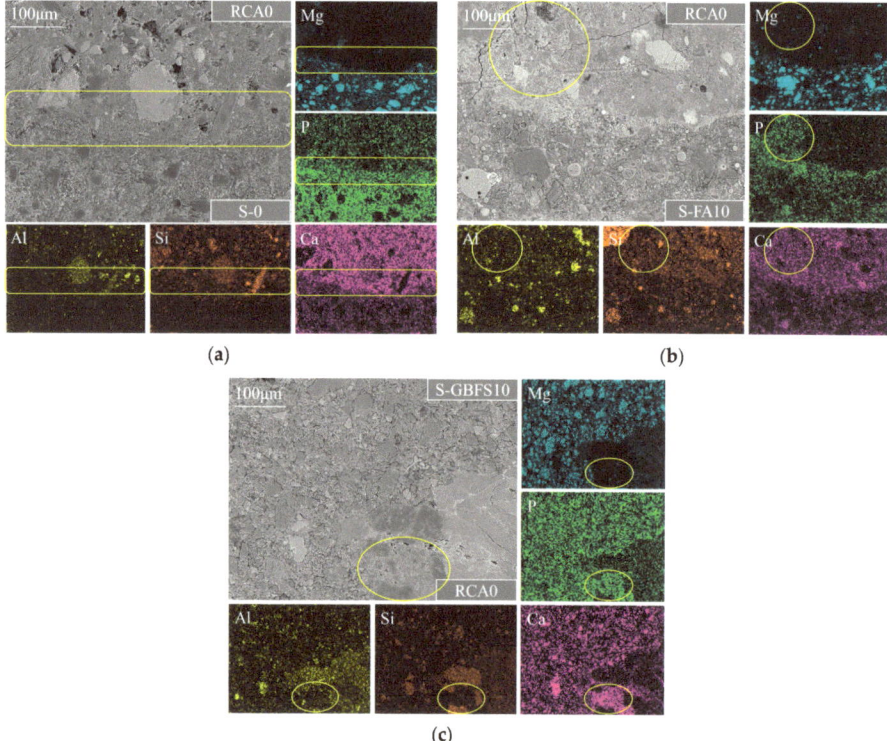

Figure 10. Typical BSE images and elemental maps of the bond interfaces in (**a**) C-1, (**b**) C-2, and (**c**) C-3.

For preventing interference from elements present in mineral admixtures, C-1 was used as an example to clarify the bonding mechanism of the strengthening paste to the new mortar, as well as its reinforcing mechanism on the new ITZ based on BSE-EDS analysis, as displayed in Figure 11a. For comparison, a typical BSE image of the new ITZ between RCA0 and the new mortar in the untreated C-0 sample is shown in Figure 11b. As seen in Figure 11b, it was evident that there were sizeable cracks in C-0's new ITZ, probably due to interfacial debonding. Large pores can also be observed within the new ITZ due to the wall effect and increased moisture content. Consequently, C-0's final performance was significantly weakened. Figure 11a shows a significant reduction of pores and microcracks in the new ITZ of C-1 compared to C-0. The distributions of Ca, P, and Mg elements within the area circled in yellow in Figure 11a indicated that Ca ions from the new mortar permeated into the MPC matrix near the interface, reacting with struvite or unhydrated phosphates to generate new cementitious materials, leading to a denser hardened MPC matrix near the interface and promoting hydration reactions in the new ITZ. The same phenomenon can also be observed in C-2 and C-3. These findings indicated that the MPC pastes exhibited excellent chemical bonding with the new mortar, significantly lowering the likelihood of shrinkage-induced debonding cracks and the appearance of large pores in the modified RAC's new ITZ. This contributed to the new ITZ's more compact and superior microstructure, further enhanced through the beneficial effects of the double mixing procedure.

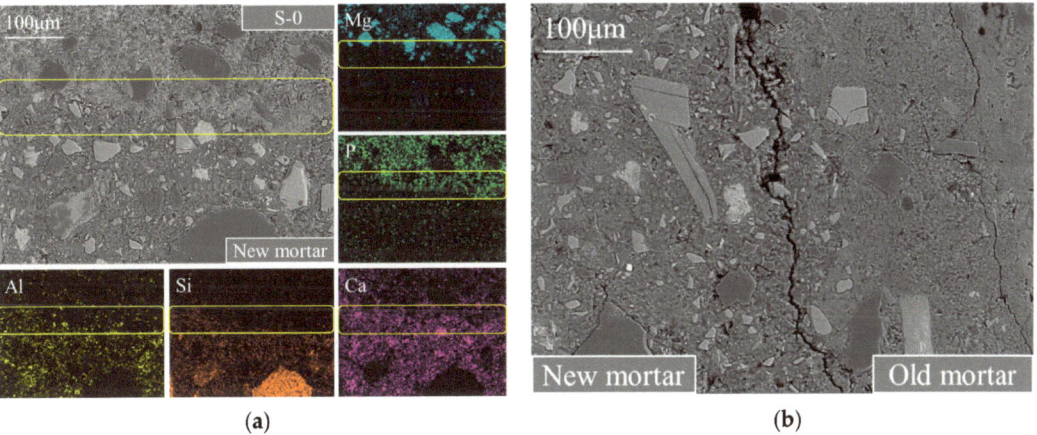

Figure 11. Typical new ITZs' (**a**) BSE-EDS image in C-1 and (**b**) BSE image in C-0.

Overall, the MPC paste precoating treatment yielded excellent interfacial bonding properties and microstructure, which exposed the mechanism for obtaining improved mechanical strengths and chloride ion permeability resistance of the modified RAC. Meanwhile, these findings also provided evidence to support the inferences in Section 3.3.

3.5. ITZ Microhardness Analysis

Microhardness analysis was used to quantitatively evaluate the micro-mechanical characteristics of the new ITZ in RAC to define the improving effectiveness of the surface treatment method employing MPC paste. Figures 12 and 13 present the typical microhardness distribution maps and the average microhardness values for the new ITZs of C-0, C-1, C-2, and C-3. The boundaries of the new ITZ in the microhardness distribution maps were depicted with red dashed lines. In all samples, the microhardness values were relatively low when located within the ITZs (with a width of approximately 85 μm to 150 μm), but they rose and stayed steady as one moved away from the ITZs, as shown in Figure 12. Figure 12a–d shows that the new ITZs' widths (approximately between 85 μm and 110 μm) following enhancement with various MPC pastes dramatically decreased in comparison to the width (approximately between 110 μm and 150 μm) of the untreated C-0, accompanied with higher microhardness values. The average microhardness values of the new ITZs in C-1, C-2, and C-3 rose by 30.2%, 30.4%, and 33.4%, respectively, compared to C-0, as shown in Figure 13. These findings indicated that the new ITZ of the modified RAC has been effectively strengthened, aligning with the BSE observations and providing further evidence for the effectiveness of the proposed strengthening method in this study.

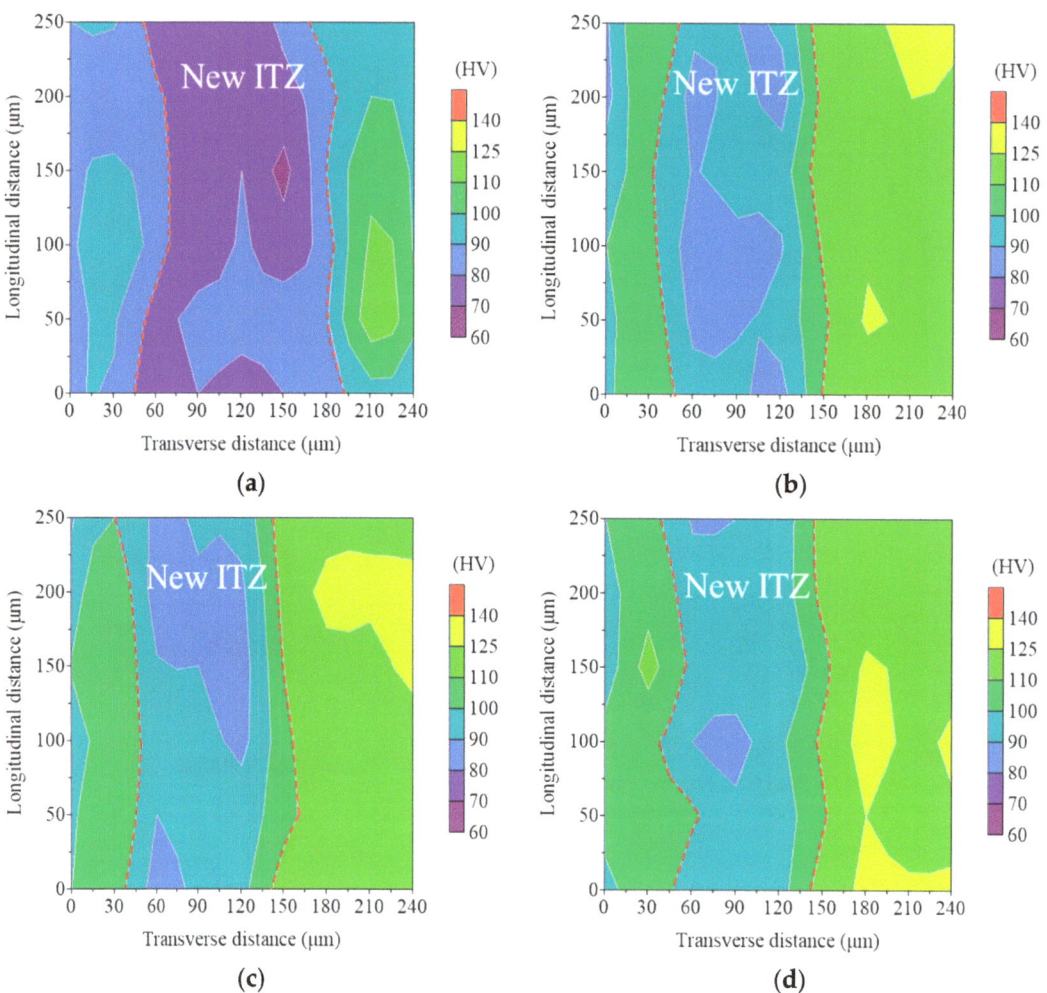

Figure 12. Typical microhardness distribution maps within indent areas of (**a**) C-0, (**b**) C-1, (**c**) C-2, and (**d**) C-3.

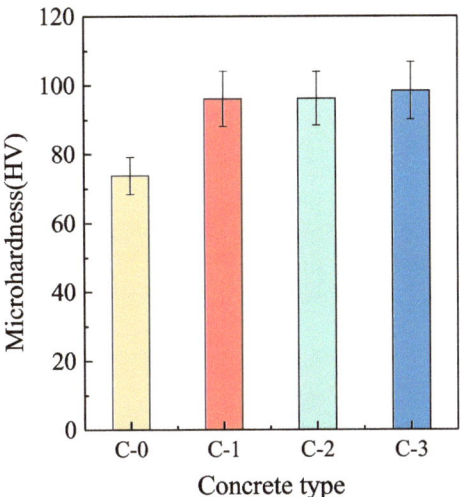

Figure 13. The average microhardness values of the new ITZs in C-0, C-1, C-2, and C-3.

4. Conclusions

In this work, seven surface-strengthening pastes were prepared; then, compressive strength and chloride ion penetration resistance were comparatively studied; afterward, the most suitable FA–MPC and GBFS–MPC hybrid slurries with the best comprehensive performance were used to coat RCA0, followed by 3 days of maintenance, and MPC slurries without mineral admixtures were also selected for comparison purposes; lastly, the physical properties of RCA before and after the surface reinforcement were compared, and the macroscopic and microscopic properties of target concrete were evaluated. The conclusions are summarized as follows:

(1) S-FA10 and S-GBFS10 were most suitable to coat RCA0 due to the higher strength and chloride ion penetration resistance, and the reason was that the mineral admixtures facilitated secondary reactions and enhanced the integrity of the hardened pastes.

(2) After the surface reinforcement with S-0, S-FA10, and S-GBFS10, the crushing value of RCA decreased from 19.52% to 19.14%, 18.86%, and 17.93%, respectively. Nevertheless, surface-strengthening pastes had little effect on the water absorption of RCA. Furthermore, the RAC prepared from R-1, R-2, and R-3 performed better than that from RCA0 regarding mechanical properties and durability. The enhancement efficiencies on the performance of RCA and RAC improved with strengthening paste performance.

(3) The BSE-EDS observations of the modified RAC showed the presence of mechanical and chemical interlocking between the strengthening paste and RCA0 or new mortar, which led to the effective filling of micro-defects near the RCA0 surface and the well-bonded interfaces between MPC pastes and RCA0 or new mortar. Furthermore, a denser microstructure within the new ITZ was observed to further improve the strengths and durability of the RAC under the combined effect of the precoating treatment and the double mixing method.

(4) Based on the microhardness test results of the ITZs, it can be seen that the breadths of the new ITZs were reduced, and the average microhardness values were improved after modification with MPC pastes, showing an improvement in its micromechanical properties, which further confirmed the effectiveness of the surface-strengthening treatment using MPC pastes.

Overall, the surface treatment method proposed in this study was considered effective and applicable. Furthermore, the construction duration was shortened compared to previous studies [34], thus improving the operability of RCA for applications in practice. In the future, the effects of MPC slurry precoating treatment on the morphology of RCA [71] as well as on heat resistance [72] and the internal moisture content [73] of RAC need further investigation, which has essential implications for the application of MPC in the surface modification of RCA.

Author Contributions: S.W., Y.G., X.X. and X.Y. conceived and designed the experiments; S.W., J.H., Z.S. and X.X. performed the experiments; S.W., J.H., Z.S. and X.X. analyzed the data; S.W. and X.X. wrote the paper; and all authors revised the paper. All authors have read and agreed to the published version of the manuscript.

Funding: This study was supported by the Natural Science Foundation of the Jiangsu Higher Education Institutions of China (No. 22KJB560010) and the Nantong Basic Science Research Program of China (No. JC12022098).

Institutional Review Board Statement: Not applicable.

Informed Consent Statement: Not applicable.

Data Availability Statement: The data presented in this study are available on request from the corresponding author.

Conflicts of Interest: The authors declare no conflict of interest.

References

1. Ryu, J.S. Improvement on Strength and Impermeability of Recycled Concrete Made from Crushed Concrete Coarse Aggregate. *J. Mater. Sci. Lett.* **2002**, *21*, 1565–1567. [CrossRef]
2. Poon, C.S.; Shui, Z.H.; Lam, L. Effect of Microstructure of ITZ on Compressive Strength of Concrete Prepared with Recycled Aggregates. *Constr. Build. Mater.* **2004**, *18*, 461–468. [CrossRef]
3. Kou, S.C.; Poon, C.S.; Chan, D. Influence of Fly Ash as Cement Replacement on the Properties of Recycled Aggregate Concrete. *J. Mater. Civ. Eng.* **2007**, *19*, 709–717. [CrossRef]
4. Ferreira, R.L.S.; Anjos, M.A.S.; Maia, C.; Pinto, L.; de Azevedo, A.R.G.; de Brito, J. Long-Term Analysis of the Physical Properties of the Mixed Recycled Aggregate and Their Effect on the Properties of Mortars. *Constr. Build. Mater.* **2021**, *274*, 121796. [CrossRef]
5. Makul, N.; Fediuk, R.; Amran, M.; Zeyad, A.M.; Klyuev, S.; Chulkova, I.; Ozbakkaloglu, T.; Vatin, N.; Karelina, M.; Azevedo, A. Design Strategy for Recycled Aggregate Concrete: A Review of Status and Future Perspectives. *Crystals* **2021**, *11*, 695. [CrossRef]
6. González-Fonteboa, B.; Martínez-Abella, F. Concretes with Aggregates from Demolition Waste and Silica Fume. Materials and Mechanical Properties. *Build. Environ.* **2008**, *43*, 429–437. [CrossRef]
7. Otsuki, N.; Miyazato, S.; Yodsudjai, W. Influence of Recycled Aggregate on Interfacial Transition Zone, Strength, Chloride Penetration and Carbonation of Concrete. *J. Mater. Civ. Eng.* **2003**, *15*, 443–451. [CrossRef]
8. Ahmed, W.; Lim, C.W. Production of Sustainable and Structural Fiber Reinforced Recycled Aggregate Concrete with Improved Fracture Properties: A Review. *J. Clean. Prod.* **2021**, *279*, 123832. [CrossRef]
9. Belén, G.-F.; Fernando, M.-A.; Diego, C.L.; Sindy, S.-P. Stress–strain Relationship in Axial Compression for Concrete Using Recycled Saturated Coarse Aggregate. *Constr. Build. Mater.* **2011**, *25*, 2335–2342. [CrossRef]
10. Butler, L.; West, J.S.; Tighe, S.L. The Effect of Recycled Concrete Aggregate Properties on the Bond Strength between RCA Concrete and Steel Reinforcement. *Cem. Concr. Res.* **2011**, *41*, 1037–1049. [CrossRef]
11. Domingo-Cabo, A.; Lázaro, C.; López-Gayarre, F.; Serrano-López, M.A.; Serna, P.; Castaño-Tabares, J.O. Creep and Shrinkage of Recycled Aggregate Concrete. *Constr. Build. Mater.* **2009**, *23*, 2545–2553. [CrossRef]
12. Golafshani, E.M.; Behnood, A. Automatic Regression Methods for Formulation of Elastic Modulus of Recycled Aggregate Concrete. *Appl. Soft Comput.* **2018**, *64*, 377–400. [CrossRef]
13. Liang, C.; Pan, B.; Ma, Z.; He, Z.; Duan, Z. Utilization of CO_2 Curing to Enhance the Properties of Recycled Aggregate and Prepared Concrete: A Review. *Cem. Concr. Compos.* **2020**, *105*, 103446. [CrossRef]
14. Golafshani, E.M.; Behnood, A. Application of Soft Computing Methods for Predicting the Elastic Modulus of Recycled Aggregate Concrete. *J. Clean. Prod.* **2018**, *176*, 1163–1176. [CrossRef]
15. Djerbi, A. Effect of Recycled Coarse Aggregate on the New Interfacial Transition Zone Concrete. *Constr. Build. Mater.* **2018**, *190*, 1023–1033. [CrossRef]
16. Wang, R.; Yu, N.; Li, Y. Methods for Improving the Microstructure of Recycled Concrete Aggregate: A Review. *Constr. Build. Mater.* **2020**, *242*, 118164. [CrossRef]
17. Kou, S.-C.; Poon, C.S. Properties of Concrete Prepared with PVA-Impregnated Recycled Concrete Aggregates. *Cem. Concr. Compos.* **2010**, *32*, 649–654. [CrossRef]
18. Tsujino, M.; Noguchi, T.; Tamura, M.; Kanematsu, M.; Maruyama, I. Application of Conventionally Recycled Coarse Aggregate to Concrete Structure by Surface Modification Treatment. *J. Adv. Concr. Technol.* **2007**, *5*, 13–25. [CrossRef]

19. Spaeth, V.; Tegguer, A.D. Improvement of Recycled Concrete Aggregate Properties by Polymer Treatments. *Int. J. Sustain. Built Environ.* **2013**, *2*, 143–152. [CrossRef]
20. Abbas, A.; Fathifazl, G.; Fournier, B.; Isgor, O.B.; Zavadil, R.; Razaqpur, A.G.; Foo, S. Quantification of the Residual Mortar Content in Recycled Concrete Aggregates by Image Analysis. *Mater. Charact.* **2009**, *60*, 716–728. [CrossRef]
21. Shayan, A.X.A. Performance and Properties of Structural Concrete Made with Recycled Concrete Aggregate. *ACI Mater. J.* **2003**, *100*, 371–380. [CrossRef]
22. Shaban, W.M.; Yang, J.; Su, H.; Liu, Q.; Tsang, D.C.W.; Wang, L.; Xie, J.; Li, L. Properties of Recycled Concrete Aggregates Strengthened by Different Types of Pozzolan Slurry. *Constr. Build. Mater.* **2019**, *216*, 632–647. [CrossRef]
23. Ouyang, K.; Shi, C.; Chu, H.; Guo, H.; Song, B.; Ding, Y.; Guan, X.; Zhu, J.; Zhang, H.; Wang, Y.; et al. An Overview on the Efficiency of Different Pretreatment Techniques for Recycled Concrete Aggregate. *J. Clean. Prod.* **2020**, *263*, 121264. [CrossRef]
24. Zhang, H.; Zhao, Y.; Meng, T.; Shah, S.P. Surface Treatment on Recycled Coarse Aggregates with Nanomaterials. *J. Mater. Civ. Eng.* **2016**, *28*, 04015094. [CrossRef]
25. Zhang, H.; Ji, T.; Liu, H.; Su, S. Modifying Recycled Aggregate Concrete by Aggregate Surface Treatment Using Sulphoaluminate Cement and Basalt Powder. *Constr. Build. Mater.* **2018**, *192*, 526–537. [CrossRef]
26. Choi, H.; Choi, H.; Lim, M.; Inoue, M.; Kitagaki, R.; Noguchi, T. Evaluation on the Mechanical Performance of Low-Quality Recycled Aggregate through Interface Enhancement between Cement Matrix and Coarse Aggregate by Surface Modification Technology. *Int. J. Concr. Struct. Mater.* **2016**, *10*, 87–97. [CrossRef]
27. Yew, M.K.; Yew, M.C.; Beh, J.H.; Saw, L.H.; Lim, S.K. Effects of Pre-Treated on Dura Shell and Tenera Shell for High Strength Lightweight Concrete. *J. Build. Eng.* **2021**, *42*, 102493. [CrossRef]
28. Mistri, A.; Dhami, N.; Bhattacharyya, S.K.; Barai, S.V.; Mukherjee, A.; Biswas, W.K. Environmental Implications of the Use of Bio-Cement Treated Recycled Aggregate in Concrete. *Resour. Conserv. Recycl.* **2021**, *167*, 105436. [CrossRef]
29. Zeng, W.; Zhao, Y.; Poon, C.S.; Feng, Z.; Lu, Z.; Shah, S.P. Using Microbial Carbonate Precipitation to Improve the Properties of Recycled Aggregate. *Constr. Build. Mater.* **2019**, *228*, 116743. [CrossRef]
30. Tam, V.W.Y.; Butera, A.; Le, K.N.; Li, W. Utilising CO_2 Technologies for Recycled Aggregate Concrete: A Critical Review. *Constr. Build. Mater.* **2020**, *250*, 118903. [CrossRef]
31. Hosseini Zadeh, A.; Mamirov, M.; Kim, S.; Hu, J. CO_2-Treatment of Recycled Aggregates to Improve Mechanical and Environmental Properties for Unbound Applications. *Constr. Build. Mater.* **2021**, *275*, 122180. [CrossRef]
32. Qiu, J.; Tng, D.Q.S.; Yang, E.-H. Surface Treatment of Recycled Concrete Aggregates through Microbial Carbonate Precipitation. *Constr. Build. Mater.* **2014**, *57*, 144–150. [CrossRef]
33. De Muynck, W.; De Belie, N.; Verstraete, W. Microbial Carbonate Precipitation in Construction Materials: A Review. *Ecol. Eng.* **2010**, *36*, 118–136. [CrossRef]
34. Chen, X.; Xiao, X.; Wu, Q.; Cheng, Z.; Xu, X.; Cheng, S.; Zhao, R. Effect of Magnesium Phosphate Cement on the Mechanical Properties and Microstructure of Recycled Aggregate and Recycled Aggregate Concrete. *J. Build. Eng.* **2022**, *46*, 103611. [CrossRef]
35. Qiao, F.; Chau, C.K.; Li, Z. Property Evaluation of Magnesium Phosphate Cement Mortar as Patch Repair Material. *Constr. Build. Mater.* **2010**, *24*, 695–700. [CrossRef]
36. Opara, E.U.; Karthäuser, J.; Köhler, R.; Kowald, T.; Koddenberg, T.; Mai, C. Low-Carbon Magnesium Potassium Phosphate Cement (MKPC) Binder Comprising Caustic Calcined Magnesia and Potassium Hydroxide Activated Biochar from Softwood Technical Lignin. *Constr. Build. Mater.* **2023**, *398*, 132475. [CrossRef]
37. Du, Y.; Gao, P.; Yang, J.; Shi, F. Research on the Chloride Ion Penetration Resistance of Magnesium Phosphate Cement (MPC) Material as Coating for Reinforced Concrete Structures. *Coatings* **2020**, *10*, 1145. [CrossRef]
38. Lahalle, H.; Patapy, C.; Glid, M.; Renaudin, G.; Cyr, M. Microstructural Evolution/Durability of Magnesium Phosphate Cement Paste over Time in Neutral and Basic Environments. *Cem. Concr. Res.* **2019**, *122*, 42–58. [CrossRef]
39. Xing, S.; Wu, C. Preparation of Magnesium Phosphate Cement and Application in Concrete Repair. *MATEC Web Conf.* **2018**, *142*, 02007. [CrossRef]
40. Wagh, A.S. *Chemically Bonded Phosphate Ceramics: Twenty-First Century Materials with Diverse Applications*; Elsevier: Amsterdam, The Netherlands, 2004.
41. Haque, M.A.; Chen, B.; Li, S. Water-Resisting Performances and Mechanisms of Magnesium Phosphate Cement Mortars Comprising with Fly-Ash and Silica Fume. *J. Clean. Prod.* **2022**, *369*, 133347. [CrossRef]
42. He, H.; Shuang, E.; Wen, T.; Yao, J.; Wang, X.; He, C.; Yu, Y. Employing Novel N-Doped Graphene Quantum Dots to Improve Chloride Binding of Cement. *Constr. Build. Mater.* **2023**, *401*, 132944. [CrossRef]
43. Tan, H.; Yang, Z.; Deng, X.; Guo, H.; Zhang, J.; Zheng, S.; Li, M.; Chen, P.; He, X.; Yang, J.; et al. Surface Reinforcement of Recycled Aggregates by Multi-Diameter Recycled Powder Blended Cement Paste. *J. Build. Eng.* **2023**, *64*, 105609. [CrossRef]
44. Liu, T.; Wang, Z.; Zou, D.; Zhou, A.; Du, J. Strength Enhancement of Recycled Aggregate Pervious Concrete Using a Cement Paste Redistribution Method. *Cem. Concr. Res.* **2019**, *122*, 72–82. [CrossRef]
45. Le, T.; Le Saout, G.; Garcia-Diaz, E.; Betrancourt, D.; Rémond, S. Hardened Behavior of Mortar Based on Recycled Aggregate: Influence of Saturation State at Macro- and Microscopic Scales. *Constr. Build. Mater.* **2017**, *141*, 479–490. [CrossRef]
46. Covill, A.; Hyatt, N.C.; Hill, J.; Collier, N.C. Development of Magnesium Phosphate Cements for Encapsulation of Radioactive Waste. *Adv. Appl. Ceram.* **2011**, *110*, 151–156. [CrossRef]

47. Bernasconi, D.; Viani, A.; Zárybnická, L.; Mácová, P.; Bordignon, S.; Das, G.; Borfecchia, E.; Štefančič, M.; Caviglia, C.; Destefanis, E.; et al. Reactivity of MSWI-Fly Ash in Mg-K-Phosphate Cement. *Constr. Build. Mater.* **2023**, *409*, 134082. [CrossRef]
48. Xu, B.; Ma, H.; Shao, H.; Li, Z.; Lothenbach, B. Influence of Fly Ash on Compressive Strength and Micro-Characteristics of Magnesium Potassium Phosphate Cement Mortars. *Cem. Concr. Res.* **2017**, *99*, 86–94. [CrossRef]
49. Mo, L.; Lv, L.; Deng, M.; Qian, J. Influence of Fly Ash and Metakaolin on the Microstructure and Compressive Strength of Magnesium Potassium Phosphate Cement Paste. *Cem. Concr. Res.* **2018**, *111*, 116–129. [CrossRef]
50. Gardner, L.J.; Bernal, S.A.; Walling, S.A.; Corkhill, C.L.; Provis, J.L.; Hyatt, N.C. Response to the Discussion by Hongyan Ma and Ying Li of the Paper "Characterization of Magnesium Potassium Phosphate Cement Blended with Fly Ash and Ground Granulated Blast Furnace Slag". *Cem. Concr. Res.* **2018**, *103*, 249–253. [CrossRef]
51. Gardner, L.J.; Bernal, S.A.; Walling, S.A.; Corkhill, C.L.; Provis, J.L.; Hyatt, N.C. Characterisation of Magnesium Potassium Phosphate Cements Blended with Fly Ash and Ground Granulated Blast Furnace Slag. *Cem. Concr. Res.* **2015**, *74*, 78–87. [CrossRef]
52. Ma, H.; Li, Y. Discussion of the Paper "Characterisation of Magnesium Potassium Phosphate Cement Blended with Fly Ash and Ground Granulated Blast Furnace Slag" by L.J. Gardner et Al. *Cem. Concr. Res.* **2018**, *103*, 245–248. [CrossRef]
53. *GB/T 14685-2022*; Pebble and Crushed Stone for Construction. Chinese National Standard: Beijing, China, 2022. (In Chinese)
54. Xie, X.; Zhang, T.; Yang, Y.; Lin, Z.; Wei, J.; Yu, Q. Maximum Paste Coating Thickness without Voids Clogging of Pervious Concrete and Its Relationship to the Rheological Properties of Cement Paste. *Constr. Build. Mater.* **2018**, *168*, 732–746. [CrossRef]
55. Tam, V.W.Y.; Gao, X.F.; Tam, C.M. Microstructural Analysis of Recycled Aggregate Concrete Produced from Two-Stage Mixing Approach. *Cem. Concr. Res.* **2005**, *35*, 1195–1203. [CrossRef]
56. Tam, V.W.Y.; Gao, X.F.; Tam, C.M. Environmental Enhancement through Use of Recycled Aggregate Concrete in a Two-Stage Mixing Approach. *Hum. Ecol. Risk Assess. An Int. J.* **2006**, *12*, 277–288. [CrossRef]
57. Tam, V.W.Y.; Tam, C.M.; Wang, Y. Optimization on Proportion for Recycled Aggregate in Concrete Using Two-Stage Mixing Approach. *Constr. Build. Mater.* **2007**, *21*, 1928–1939. [CrossRef]
58. Tam, V.W.Y.; Tam, C.M. Assessment of Durability of Recycled Aggregate Concrete Produced by Two-Stage Mixing Approach. *J. Mater. Sci.* **2007**, *42*, 3592–3602. [CrossRef]
59. *GB/T 17671-2021*; Test Method of Cement Mortar Strength (ISO Method). Chinese National Standard: Beijing, China, 2021. (In Chinese)
60. *ASTM C1202-19*; Test Method for Electrical Indication of Concrete's Ability to Resist Chloride Ion Penetration. American Society for Testing and Materials Standard: West Conshohocken, PA, USA, 2019.
61. *GB/T 50081-2019*; Standard for Test Methods of Concrete Physical and Mechanical Properties. Chinese National Standard: Beijing, China, 2019. (In Chinese)
62. Igarashi, S.; Bentur, A.; Mindess, S. Microhardness Testing of Cementitious Materials. *Adv. Cem. Based Mater.* **1996**, *4*, 48–57. [CrossRef]
63. Letelier, V.; Hott, F.; Bustamante, M.; Wenzel, B. Effect of Recycled Coarse Aggregate Treated with Recycled Binder Paste Coating and Accelerated Carbonation on Mechanical and Physical Properties of Concrete. *J. Build. Eng.* **2024**, *82*, 108311. [CrossRef]
64. Zaidi, S.A.; Khan, M.A.; Naqvi, T. A Review on the Properties of Recycled Aggregate Concrete (RAC) Modified with Nano-Silica. *Mater. Today Proc.* **2023**. [CrossRef]
65. Liu, H.; Feng, Q.; Yang, Y.; Zhang, J.; Zhang, J.; Duan, G. Experimental Research on Magnesium Phosphate Cements Modified by Fly Ash and Metakaolin. *Coatings* **2022**, *12*, 1030. [CrossRef]
66. Fernández-Jiménez, A.; Palomo, A. Nanostructure/Microstructure of Fly Ash Geopolymers. In *Geopolymers: Structures, Processing, Properties and Industrial Applications*; Elsevier: Cambridge, UK, 2009; pp. 89–117.
67. Haque, M.A.; Chen, B.; Liu, Y. The Role of Bauxite and Fly-Ash on the Water Stability and Microstructural Densification of Magnesium Phosphate Cement Composites. *Constr. Build. Mater.* **2020**, *260*, 119953. [CrossRef]
68. Lu, S.; Bai, S.; Shan, H. Mechanisms of Phosphate Removal from Aqueous Solution by Blast Furnace Slag and Steel Furnace Slag. *J. Zhejiang Univ. A* **2008**, *9*, 125–132. [CrossRef]
69. Olofinnade, O.M.; Osoata, O.P. Performance Assessment of Mechanical Properties of Green Normal Strength Concrete Produced with Metakaolin-Cement Coated Recycled Concrete Aggregate for Sustainable Construction. *Constr. Build. Mater.* **2023**, *407*, 133508. [CrossRef]
70. Raman, J.V.M.; Ramasamy, V. Various Treatment Techniques Involved to Enhance the Recycled Coarse Aggregate in Concrete: A Review. *Mater. Today Proc.* **2021**, *45*, 6356–6363. [CrossRef]
71. Wang, M.; Yang, X.; Wang, W. Establishing a 3D Aggregates Database from X-ray CT Scans of Bulk Concrete. *Constr. Build. Mater.* **2022**, *315*, 125740. [CrossRef]
72. Tang, Y.; Wang, Y.; Wu, D.; Chen, M.; Pang, L.; Sun, J.; Feng, W.; Wang, X. Exploring Temperature-Resilient Recycled Aggregate Concrete with Waste Rubber: An Experimental and Multi-Objective Optimization Analysis. *Rev. Adv. Mater. Sci.* **2023**, *62*, 20230347. [CrossRef]
73. Zhou, F.; Li, W.; Hu, Y.; Huang, L.; Xie, Z.; Yang, J.; Wu, D.; Chen, Z. Moisture Diffusion Coefficient of Concrete under Different Conditions. *Buildings* **2023**, *13*, 2421. [CrossRef]

Disclaimer/Publisher's Note: The statements, opinions and data contained in all publications are solely those of the individual author(s) and contributor(s) and not of MDPI and/or the editor(s). MDPI and/or the editor(s) disclaim responsibility for any injury to people or property resulting from any ideas, methods, instructions or products referred to in the content.

Article

Prediction of Hydration Heat for Diverse Cementitious Composites through a Machine Learning-Based Approach

Liqun Lu [1,2,3], **Yingze Li** [1,4], **Yuncheng Wang** [1,4,*], **Fengjuan Wang** [1,4], **Zeyu Lu** [1,4], **Zhiyong Liu** [1,4] and **Jinyang Jiang** [1,4]

1. School of Materials Science and Engineering, Southeast University, Nanjing 211189, China; 230209138@seu.edu.cn (L.L.); 230238681@seu.edu.cn (Y.L.); fjwang1118@163.com (F.W.); 101012819@seu.edu.cn (Z.L.); liuzhiyong0728@163.com (Z.L.); jiangjinyang16@163.com (J.J.)
2. State Key Laboratory of High Performance Civil Engineering Materials, Jiangsu Research Institute of Building Science Co., Ltd., Nanjing 210008, China
3. Jiangsu Sobute New Materials Co., Ltd., Nanjing 211103, China
4. Jiangsu Key Laboratory for Construction Materials, Southeast University, Nanjing 211189, China
* Correspondence: wangyc950902@foxmail.com

Abstract: Hydration plays a crucial role in cement composites, but the traditional methods for measuring hydration heat face several limitations. In this study, we propose a machine learning-based approach to predict hydration heat at specific time points for three types of cement composites: ordinary Portland cement pastes, fly ash cement pastes, and fly ash–metakaolin cement composites. By adjusting the model architecture and analyzing the datasets, we demonstrate that the optimized artificial neural network model not only performs well during the learning process but also accurately predicts hydration heat for various cement composites from an extra dataset. This approach offers a more efficient way to measure hydration heat for cement composites, reducing the need for labor- and time-intensive sample preparation and testing. Furthermore, it opens up possibilities for applying similar machine learning approaches to predict other properties of cement composites, contributing to efficient cement research and production.

Keywords: cementitious composites; hydration; characterization; machine learning; prediction

1. Introduction

Cement hydration is a chemical reaction that occurs between cement minerals and water, ultimately resulting in the formation of key hydration products crucial for cement hardening. A key characteristic of this process comes in the form of hydration heat, the heat released during this chemical reaction. It is widely accepted that the comprehension and study of this hydration heat is vitally significant to the production, design, and application of cement materials [1–3].

The primary relevance of hydration heat in cement investigations chiefly arises from its critical role in ascertaining structural characterization and evolution. Hydration heat is inextricably linked to the hydration kinetics of cement compositions and plays a pivotal role in defining distinct stages of cement hydration—namely, the initial reaction, induction, acceleration, and deceleration periods. This key identifying feature is widely recognized within the domain of cement-based studies [4,5].

Typically, such distinctions are made noticeable through instances like the notable hydration peak evident close to the 10 h mark, which can be specifically associated with the hydration of alite/C3S [6,7]. This phase forms a significant part of the formation of primary hydration products (C-S-H). By drawing upon the hydration mechanism, hydration heat is perceived as an effective pointer indicative of modifications in the crystal morphology and crystallinity within cement minerals. These changes, reciprocally, have a direct impact on the microstructure and overall performance of cement-based materials [8]. Numerous other studies have explored the variance in heat across different stages to demonstrate the ubiquitous mechanism encompassing various parameters, including but not limited to the water–cement (w/c) ratio [9,10], supplementary cementitious materials (SCMs) [11–14], and nanomaterials [15–17], among others [18–21]. A thorough understanding of the variations in hydration heat can substantially contribute to the refined design and modification of cementitious materials, which can in turn enhance their mechanical durability and prolong their lifespan.

However, accurately characterizing the hydration heat of cement poses a complex and labor-intensive challenge, largely due to a plethora of influencing factors such as the equipment used, environmental conditions, and the specific methods implemented. For instance, the selection and calibration of instruments for the measurement of hydration heat are vitally important, with isothermal calorimeters and heat flow calorimeters being traditionally used [1,22,23]. These devices require dedicated calibration and consistent maintenance to assure accurate and reliable measurements. Any disruptions or inconsistencies in equipment performance might inevitably result in flawed data, subsequently undermining the reliability of the characterization [3,24,25]. Additionally, the influence of external conditions cannot be overlooked in the process of hydration heat measurement. Environmental elements such as ambient temperature, humidity, and air circulation need to be persistently monitored and controlled to minimize their impact on the measurement. In addition, the hydration process of cement occurs at a relatively slow pace, causing the heat release to span over longer periods. Depending on the type of cementitious material and its composition, the duration of testing can range from mere hours to days or even extend to weeks. This factor makes the task of acquiring comprehensive hydration heat data inherently time-consuming.

Recently, machine learning techniques have surfaced as a promising tool for predicting the properties [26,27] and characterizing the microstructures [28–30] of cement composites. Machine learning employs computational algorithms to detect patterns in existing data and to develop predictive models [31]. This data-driven approach allows these algorithms to learn from extensive datasets derived from past experiments, aptly capturing the intricate relationships between cement composition and its properties. There have been various attempts to investigate cement hydration by employing these techniques [32–34]. For example, estimates of concrete hydration degrees have been provided through automated machine learning-based microstructure analyses [35], and heat flow rate profiles for cementitious binders containing fly ash have been predicted using deep forest models [36]. Nevertheless, there have yet to be any reported cases of direct hydration heat predictions for various cementitious composites within specific timeframes.

This study presents the development and implementation of a machine learning model specifically designed to predict cement hydration heat. The dataset design included three separate types of cementitious composites, each with distinct mix designs to promote its diversity. Measurements of heat release, gathered experimentally over 24.5 h, were splintered into 18 time intervals to collect data points for each composite. Artificial neural networks, a typical machine learning algorithm, were employed with its network architecture and parameters optimized for maximum prediction accuracy. The performance of the finalized model was validated by comparison with an extra experimental dataset to confirm its precision and efficacy. The model excelled in both accuracy and efficiency, outperforming traditional testing methods. With these trained models, researchers can obtain immediate results, eliminating waits for experiment completion, thereby facilitating significant time and resource savings. Furthermore, potential limitations resulting from environmental sensitivities can be minimized.

2. Methods

Figure 1 outlines the workflow used in this study. The main objective was to accurately predict the hydration heat for various types of cementitious composites at specific time intervals. The term "ground truth data" refers to data acquired from experimental measurements of hydration heat. This study leveraged two types of datasets: the standard dataset, which was used for model training and evaluation, and an extra dataset, which was utilized to demonstrate the broader applicability of the predictive model. Each of these stages is further detailed in the following Sections.

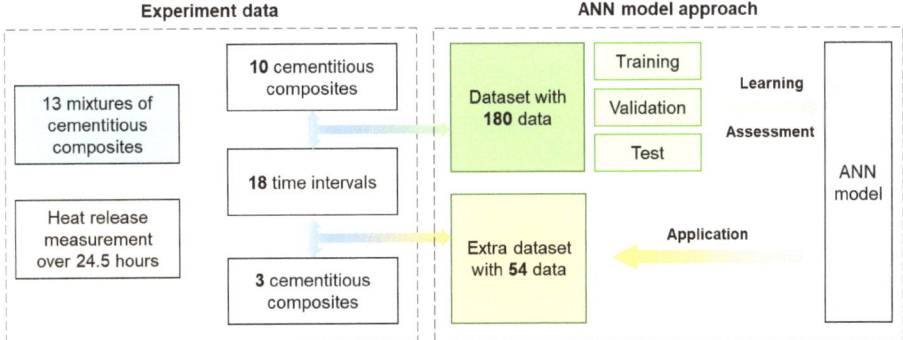

Figure 1. Schematic of the proposed machine learning-based approach for the prediction of hydration heat for diverse cementitious composites.

2.1. Design of Cementitious Composites

To bolster the robustness and broad applicability of the model, 13 distinct cementitious composites were designed, each with varying weight proportions of its components. The cumulative weight of the binders, which includes cement, fly ash, and metakaolin, was kept constant at 1. As denoted in Table 1, samples 1–5 comprised varying water-to-binder (w/b) ratios of ordinary Portland cement pastes. Samples 6–9 were made up of fly ash cement pastes, featuring a consistent w/b ratio but varying amounts of fly ash. Lastly, samples 10–13 encompassed a mixture of Portland cement, fly ash, and metakaolin, all of which consisted of the same w/b ratios and fly ash quantities while differing in the proportions of metakaolin included.

Table 1. Design of cementitious composites in terms of weight ratio.

Sample No.	Water	Cement	Fly Ash	Metakaolin
1	0.32	1	0	0
2	0.34	1	0	0
3	0.36	1	0	0
4	0.38	1	0	0
5	0.4	1	0	0
6	0.38	0.95	0.05	0
7	0.38	0.9	0.1	0
8	0.38	0.85	0.15	0
9	0.38	0.8	0.2	0
10	0.38	0.85	0.1	0.05
11	0.38	0.8	0.1	0.1
12	0.38	0.75	0.1	0.15
13	0.38	0.7	0.1	0.2

2.2. Hydration Heat Data Collection

Heat release from the hydration of cement composites was evaluated using a TAM-Air isothermal calorimeter (TA Instruments, Newcastle, DE, USA). The experiment was performed at a constant temperature of 20 °C, equivalent to room temperature, to maintain consistency given the study's primary objective of predicting hydration heat at specific time intervals for different cement composites. Measurements of heat release were recorded at 30 s intervals over a duration of 24.5 h. The overall heat release was determined by calculating the area under the curve that plots heat release against time. For subsequent analysis, data points at 18 specific time intervals—1500, 3000, 4500, 6000, 12,000, 18,000, 24,000, 30,000, 36,000, 42,000, 48,000, 54,000, 60,000, 66,000, 72,000, 78,000, 84,000, and 87,000 s—were collected and added to a database.

2.3. Data Preprocessing

As shown in Figure 1, the size of the dataset in this study equals the product of the number of composites (13) and the time intervals of hydration heat (18), resulting in 234 total data points. Three composites, samples no. 3, no. 7, and no. 12, were selectively chosen to create an extra dataset of 54 data points to demonstrate the predictive capabilities of the trained model in estimating cement hydration heat. The remaining 10 composites made up a separate dataset with 180 data points for fine-tuning the machine learning model. This dataset was randomly divided into training, validation, and test sets in a specific ratio.

In the traditional machine learning approach, the training set is used for the learning phase, during which parameters such as weights and biases are adjusted to enhance the model's performance. The validation set is used to identify any potential model overfitting, and based on these results hyperparameters such as the learning rate can be adjusted to avoid overfitting. The test set is used to evaluate the performance of the fully trained model, without any further model training.

In this study, two division scenarios were analyzed: 7:2:1 and 8:1:1. In the 7:2:1 division, the proportion of training, validation, and test data was 70%, 20%, and 10%, respectively. In the 8:1:1 division, 80% of the data were allocated for training, with 10% used for validation and the remaining 10% used for testing.

2.4. ANN Model Design and Adjustment

Figure 2 shows the schematic architecture of the artificial neural network (ANN). The network included five input parameters: water, cement, fly ash, metakaolin, and time. The network's output was the hydration heat for particular composites at specified times. Preliminary investigations indicated that additional hidden layers could lead to a decline in model performance. Therefore, this study employed a single hidden layer, testing three different neuron counts (10, 20, and 30) to optimize performance. These three ANN models are referred to as NN 10, NN 20, and NN 30, respectively, in subsequent Sections. The Levenberg–Marquardt algorithm was used, paired with a sigmoid activation function.

To prevent overfitting, the epoch was capped at 100 for all scenarios, and early stopping was implemented to halt training if the validation error ceased to decrease. An epoch denotes a full run-through of the entire training dataset in a single cycle when the machine learning model is trained. During an epoch, every training sample in the dataset is processed, and the model's parameters, such as weight and bias, are updated based on the calculated error. This is a self-learning process that allows the ANN model to adjust its parameters for optimized performance. Also, a 5-fold cross-validation was utilized, and other hyperparameters, like the learning rate and regularization, were adjusted as needed to enhance performance across all scenarios.

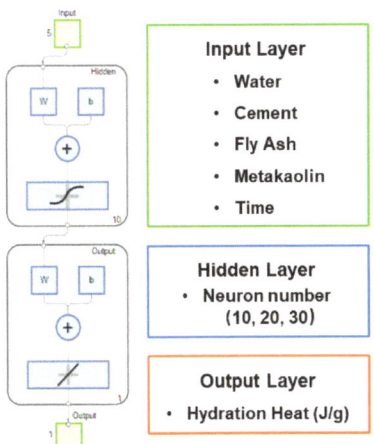

Figure 2. Schematic architecture of the ANN model and parameters.

In this study, the performance of the model was assessed based on two criteria. One of them is the mean square error (MSE), which is defined by Equation (1), where Y_i represents the actual output value and $\hat{Y}_i$ represents the predicted value. MSE, or the root mean square error (RMSE), is a widely employed metric for evaluating the performance of machine learning models.

$$\text{MSE} = \frac{1}{n}\sum_{i=1}^{n}(Y_i - \hat{Y}_i)^2 \tag{1}$$

The other criterion used involves the following steps:

(1) Plot a scatter diagram to compare the predicted output values $\hat{Y}_i$ (represented by Y) against the true output values Y_i (represented by T);
(2) Perform a linear fitting for the scatter plot points;
(3) Generate a regression (Equation (2)) relating the predicted output values $\hat{Y}_i$ (Y) and the true output values Y_i (T).

$$Y = k \times T + b \tag{2}$$

The regression equation was further evaluated by calculating the coefficient of determination, denoted as R^2, using Equation (3). In this equation, $\overline{Y}$ represents the average value of the true output.

$$R^2 = 1 - \frac{\frac{1}{n}\sum_{i=1}^{n}(Y_i - \hat{Y}_i)^2}{\frac{1}{n}\sum_{i=1}^{n}(Y_i - \overline{Y})^2} \tag{3}$$

$$R = \sqrt{1 - \frac{\frac{1}{n}\sum_{i=1}^{n}(Y_i - \hat{Y}_i)^2}{\frac{1}{n}\sum_{i=1}^{n}(Y_i - \overline{Y})^2}} \tag{4}$$

The performance of the model was evaluated based on the values of k, b, and R. For k and b, perfect prediction is indicated by their values being 1 and 0, respectively. These values represent an exact linear relationship between the predicted and true output values. R is a coefficient of determination ranging between 0 and 1. A higher value of R indicates that the scatter plot points are better aligned with the fitting line and show less fluctuation around it. This suggests a stronger correlation and better fit between the predicted and true output values.

3. Results and Discussion

3.1. Machine Learning Process

Figure 3 presents an example of the loss function for the training, validation, and test sets during the learning process. The graph demonstrates that the error drops rapidly until approximately epoch 10, after which it begins decreasing at a slower rate. As the training progresses, the trajectories for the three groups diverge. At around epoch 70, the training error continues to diminish, while the validation error stabilizes, suggesting that further learning could lead to overfitting. Hence, the termination point denoted by the green circle was chosen as the model selection criterion. This approach of stopping the model at this termination point was consistently implemented during the learning process in this study.

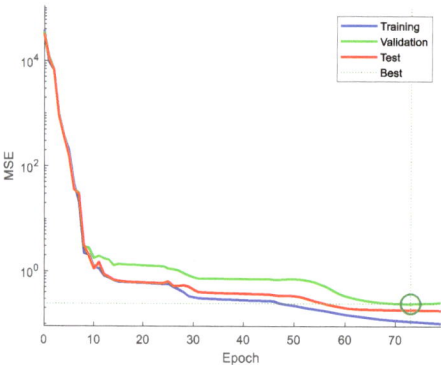

Figure 3. Example of learning termination and model selection.

3.2. ANN Model Performance Analysis

In commonly reported literature, it is expected that the performance of the training and validation sets would be similar. If the validation performance is significantly worse than that of the training set, it indicates overfitting of the model during the learning process. It is generally acceptable for the performance of the test set to be slightly worse than that of the training set. The performance of the entire dataset serves as an indication of how well the ANN model can predict hydration heat for the studied cement composites.

Figure 4 depicts the performance of ANN models with different numbers of hidden neurons (NNs)—specifically, 10, 20, and 30. Each model was evaluated using four different sets: training, validation, test, and the entire dataset. The division ratio was 8:1:1, which represents that the training, validation, and test set comprised 80%, 10%, and 10% of the entire dataset, respectively. As introduced in Section 2.4, the X axis and Y axis for the scattered data points are real heat value and predicted heat value, respectively.

Through observation, it was noted that all three models exhibited excellent performance across all sets. The scattered points in the scatter plots align closely with the fitting line, as evidenced by the high coefficient of determination (R) values exceeding 0.999. Furthermore, in all cases, the coefficient k is equal to 1, and b is less than 1. These findings suggest that the ANN model is adept at accurately predicting hydration heat at different specific times.

It is important to consider the results in the four sets as they can provide insights into the learning process of the model. For the NN 10 case, the coefficient b in the validation set (0.3) is similar to that in the training set (0.39), and even smaller in the test sets (0.15). This indicates that the trained model captures similarities across the entire dataset. However, the relatively high coefficient suggests that the model has room for improvement.

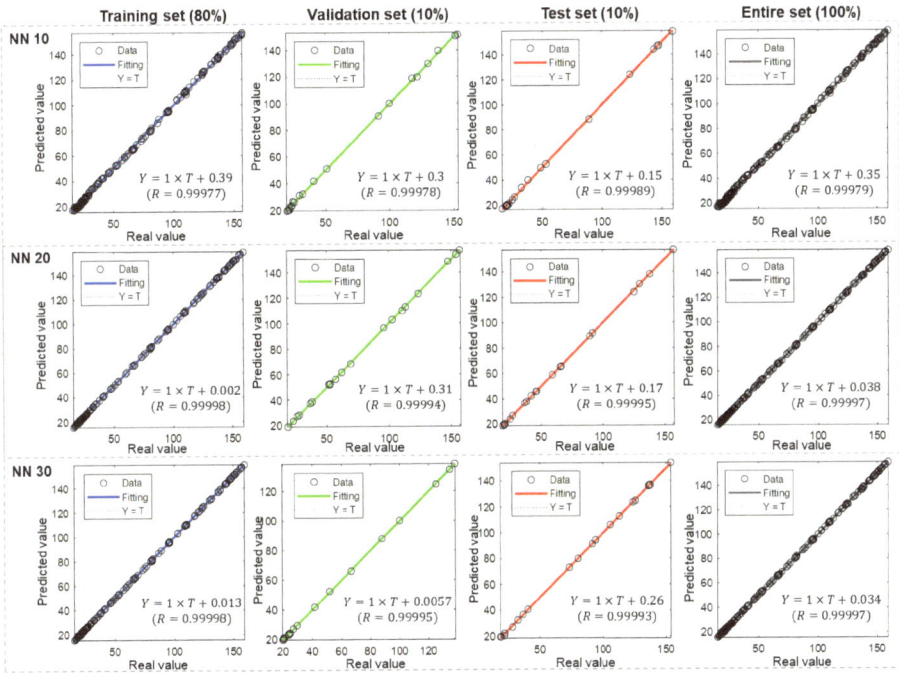

Figure 4. Performance of ANN models with different hidden neuron numbers under an 8:1:1 division ratio.

In contrast, for the NN 20 case, the coefficient b in the training set significantly decreases to 0.002, while it remains relatively unchanged in the other two sets. As a result, the values in the validation and test sets becomes noticeably larger than that in the training set. This suggests the potential occurrence of overfitting during the learning process. In the NN 30 case, the coefficient b has a very small value in the training and validation sets, but it increases to 0.26 in the test set. This situation highlights the importance of dataset division. The model is expected to learn specific features from the training set and may lose accuracy when applied to data outside the set it learned from.

It is worth noting that, in most of the reported literature, performance is evaluated by calculating the R of points with the X = Y line. However, in this study, the performance would not be easily distinguished because almost all points lie on the X = Y line. Therefore, we applied another method as introduced in the Methods Section. Based on the coefficients of regression equation k, b, and R, the slight difference could also be revealed.

Figure 5 presents the performance of ANN models with a division ratio of 7:2:1. Similarly, each model was evaluated using four different sets while the proportion of training, validation, and test sets was 70%, 20%, and 10%, respectively.

The overall performance was satisfactory with changes in the proportion of the training and validation sets. In the cases of NN 10 and NN 20, the performance in the validation set improves as the coefficient b decreases from 0.3 to 0.18 and from 0.31 to 0.16, respectively. Meanwhile, the coefficient b for the training set is only 0.056 and 0.0047, respectively. This indicates that the model benefits from having more validation data, leading to improved learning. However, in the NN 30 case, the adjustment in the ratio has a different effect. With a higher validation ratio, both the coefficient b in the training set and the validation set increase. This suggests that the model is not effectively trained based on the current dataset. Overall, the changes in the division ratio highlight the importance of dataset selection and the impact it can have on the model's performance.

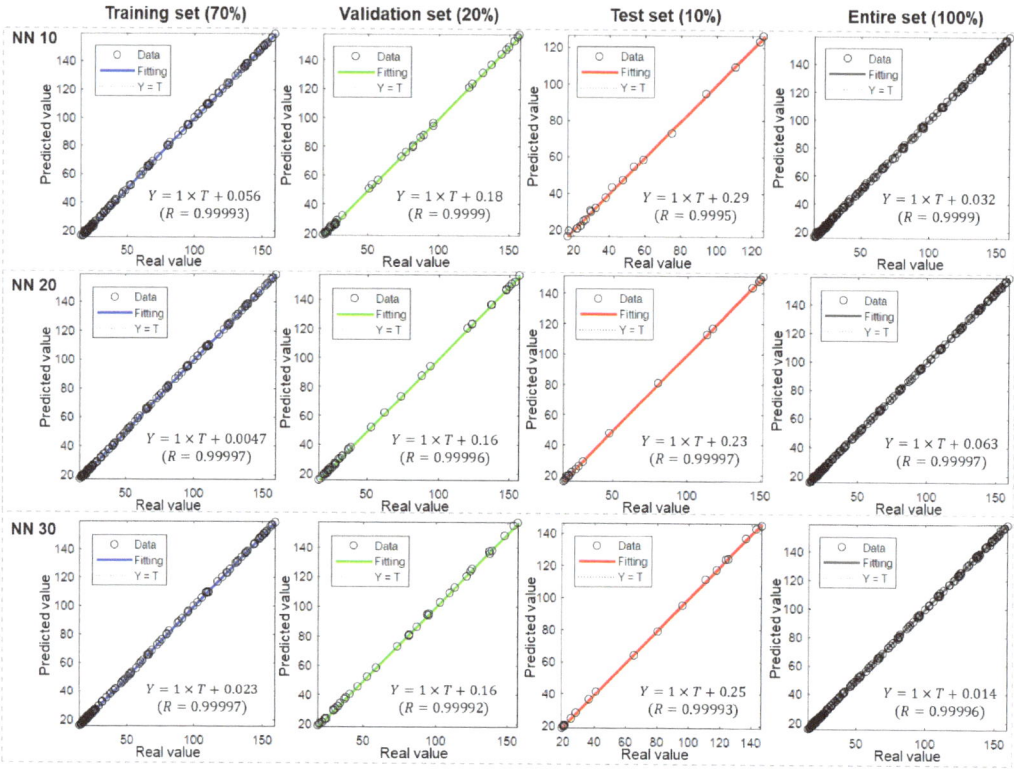

Figure 5. Performance of ANN models with different hidden neuron numbers under a 7:2:1 division ratio.

Based on the observations, there was a suspicion that with a hidden neuron number of 10 or 20, the model is capable of learning effectively when the training dataset constitutes 70% or more of the total data. Additionally, having more validation data appears to aid in model learning and helps prevent overfitting. However, as the neuron number increases to 30, the model becomes more complex and demands a larger amount of training data to achieve optimal performance. This suggests that the increased complexity and representation power of the model may require a greater amount of data for training and generalization.

Figure 6 provides the MSE error distribution for all division sets across six cases, offering further assessment of the model's performance. The X axis is the error value while the Y axis is the count number of samples. Evaluation can be based on the error boundary values and the distribution of samples among 20 bins.

For instance, in the NN 10 case with an 8:1:1 ratio, the error boundary is -2.704 and 1.949. Although the sample distribution is relatively even with more than 10 samples in 13 bins, indicating a lack of strong bias, the model could be considered as not well trained due to this distribution.

In comparison, the NN 10 case with a 7:2:1 ratio has a similar error boundary (-2.515 and 1.653), but the sample distribution is closer to a normal distribution. Notably, the samples located on the boundary mainly belong to the test set, indicating that the model is relatively well trained based on the training and validation sets. However, its performance on other data may be weaker.

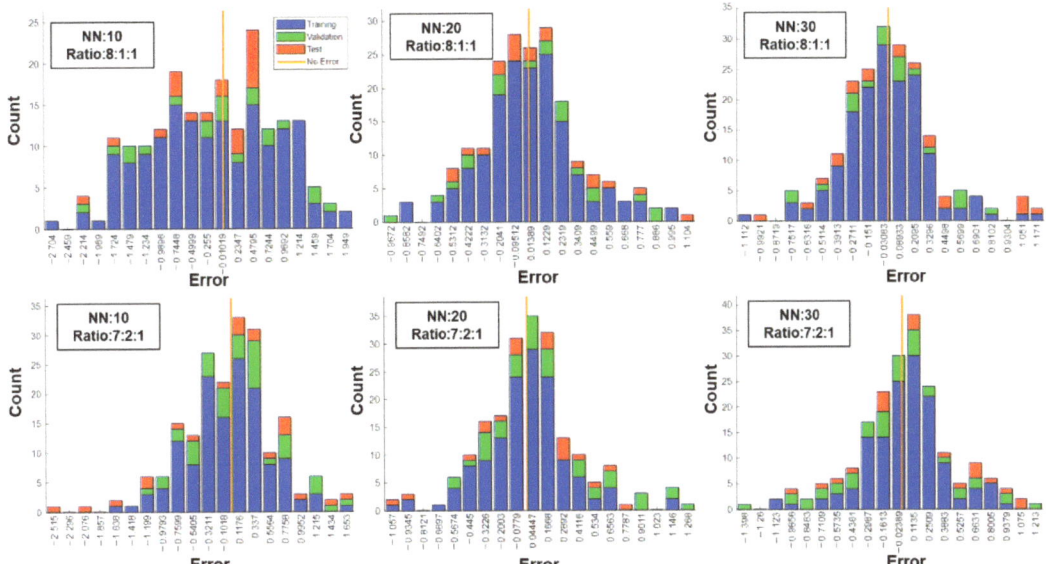

Figure 6. MSE error distribution for all division sets across six cases.

Among all cases, the NN 20 case with a 7:2:1 ratio was considered the best model in this study. It has a relatively small error boundary (−1.057 and 1.268), suggesting that the entire dataset has been included and no specific data were excluded during the learning process. Additionally, the distribution can be characterized as a normal distribution with a higher peak and smaller standard deviation.

3.3. ANN Model Application

The finalized model was applied to predict an extra dataset, which was completely distinct from the test set. Referring to Table 1, the extra dataset consisted of samples no. 3, no. 7, and no. 12. The exact values of w/c ratio, fly ash ratio, and metakaolin ratio in the extra dataset were not included in the learning dataset. However, it is important to note that the values in the extra dataset fall within the range covered by the learning dataset. Therefore, it can be expected that the constructed ANN model is capable of predicting the hydration heat for the extra dataset as well.

Figure 7 displays a scatter diagram and fitting equation of the prediction results for all cases. Observations show that, with a division ratio of 8:1:1, models with more hidden neurons perform worse. For NN 10, the coefficients k and b are 1 and −3.9, respectively, and the R-value is 0.99438. This indicates that the predicted points follow the same trend as the true values, but the exact values are underestimated.

As for NN 20 and NN 30, the coefficient k is 1.1 and 0.98, respectively, and the coefficient b is −1.2 and 1.8, respectively. In these cases, the predicted points no longer exhibit the same trend as the true values, and the fitting line shifts. The corresponding R-squared values are 0.98286 and 0.97185, indicating that the points deviate significantly from the fitting line. It is evident that these models fail to accurately predict the extra dataset. In the case of NN 30, the scatter points form three distinct lines, suggesting that the trained model is only applicable to the learning dataset and lacks generalizability to other data. A similar phenomenon is observed when the division ratio is 7:2:1, although the performance improves slightly with an R-value of 0.98644.

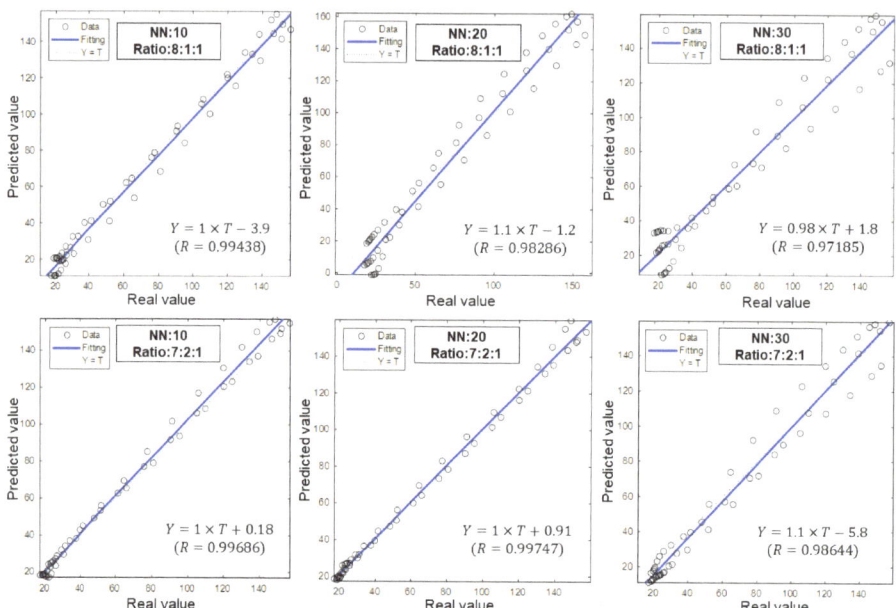

Figure 7. Performance of different ANN models cases on predicting the extra dataset.

In comparison, NN 10 and NN 20 show more promising results in terms of their predictive performance. Both cases exhibit a coefficient (k) value of 1, while NN 10 has a coefficient (b) of 0.18 and NN 20 has a coefficient (b) of 0.91. These coefficients indicate a desirable linear relationship between the predicted and true output values. Notably, both cases possess coefficients greater than 0.995, suggesting an acceptable variance for the correlation. As a result, the trained model aptly predicts the extra dataset, highlighting the robustness and efficacy of the approach.

Table 2 presents a comparison of MSE results for all datasets examined in this study. The MSE values for the learning datasets (training, validation, and test) are relatively low when compared to the MSE value for the extra dataset, which is expected. These findings align with the detailed analysis provided in the previous sections. Considering the MSE metric, the optimal model for predicting the extra application is the one with 20 hidden neurons, trained using a dataset division ratio of 7:2:1.

Table 2. Comparison of MSE results for all datasets.

	Ratio 8:1:1			Ratio 7:2:1		
	NN10	NN20	NN30	NN10	NN20	NN30
Training	1.07	0.11	0.10	0.31	0.11	0.13
Validation	1.17	0.24	0.18	0.49	0.27	0.36
Test	0.62	0.19	0.38	1.26	0.21	0.39
Extra	34.23	153.66	129.68	20.35	12.06	81.33

It is important to recognize that while the MSE value can provide an overall assessment of a model's performance during the learning process, it may not directly indicate its suitability for the extra application. For instance, in the case of NN 20 with a ratio of 8:1:1, the MSE values for the learning datasets are lower compared to those of NN 20 with a ratio of 7:2:1. However, the MSE value for extra prediction in the former case is significantly larger than the latter case.

Figure 8 shows a more direct way to compare the model's performance by plotting the predicted hydration heat value against the real value for each cementitious composite. The best and the worst models under two division ratios are presented. As for the performance on the learning dataset including training, validation, and test, almost all data are well predicted while no obvious differences can be distinguished among those models. However, the performance on the extra dataset is apparently different. In comparison to the conventional MSE assessment, the criteria employed in this study enabled us to identify variations across different learning scenarios, thus providing valuable insights for optimizing the neural network model.

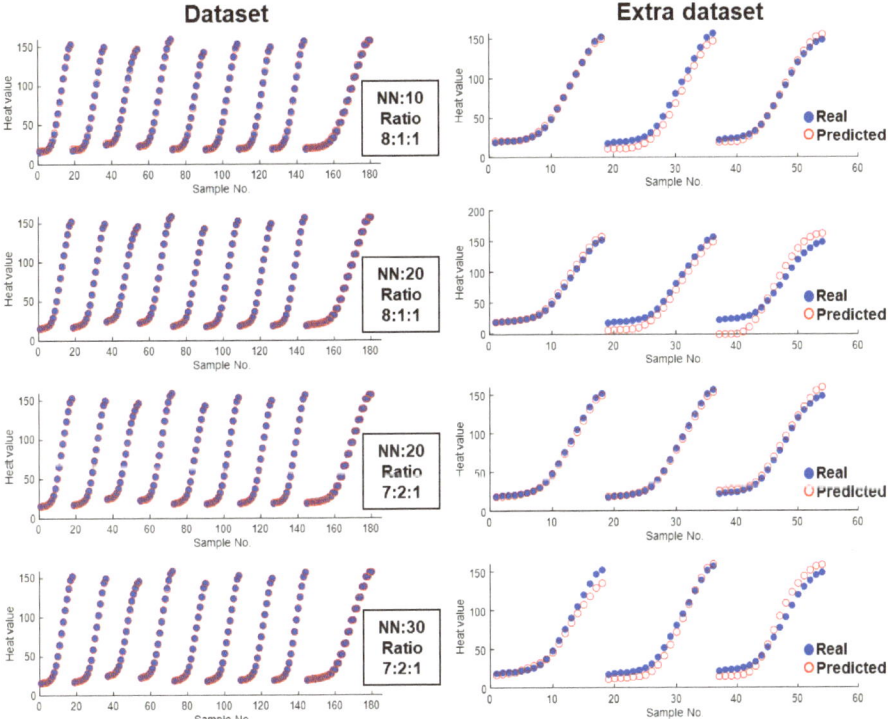

Figure 8. Direct comparison between predicted value and real value for different datasets.

The optimized ANN model exhibits tremendous potential in predicting the hydration heat of diverse cement composites. The results indicate that accurate predictions of hydration heat can be achieved for cement composites falling within the range of the training dataset, for the given composition and time. This breakthrough has the potential to substantially diminish the time and labor required for sample preparation and testing. Furthermore, the proposed machine learning approach holds promise for predicting other properties of various cement composites, extending its applicability beyond hydration heat prediction.

4. Conclusions

The characterization of hydration heat during the hydration process of cement composites is crucial in cement material production, design, and usage. However, traditional methods for measuring hydration heat are labor-intensive, time-consuming, and require specialized equipment and controlled environments.

In response to these challenges, this study introduces a machine learning-based approach to predict hydration heat at 18 precise time intervals for 13 different cement composites of varying composition ratios. There are a total of 234 data points, with 180 used for model training and evaluation and 54 used as an additional dataset to demonstrate the model's robustness. Three diverse model architectures and two dataset division scenarios were compared, and the ANN models' performances during the learning process were thoroughly analyzed.

The results indicate that the optimized ANN model not only delivers excellent performance on the datasets used during training but also accurately predicts the hydration heat of various cement composites from the extra dataset. This approach has the potential to significantly improve the efficiency of hydration heat measurements for cement composites. Additionally, it underscores the potential that this method may be extended to predict other properties of cement composites.

Author Contributions: L.L.: Conceptualization, Methodology, Software, Visualization, Investigation, Writing—Original Draft Preparation; Y.L.: Experiment, Software, Writing—Reviewing and Editing; Y.W.: Software, Supervision, Writing—Reviewing and Editing; F.W.: Writing—Reviewing and Editing; Z.L. (Zeyu Lu): Writing—Reviewing and Editing; Z.L. (Zhiyong Liu): Writing—Reviewing and Editing; J.J.: Supervision, Writing—Reviewing and Editing. All authors have read and agreed to the published version of the manuscript.

Funding: This research was funded by National Key R & D Program of China (2021YFF0500803), the National Outstanding Youth Science Fund Project of the National Natural Science Foundation of China (51925903), the National Natural Science Foundation of China (52350004), the Jiangsu Key Laboratory for Construction Materials (4212002203), and the State Key Laboratory of High Performance Civil Engineering Materials (2020CEM001).

Institutional Review Board Statement: Not applicable.

Informed Consent Statement: Not applicable.

Data Availability Statement: Data are contained within the article.

Conflicts of Interest: Author Liqun Lu was employed by the company Jiangsu Research Institute of Building Science Co., Ltd. and Jiangsu Sobute New Materials Co., Ltd. The remaining authors declare that the research was conducted in the absence of any commercial or financial relationships that could be construed as a potential conflict of interest.

References

1. Scrivener, K.; Ouzia, A.; Juilland, P.; Mohamed, A.K. Advances in understanding cement hydration mechanisms. *Cem. Concr. Res.* **2019**, *124*, 105823. [CrossRef]
2. Scrivener, K.L.; Nonat, A. Hydration of cementitious materials, present and future. *Cem. Concr. Res.* **2011**, *41*, 651–665. [CrossRef]
3. Scrivener, K.L.; Juilland, P.; Monteiro, P.J. Advances in understanding hydration of Portland cement. *Cem. Concr. Res.* **2015**, *78*, 38–56. [CrossRef]
4. Bullard, J.W.; Jennings, H.M.; Livingston, R.A.; Nonat, A.; Scherer, G.W.; Schweitzer, J.S.; Scrivener, K.L.; Thomas, J.J. Mechanisms of cement hydration. *Cem. Concr. Res.* **2011**, *41*, 1208–1223. [CrossRef]
5. Gartner, E.; Young, J.; Damidot, D.; Jawed, I. Hydration of Portland cement. *Struct. Perform. Cem.* **2002**, *2*, 57–113.
6. Brown, P.W.; Franz, E.; Frohnsdorff, G.; Taylor, H. Analyses of the aqueous phase during early C3S hydration. *Cem. Concr. Res.* **1984**, *14*, 257–262. [CrossRef]
7. Nicoleau, L.; Nonat, A. A new view on the kinetics of tricalcium silicate hydration. *Cem. Concr. Res.* **2016**, *86*, 1–11. [CrossRef]
8. Scherer, G.W.; Zhang, J.; Thomas, J.J. Nucleation and growth models for hydration of cement. *Cem. Concr. Res.* **2012**, *42*, 982–993. [CrossRef]
9. Kirby, D.M.; Biernacki, J.J. The effect of water-to-cement ratio on the hydration kinetics of tricalcium silicate cements: Testing the two-step hydration hypothesis. *Cem. Concr. Res.* **2012**, *42*, 1147–1156. [CrossRef]
10. Lahalle, H.; Coumes, C.C.D.; Mercier, C.; Lambertin, D.; Cannes, C.; Delpech, S.; Gauffinet, S. Influence of the w/c ratio on the hydration process of a magnesium phosphate cement and on its retardation by boric acid. *Cem. Concr. Res.* **2018**, *109*, 159–174. [CrossRef]
11. Zhang, J.; Dong, B.; Hong, S.; Teng, X.; Li, G.; Li, W.; Tang, L.; Xing, F. Investigating the influence of fly ash on the hydration behavior of cement using an electrochemical method. *Constr. Build. Mater.* **2019**, *222*, 41–48. [CrossRef]
12. Kadri, E.-H.; Duval, R. Hydration heat kinetics of concrete with silica fume. *Constr. Build. Mater.* **2009**, *23*, 3388–3392. [CrossRef]
13. Avet, F.; Scrivener, K. Investigation of the calcined kaolinite content on the hydration of Limestone Calcined Clay Cement (LC3). *Cem. Concr. Res.* **2018**, *107*, 124–135. [CrossRef]

14. Lothenbach, B.; Scrivener, K.; Hooton, R. Supplementary cementitious materials. *Cem. Concr. Res.* **2011**, *41*, 1244–1256. [CrossRef]
15. Dong, P.; Allahverdi, A.; Andrei, C.M.; Bassim, N.D. The effects of nano-silica on early-age hydration reactions of nano Portland cement. *Cem. Concr. Compos.* **2022**, *133*, 104698. [CrossRef]
16. Lin, J.; Shamsaei, E.; de Souza, F.B.; Sagoe-Crentsil, K.; Duan, W.H. Dispersion of graphene oxide–silica nanohybrids in alkaline environment for improving ordinary Portland cement composites. *Cem. Concr. Compos.* **2020**, *106*, 103488. [CrossRef]
17. Li, W.; Li, X.; Chen, S.J.; Liu, Y.M.; Duan, W.H.; Shah, S.P. Effects of graphene oxide on early-age hydration and electrical resistivity of Portland cement paste. *Constr. Build. Mater.* **2017**, *136*, 506–514. [CrossRef]
18. Myers, R.J.; Geng, G.; Rodriguez, E.D.; da Rosa, P.; Kirchheim, A.P.; Monteiro, P.J. Solution chemistry of cubic and orthorhombic tricalcium aluminate hydration. *Cem. Concr. Res.* **2017**, *100*, 176–185. [CrossRef]
19. Quennoz, A.; Scrivener, K.L. Interactions between alite and C3A-gypsum hydrations in model cements. *Cem. Concr. Res.* **2013**, *44*, 46–54. [CrossRef]
20. Pustovgar, E.; Mishra, R.K.; Palacios, M.; de Lacaillerie, J.-B.d.E.; Matschei, T.; Andreev, A.S.; Heinz, H.; Verel, R.; Flatt, R.J. Influence of aluminates on the hydration kinetics of tricalcium silicate. *Cem. Concr. Res.* **2017**, *100*, 245–262. [CrossRef]
21. Pustovgar, E.; Palacios, M.; de Lacaillerie, J.-B.d.E.; Matschei, T.; Ruffray, N.; Verel, R.; Flatt, R. New Insights into the Retarding Effect of Aluminates on C3S Hydration. *Spec. Publ.* **2017**, *320*, 14.11–14.12.
22. Xu, Q.; Hu, J.; Ruiz, J.M.; Wang, K.; Ge, Z. Isothermal calorimetry tests and modeling of cement hydration parameters. *Thermochim. Acta* **2010**, *499*, 91–99. [CrossRef]
23. Schöler, A.; Lothenbach, B.; Winnefeld, F.; Haha, M.B.; Zajac, M.; Ludwig, H.-M. Early hydration of SCM-blended Portland cements: A pore solution and isothermal calorimetry study. *Cem. Concr. Res.* **2017**, *93*, 71–82. [CrossRef]
24. Thomas, J.J. The instantaneous apparent activation energy of cement hydration measured using a novel calorimetry-based method. *J. Am. Ceram. Soc.* **2012**, *95*, 3291–3296. [CrossRef]
25. Jansen, D.; Naber, C.; Ectors, D.; Lu, Z.; Kong, X.-M.; Goetz-Neunhoeffer, F.; Neubauer, J. The early hydration of OPC investigated by in-situ XRD, heat flow calorimetry, pore water analysis and 1H NMR: Learning about adsorbed ions from a complete mass balance approach. *Cem. Concr. Res.* **2018**, *109*, 230–242. [CrossRef]
26. Kazemi, F.; Asgarkhani, N.; Jankowski, R. Machine learning-based seismic fragility and seismic vulnerability assessment of reinforced concrete structures. *Soil Dyn. Earthq. Eng.* **2023**, *166*, 107761. [CrossRef]
27. Albert, C.; Isgor, O.B.; Angst, U. Exploring machine learning to predict the pore solution composition of hardened cementitious systems. *Cem. Concr. Res.* **2022**, *162*, 107001. [CrossRef]
28. Guo, P.; Meng, W.; Bao, Y. Automatic identification and quantification of dense microcracks in high-performance fiber-reinforced cementitious composites through deep learning-based computer vision. *Cem. Concr. Res.* **2021**, *148*, 106532. [CrossRef]
29. Song, Y.; Huang, Z.; Shen, C.; Shi, H.; Lange, D.A. Deep learning-based automated image segmentation for concrete petrographic analysis. *Cem. Concr. Res.* **2020**, *135*, 106118. [CrossRef]
30. Lin, J.; Liu, Y.; Sui, H.; Sagoe-Crentsil, K.; Duan, W. Microstructure of graphene oxide–silica-reinforced OPC composites: Image-based characterization and nano-identification through deep learning. *Cem. Concr. Res.* **2022**, *154*, 106737. [CrossRef]
31. Jordan, M.I.; Mitchell, T.M. Machine learning: Trends, perspectives, and prospects. *Science* **2015**, *349*, 255–260. [CrossRef]
32. Lapeyre, J.; Han, T.; Wiles, B.; Ma, H.; Huang, J.; Sant, G.; Kumar, A. Machine learning enables prompt prediction of hydration kinetics of multicomponent cementitious systems. *Sci. Rep.* **2021**, *11*, 3922. [CrossRef]
33. Guo, J.; Chen, C.P.; Wang, L.; Yang, B.; Zhang, T.; Zhang, L. Constructing Microstructural Evolution System for Cement Hydration From Observed Data Using Deep Learning. *IEEE Trans. Syst. Man Cybern. Syst.* **2023**, *53*, 4576–4589. [CrossRef]
34. Tong, Z.; Wang, Z.; Wang, X.; Ma, Y.; Guo, H.; Liu, C. Characterization of hydration and dry shrinkage behavior of cement emulsified asphalt composites using deep learning. *Constr. Build. Mater.* **2021**, *274*, 121898. [CrossRef]
35. Bangaru, S.S.; Wang, C.; Hassan, M.; Jeon, H.W.; Ayiluri, T. Estimation of the degree of hydration of concrete through automated machine learning based microstructure analysis–A study on effect of image magnification. *Adv. Eng. Inform.* **2019**, *42*, 100975. [CrossRef]
36. Han, T.; Bhat, R.; Ponduru, S.A.; Sarkar, A.; Huang, J.; Sant, G.; Ma, H.; Neithalath, N.; Kumar, A. Deep learning to predict the hydration and performance of fly ash-containing cementitious binders. *Cem. Concr. Res.* **2023**, *165*, 107093. [CrossRef]

Disclaimer/Publisher's Note: The statements, opinions and data contained in all publications are solely those of the individual author(s) and contributor(s) and not of MDPI and/or the editor(s). MDPI and/or the editor(s) disclaim responsibility for any injury to people or property resulting from any ideas, methods, instructions or products referred to in the content.

Article

Synthesis and Investigation of the Hydration Degree of CA$_2$ Phase Modified with Boron and Fluorine Compounds

Michał Pyzalski [1,*], Karol Durczak [2], Agnieszka Sujak [2], Michał Juszczyk [3], Tomasz Brylewski [1] and Mateusz Stasiak [4]

[1] Faculty of Materials Science and Ceramics, AGH University of Krakow, Al. Mickiewicza 30, 30-059 Cracow, Poland; brylew@agh.edu.pl

[2] Department of Biosystems Engineering, Faculty of Environmental and Mechanical Engineering, Poznan University of Life Sciences, Wojska Polskiego 50, 60-627 Poznan, Poland; karol.durczak@up.poznan.pl (K.D.); agnieszka.sujak@up.poznan.pl (A.S.)

[3] Faculty of Civil Engineering, Cracow University of Technology, 31-155 Cracow, Poland; michal.juszczyk@pk.edu.pl

[4] Institute of Agrophysics, Polish Academy of Sciences, Doświadczalna 4, 20-290 Lublin, Poland; m.stasiak@ipan.lublin.pl

* Correspondence: michal.pyzalski@agh.edu.pl; Tel./Fax: +48-12-617-45-52

Abstract: This study investigated the effect of fluoride and boron compound additives on the synthesis and hydration process of calcium aluminate (CA$_2$). The analysis showed that the temperature of the full synthesis of CA$_2$ without mineralizing additives was 1500 °C. However, the addition of fluorine and boron compounds at 1% and 3% significantly reduced the synthesis temperature to a range of 1100–1300 °C. The addition of fluoride compounds did not result in the formation of fluoride compounds from CaO and Al$_2$O$_3$, except for the calcium borate phase (Ca$_3$(BO$_3$)$_2$) under certain conditions. In addition, the cellular parameters of the synthesized calcium aluminate phases were not affected by the use of these additives. Hydration studies showed that fluoride additives accelerate the hydration process, potentially improving mechanical properties, while boron additives slow down the reaction with water. These results highlight the relevance of fluoride and boron additives to the synthesis process and hydration kinetics of calcium aluminate, suggesting the need for further research to optimize their application in practice. TG studies confirmed the presence of convergence with respect to X-ray determinations made. SEM, EDS and elemental concentration maps confirmed the presence of a higher Al/Ca ratio in the samples and also showed the presence of hexagonal and regular hydration products.

Keywords: calcium aluminate; fluoride compound; mineralizing additives; boron compound; hydration degree; synthesis temperature

1. Introduction

Aluminate cement is a rapidly setting, high-strength hydraulic material, consisting mainly of calcium aluminates. The main crystalline phases in aluminate cements include monocalcium aluminate CaO·Al$_2$O$_3$ (CA) and dicalcium aluminate CaO·2Al$_2$O$_3$ (CA$_2$), as well as calcium ferrite and calcium aluminoferrite of the type 2CaO·Fe$_2$O$_3$ (C$_2$F), 4CaO·2Al$_2$O$_3$·Fe$_2$O$_3$ (C$_4$AF). Additionally, in smaller quantities, phases such as 12CaO·7Al$_2$O$_3$ (C$_{12}$A$_7$), 2CaO·SiO$_2$ (C$_2$S), or 2CaO·2Al$_2$O$_3$·SiO$_2$ (C$_2$AS) are present [1–4]. Literature review [5,6] indicates that the physicochemical characteristics of aluminate cements, such as strength and setting times, mainly depend on the content of the CA phase and its properties. The dicalcium aluminate phase typically exhibits lower reactivity with water compared to the CA phase. Increasing the content of the CA$_2$ phase leads to improved refractoriness of the material and the final strength of concretes prepared with such cement [5].

These types of cement are among the high-performance binders characterized by a very rapid increase in initial strength. After only a dozen hours the strength for serviceability is reached. A major advantage of these cements is also their resistance to high temperatures and ability to be used in operations carried out in winter conditions due to their significant heat of hydration. They also show satisfactory durability in waters rich in various mineral salts. They are relatively resistant to weak acids, CO_2-rich waters, and industrial effluents. Aluminous cements are used in the production of expansive cement, in the metallurgy, energy and chemical industries for various types of concrete and concrete masses operating at high temperatures (even above 1700 °C). In the mining industry, they are used for repair and construction work, injection and pit protection. The disadvantage of these cements is their poor resistance to alkaline solutions. Unfortunately, the production cost of aluminous types of cement is many times that of Portland cement [5].

The physicochemistry of aluminous cements is integrally linked to the $CaO-Al_2O_3$ system and requires detailed analysis. Work on this system has been the subject of numerous studies and the results obtained often show inconsistencies. It should be noted that this system is not yet fully understood and described. There are several calcium aluminates in this system including C_3A, $C_{12}A_7$, CA, CA_2 and CA_6. $C_{12}A_7$ was previously referred to as C_5A_3, while CA_2 as C_3A_5 [5].

Most researchers believe that C_3A and CA_6 melt incongruently, but opinions differ on the melting of the other aluminates. Nurse's research has shown that calcium aluminates melt incongruently, while Kravchenko reports that $C_{12}A_7$ occurs in cement as a unstable variety, and that stability is reached upon heating. Other authors claim that calcium dodecaluminate has polymorphic varieties, the transition of which is observed in the temperature range 1185–1305 °C. Some researchers believe that CA melts incongruently, forming a liquid similar in composition to the theoretical chemical composition of $CaO-Al_2O_3$. Others, however, suggest that calcium monoaluminate melts congruently, forming a liquid with a composition corresponding to the theoretical chemical composition of calcium aluminate [5,6]. The most important physicochemical properties of the phases of the $CaO-Al_2O_3$ system are shown in Tables 1 and 2.

Table 1. Physicochemical properties of the phases occurring in the $CaO-Al_2O_3$ system.

Phase	Physicochemical Properties
C_3A	Melts incongruently at 1639 °C (5) into CaO and liquid phase with a composition of 57.2% CaO, 42.8% Al_2O_3, and forms an eutectic with CA at 1360 °C (5) with a composition of 49.35% CaO and 50.65% of Al_2O_3.
CA	Melts incongruently at 1602 °C (5) into CA_2 and liquid phase with a composition of 36% CaO and 64% Al_2O_3. Forms an eutectic with C_3A.
CA_2	Melts incongruently at 1762 °C (±5) into CA_6 and liquid phase consisted of 22% of CaO and 78% of Al_2O.
CA_6	Melts incongruently at 1830 °C into a corund and liquid phase consisted of 16% of CaO and 84% of Al_2O_3.
$C_{12}A_7$	In dry air melts incongruently at. 1374 °C into CA and liquid phase. In the air of normal humidity incongruently at 1392 °C.

Table 2. The selected synthesis temperatures for samples containing boron and fluoride compounds.

Sample [1]	Synthesis Temperature					
	1100 [°C]	1200 [°C]	1250 [°C]	1300 [°C]	1400 [°C]	1500 [°C]
No additive	X	X		X	X	X
1% B		X	X	X		
3% B	X	X				
1% F		X		X		
3% F	X	X		X		

[1] Percentages stand for weight content of pure boron (B) or fluoride (F) in the samples. X—Sample selected for synthesis in given temperature

The properties of aluminate cements depend on their chemical and phase composition and the individual properties of each phase. Understanding the processes occurring between water and the dicalcium aluminate compound CA_2 is crucial for comprehending the essential physicochemical properties of refractory concrete. Comparative studies regarding the speed and dynamics of the hydration processes of CA_2 compound preparations, with the addition of temperature synthesis-reducing mineralizers, are particularly interesting and not yet fully explored [7,8]. The addition of various types of mineralizers (additives) typically alters the characteristic temperatures of mineral formation or melting during their thermal processing. These additives may lead to the formation of solid solutions, defecting the structure, the appearance of polymorphic varieties, or the creation of new phases, which consequently affect the changes in the physicochemical properties of the newly formed compounds [9]. Understanding the structure of minerals, especially those contained in aluminate clinkers with additives such as boron and fluorine compounds, is significant from the perspective of cement chemistry, mainly due to the reduction in the temperature of their formation, which will also influence sustainable development phenomena [10].

The addition of boron compounds to calcium aluminates, such as $C_{12}A_7$ and CA, results in the decomposition of these phases and the formation of new minerals [11,12]. In the case of $C_{12}A_7$ samples, the formation of calcium borate $Ca_3(BO_3)_2$ and the CA phase adjacent to the parent phase is observed. The content of $Ca_3(BO_3)_2$ and CA phases varies with the amount of added boron. Therefore, it can be inferred that the addition of boron to the $C_{12}A_7$ phase results in its decomposition into CA and unbound CaO, which then reacts with B_2O_3 to form $Ca_3(BO_3)_2$ [13]. The course of the reaction likely proceeds as follows:

$$3(12CaO \cdot 7Al_2O_3) + 5(B_2O_3) \rightarrow 5[Ca_3(BO_3)_2] + 21(CaO \cdot Al_2O_3) \qquad (1)$$

In the case of doping monocalcium aluminate with boron compounds, the coexistence of the CA_2 phase and $Ca_3(BO_3)_2$ was observed. The probable course of the reaction is as follows:

$$6(CaO \cdot Al_2O_3) + B_2O_3 \rightarrow Ca_3(BO_3)_2 + 3(CaO \cdot 2Al_2O_3) \qquad (2)$$

Analysis of the binary systems CaO-B_2O_3 and Al_2O_3-B_2O_3 in the temperature range from 1100 °C to 1500 °C has shown that with the addition of boron in the range of 1% to 3% (by weight), compounds such as $3CaO \cdot B_2O_3 + CaO$ and liquid phase can coexist. Similarly, in the Al_2O_3-B_2O_3 system, under the same boron content and temperature range, compounds like $9Al_2O_3 \cdot 2B_2O_3$ and liquid phase are observed [13,14].

So far, there are no relevant scientific publications regarding the doping of calcium aluminate with fluorine compounds at levels from 1% to 3% (by weight), synthesized in the temperature range from 1100 °C to 1500 °C. There is only literature data concerning the relationship between CaF_2 and AlF_3 compounds, which has limited significance for the present study [13,15].

This research paper discusses the influence of mineralizers (additives) in the form of boron or fluorine compounds on the synthesis temperature of the calcium aluminate phase. The study aims to present, analyze, and discuss preliminary experimental investigations concerning the hydraulic activity of calcium aluminate compounds modified with selected boron or fluorine compounds. The obtained results serve as a starting point for further research on the crystal chemistry of calcium aluminate compounds modified with boron oxide and fluorine oxide, as well as their impact on hydraulic activity and refractoriness.

2. Materials and Methods

2.1. Samples Preparation

It was assumed that calcium aluminate would be obtained from chemically pure compounds (CP) $CaCO_3$ and Al_2O_3 (POCH, Poland S.A., Gliwice, Poland). Then, the calcination losses of the used compounds were determined, which amounted to −43.57% for $CaCO_3$ and −3.29% for Al_2O_3, respectively. From the weighted and combined compounds, taking into account their calcination losses, a raw material set was prepared. This set was then subjected to homogenization in a rotary mixer prepared for this purpose for a period of 20 h [16,17].

The next stage of homogenization was continued in spherical agate mills with the addition of distilled water (to the consistency of a slurry). After the homogenization process was completed and the mixture was dried, the obtained raw material set was subjected to preliminary heat treatment at 100 °C for 30 min [3,18].

It was also decided that the mineralizing additives would be boron and fluorine compounds in the form of H_3BO_3 and NH_4F, (POCH, Poland S.A., Gliwice, Poland) in such quantities that, calculated as pure boron and fluorine, their content in the dry set should be 1% and 3% by weight, respectively. The samples were soaked in the appropriate amount of distilled water, in which the proper amounts of additives were dissolved. After drying the moist samples, along with the additives, they were subjected to preliminary decarbonization at 100 °C for 30 min [17,18].

The following heating temperatures [°C] were selected: 1100, 1200, 1250, 1300, 1400, 1500, and correspondingly chosen synthesis times: 15, 30, 45, 60, and 90 min. Not all samples were subjected to the same synthesis temperatures, as shown in Table 1. The prepared samples were then subjected to the appropriate thermal treatment process at temperatures according to Table 2.

The sintered samples were subjected to the grinding process in an agate mill to ensure the appropriate grain size of the preparations. Each of the samples weighing 6 g, was moistened with water at a water-to-cement ratio W/C = 0.4 and at a constant room temperature of 20 °C (± 1 °C), then placed in closed plastic containers. For the prepared preparations, the hydration process occurred at intervals: 1 day, 3 days, 7 days, 14 days, and 28 days. After each established period, the reacted samples were ground using an agate mortar, and subsequently, they could undergo comprehensive X-ray analysis [2,18].

2.2. XRD Analysis

The primary research technique utilized in this study was qualitative and quantitative analysis performed using the XRD (X-ray diffraction) method. The process of complete analysis was divided into several stages [19]:

- Qualitative examination of the phase compositions of pure sintered materials and pastes (without mineralizing additives) of the preparations;
- Qualitative examination of the phase compositions of samples doped with boron and fluorine;
- Determination and adoption of appropriate parameters for quantitative XRD methods;
- Quantitative X-ray analysis of both types of preparations.

The samples obtained after thermal treatment and hydration, when using the powder X-ray diffraction method, were additionally ground in an agate mortar to achieve the proper degree of comminution. All tested preparations must have the same degree of comminution.

Preparations made as above were subjected to qualitative and quantitative analysis of the phase composition using XRD methods. X-ray analyses were performed on a "Philips" apparatus. Qualitative studies were conducted in the range of angles from 5 to 75° 2θ, while quantitative studies of preparations were carried out in the ranges of occurrence of selected reflexes. A Cu lamp was used with the following settings:

- lamp voltage: 45 kV;
- filament current: 25 mA.

Parameters for qualitative analysis:

- angular range: from 5 to 75° 2θ;
- step counting mode;
- step size of the diffractometer arm: 0.05° 2θ;
- counting time: 2 s;
- powder sample—rotating.

Parameters for quantitative analysis:

- step counting mode,
- step size of the diffractometer arm: 0.02° 2θ,
- counting time: 10 s,
- powder sample—rotating,
- minimum number of repetitions: ×3.

To interpret the experimental results, PANalytical software HighScore Plus (Version 2.1.0) was used. Qualitative and quantitative analysis (excluding the amorphous phase) was carried out to provide a more precise assessment of the phase composition, using predefined parameters. It was not necessary to use an internal standard. This method helped to avoid additional dilution of the analyzed samples, which contributed to increased precision in the identification of doped phases.

Wykorzystano konkretne pliki CIF do analizy, w tym: α—Al_2O_3—63648.cif; CaO—26,959.cif; $3CaO \cdot Al_2O_3$—1841.cif; $12CaO \cdot 7Al_2O_3$—6287.cif; $CaO \cdot Al_2O_3$—260.cif; $CaO \cdot 2Al_2O_3$—14,270.cif i $Ca_3(BO_3)_2$—23,664.cif [20]. Reference phase structural data available in the ICSD (International Crystal Structure Database) databases were used to determine the content of individual phases in the samples. Specific CIF files for analysis were used, including: α—Al_2O_3—63,648.cif; CaO—26,959.cif; $3CaO \cdot Al_2O_3$—1841.cif; $12CaO \cdot 7Al_2O_3$—6287.cif; $CaO \cdot Al_2O_3$—260.cif; $CaO \cdot 2Al_2O_3$—14,270.cif and $Ca_3(BO_3)_2$—23,664.cif [20].

The values obtained as the quantitative analysis were rounded to the nearest 0.5% according to the declared level of accuracy of the quantitative measurements, thus ensuring the reliability of the analysis.

Based on the information available in the help section of the programme, the reliability of the quantitative results is assessed by the GOF (Goodness of Fit) coefficient, which should range from 1 to a maximum of 5. The lower the GOF coefficient value, the more reliable the test results are. In the case of our study, GOF fit coefficients ranged from 1.5 to a maximum of 2.5, indicating a high reliability of the results obtained. For quantitative analysis, the computer program "Analyze", Rayflex Version 1.0, was used [20,21].

2.3. Quantitative and Statistical Analysis

The research results, including the percentage shares of calcium carbonate in the tested samples obtained using quantitative X-ray diffraction (XRD) analysis, were subjected to further quantitative analysis using the external standard method. This method relied on comparing the surface area of the analytical peak with the surface area of the standard peak [22].

In the case of paste analyses, it was assumed that the degree of hydration "α" of the analyzed calcium aluminate phases (CA_2) was determined as the ratio of the surface area of the selected peaks for specified values of "d" in [Å] for hydrated samples to the surface area of peaks for identical values of "d" in [Å] for samples not subjected to the hydration process (standard 100%). A preparation with a calcium carbonate content of 100% was used as standard. The surface areas of the selected peaks were determined using the computer program "Analyze" [20]. The calculation of the content of the unreacted phase was performed based on the formula:

$$X = (100\% \cdot Y/Z)\%, \qquad (3)$$

where:

Y—the surface area of the respective peak for the calculated hydrate (1 day, 3 days, 7 days, 14 and 28 days),
Z—the surface area of the respective peaks for the reference cement, not subjected to hydration.

To determine the degree of hydration "α" (the quantity of reacted phase), the differences between the content of a given phase in the reference (initial) sample (assuming 100%

content of CA_2, regardless of the actual content of this phase in the compound) and the content of the given phase in the cement subjected to hydration process were calculated:

$$\alpha = 100\% - X, \tag{4}$$

The results of the quantitative analysis were statistically processed using the computer program "Statistica v 5.0 PL" and presented in tabular form. Additionally, these results were depicted in illustrating the relationship between the degree of synthesis at assumed temperatures, the degree of reaction "α", and the duration of the hydration process. It was assumed, that when presenting this relationship as graphs, the degree of hydration "α" represents the average value of statistical calculations. According to the literature data [23], the error of the quantitative determination method using X-ray techniques was determined to be $\pm 2\%$. The mean values, medians, standard deviations, and standard errors were calculated with a confidence level of 95%. The number of measurements in the quantitative analysis is not proportional to the amount of obtained results because samples with 100% CA_2 content (i.e., full synthesis of calcium carbonate) were not included in the analysis.

2.4. Thermogravimetric Analysis

The thermogravimetric tests carried out aimed to monitor the temperature-dependent changes in the mass of the sample to accurately quantify the hydration products of the calcium aluminate phases. The mass loss analysis was carried out using a NEXTA STA200 thermogravimetric analyzer (Hitachinaka, Ibaraki, Japan). Each 3 g sample was placed in a platinum crucible and then thermally treated in an oven connected to a balance. The heating process took place in stages, where the sample was gradually heated at a constant oven temperature rise rate of 5 °C/min until the set temperature of 1000 °C was reached.

During the experiment, the temperature was measured using a thermocouple placed near the sample vessel to ensure the accuracy of the measurements. The entire measurement was carried out in a gaseous atmosphere, where air was used as the carrier gas. The use of such an environment was important to obtain reliable data on sample mass changes. The present study was crucial for the quantitative analysis of the hydration products of calcium aluminate phases, an important aspect in many scientific and industrial fields. Accurate thermogravimetric measurements provided valuable information that can be used for further research and optimisation of processes related to ceramic materials and other fields where control of thermal reactions is crucial.

2.5. Morphology Studies

The morphology of the samples was examined using an ultra-high-resolution scanning electron microscope (FEI, Nova NanoSEM 200, Philips, Eindhoven, The Netherlands), equipped with a thermal field emission electron gun (FEG-Schottky emitter, Philips, Eindhoven, The Netherlands), operating at an accelerating voltage of 18 kV. The chemical composition of the tested samples was identified using an X-ray energy dispersion analyzer (EDS) from EDAX Genesis XM, connected to the scanning electron microscope.

3. Results and Discussion

3.1. XRD Analysis Results

The results of the XRD analysis, presented in the form of diffraction patterns, were interpreted and described to interpret the present phases. The diffraction patterns are shown in Figures 1–4, where individual peaks were assigned to corresponding phases and described briefly along with their d-spacing values [Å]. The analysis of the obtained diffraction patterns also allowed for a partial assessment of the content of other mineral phases present in the samples, besides calcium aluminate, such as CaO, C_3A, $C_{12}A_7$, CA, CA_2, and Al_2O_3—PDF: 43-1001; 9-413; 34-440; 23-1037 and 46-1212.

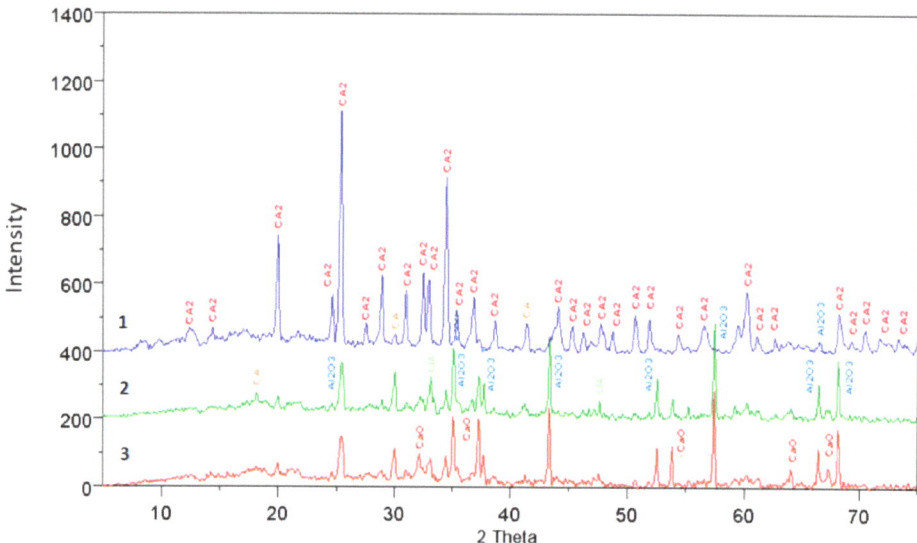

Figure 1. The diffraction pattern for a synthesis time of 15 min at temperatures of 3—1200 °C, 2—1300 °C, and 1—1500 °C, for samples without additives.

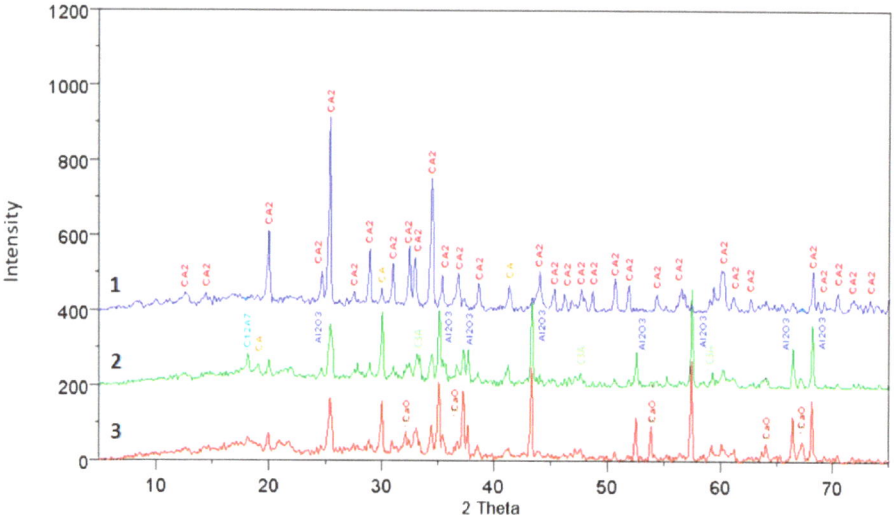

Figure 2. The diffraction pattern for a synthesis time of 90 min at temperatures of 3—1200 °C, 2—1300 °C, and 1—1500 °C, for samples without additives.

The results of this analysis are presented and compiled in Tables 3–13. Based on a commonly used, approximate method, a comparison of the intensity of reflections and the area of individual peaks (without using standards) was carried out to estimate semi-quantitatively the content of each phase.

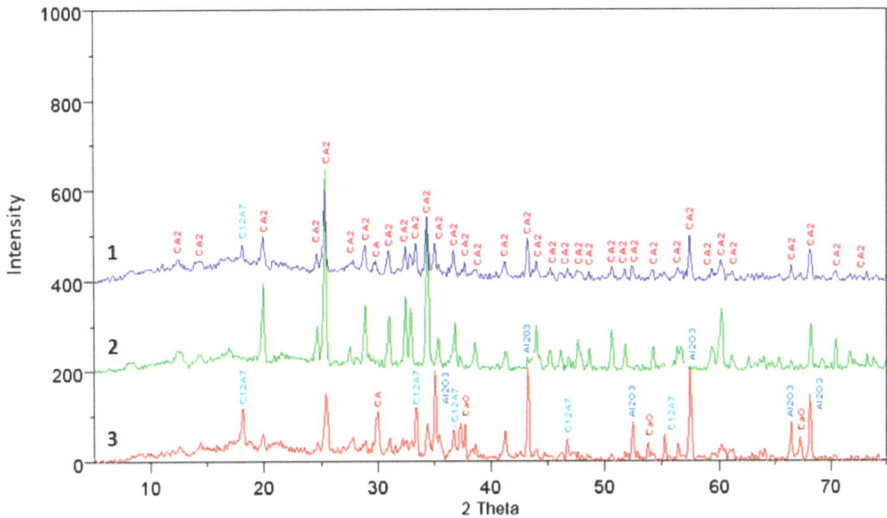

Figure 3. The diffraction pattern for a synthesis time of 15 min at temperatures of 1—1100 °C with 3% fluorine additives, 2—1200 °C with fluorine additives, and 3—1100 °C with 1% fluorine additives.

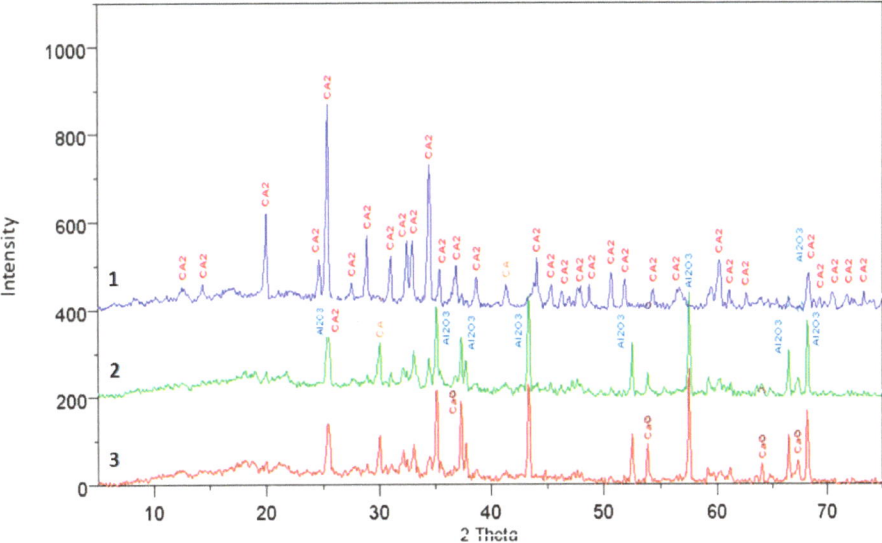

Figure 4. The diffraction pattern for synthesis times of 15, 45, and 90 min at temperatures of 3—1200 °C, 2—1250 °C, and 1—1300 °C, with a 1% boron additive.

The results of the qualitative analysis of fully reacted samples, compared to pure samples and those doped with mineralizers, show that they consist solely of the $CaO \cdot 2Al_2O_3$ phase. In the case of incomplete synthesis, in samples subjected to thermal treatment, besides the main phase, other compounds were also present, such as CaO, C_3A, $C_{12}A_7$, CA, CA_2, αAl_2O_3, and $Ca_3(BO_3)_2$.

Table 3. Identification of phases, synthesis temperature 1200 °C, no additives.

Synthesis Time [min]	CaO	C_3A	$C_{12}A_7$	CA	CA_2	α-Al_2O_3	GOF
15	8.0	6.0	3.5	4.5	18.0	50.0	1.21
30	6.5	5.5	3.0	15.5	20.5	49.0	1.55
45	6.5	5.5	3.0	15.5	20.5	49.0	1.62
60	6.5	5.5	3.5	15.5	21.0	48.0	1.75
90	5.0	5.5	4.5	17.5	22.0	45.5	1.72

Table 4. Identification of phases, synthesis temperature 1300 °C, no additives.

Synthesis Time [min]	CaO	C_3A	$C_{12}A_7$	CA	CA_2	α-Al_2O_3	GOF
15	4.5	5.0	7.0	17.0	18.5	48.0	1.55
30	3.5	4.5	7.0	18.0	18.0	49.0	1.61
45	3.0	4.5	6.0	20.5	18.5	47.5	1.57
60	2.5	4.0	6.0	21.5	19.0	47.0	1.50
90	1.5	3.5	6.0	24.5	21.5	43.0	1.45

Table 5. Identification of phases, synthesis temperature 1500 °C, no additives.

Synthesis Time [min]	CaO	C_3A	$C_{12}A_7$	CA	CA_2	α-Al_2O_3	GOF
15	0.0	1.5	1.0	2.0	93.5	2.0	2.00
30	0.0	1.0	1.0	2.5	93.5	2.0	2.08
45	0.0	0.5	0.5	1.5	96.5	1.0	2.31
60	0.0	1.0	0.5	2.0	95.5	1.0	2.07
90	0.0	0.5	0.0	1.5	97.0	1.0	1.96

Table 6. Identification of phases, synthesis temperature 1200 °C, 1% boron additive.

Synthesis Time [min]	CaO	C_3A	$C_{12}A_7$	CA	CA_2	α-Al_2O_3	GOF
15	7.0	4.5	2.5	15.5	16.0	54.5	1.70
45	4.5	5.0	4.0	20.5	17.5	48.5	1.47
90	3.0	4.5	5.0	21.0	21.5	45.0	1.44

Table 7. Identification of phases. synthesis temperature 1250 °C. 1% boron additive.

Synthesis Time [min]	CaO	C_3A	$C_{12}A_7$	CA	CA_2	α-Al_2O_3	GOF
15	5.0	5.0	4.0	17.5	21.5	47.0	1.50
45	3.5	3.0	4.5	24.0	24.5	40.5	1.60
90	1.0	2.5	4.5	26.5	37.5	28.0	1.58

Table 8. Identification of phases, synthesis temperature 1300 °C, 1% boron additive.

Synthesis Time [min]	CaO	C_3A	$C_{12}A_7$	CA	CA_2	α-Al_2O_3	GOF
15	0.0	2.0	1.0	1.5	94.0	1.5	1.61
45	0.0	1.0	1.0	2.5	94.5	1.0	1.57
90	0.0	0.5	1.0	2.0	95.5	1.0	2.08

Table 9. Identification of phases, synthesis temperature 1100 °C, 3% boron additive.

Synthesis Time [min]	CaO	C_3A	$C_{12}A_7$	CA	CA_2	$\alpha\text{-}Al_2O_3$	$Ca_3(BO_3)_2$	GOF
15	4.5	1.0	2.0	14.0	10.0	58.0	10.5	1.67
45	2.5	1.0	2.0	15.0	18.5	51.0	10.0	1.64
90	2.0	1.5	2.5	16.0	22.0	46.5	9.5	1.69

Table 10. Identification of phases, synthesis temperature 1200 °C, 3% boron additive.

Synthesis Time [min]	CaO	C_3A	$C_{12}A_7$	CA	CA_2	$\alpha\text{-}Al_2O_3$	$Ca_3(BO_3)_2$	GOF
15	0.0	0.0	0.0	1.5	88.0	10.0	0.5	1.48
30	0.0	1.0	1.0	1.5	90.5	5.5	0.5	1.51
45	0.5	0.0	1.0	2.5	90.5	5.0	0.5	1.78
90	0.5	0.0	1.5	2.5	91.5	3.0	1.0	1.56

Table 11. Identification of phases, synthesis temperature 1200 °C, 1% fluorine additive.

Synthesis Time [min]	CaO	C_3A	$C_{12}A_7$	CA	CA_2	$\alpha\text{-}Al_2O_3$	GOF
15	3.0	1.0	15.0	16.0	21.5	43.5	1.67
30	2.5	0.5	5.0	12.5	48.5	31.0	1.48
45	1.5	1.0	8.5	7.5	61.0	20.5	1.77
60	1.0	1.0	3.0	2.5	91.5	1.0	2.11
90	0.5	0.5	1.0	0.5	97.0	0.5	2.06

Table 12. Identification of phases, synthesis temperature 1300 °C, 1% fluorine additive.

Synthesis Time [min]	CaO	C_3A	$C_{12}A_7$	CA	CA_2	$\alpha\text{-}Al_2O_3$	GOF
15	0.0	0.0	1.0	1.5	96.0	1.5	2.02
45	0.0	0.0	0.0	2.5	96.5	1.0	1.77
90	0.0	0.0	0.0	2.0	97.0	1.0	2.08

Table 13. Identification of phases, synthesis temperature 1200 °C, 3% fluorine additive.

Synthesis Time [min]	CaO	C_3A	$C_{12}A_7$	CA	CA_2	$\alpha\text{-}Al_2O_3$	GOF
15	0.0	2.0	0.5	0.5	95.5	1.5	1.61
45	0.0	1.0	1.0	2.5	94.5	1.0	1.57
90	0.0	1.0	1.0	1.5	96.0	0.5	1.92

Additionally, to broaden interpretational possibilities and facilitate a direct comparison of CA_2 synthesis, selected results are presented in Figures 5–8. The charts are organized according to the convention, where the horizontal axes represent synthesis times and the vertical axes represent the percentage of synthesized CA_2 in samples. In the charts, the synthesis of CA_2 without additives in temperatures of 1200 °C and 1400 °C provides a level of reference (compare with Tables 3 and 5) for comparison with synthesis with boron compound or fluorine compound additives. In the case of synthesis with boron compound or fluorine compound additives, error bars have been added to the graphs. To make the bars visible, their values are magnified by five times.

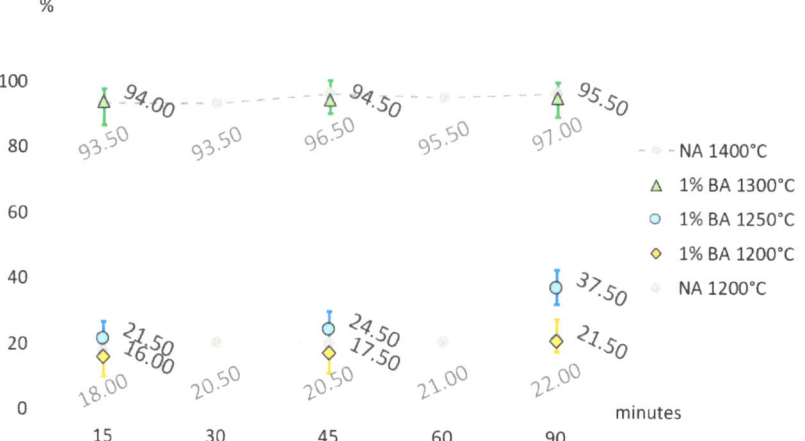

Figure 5. Comparison of the amount of CA$_2$ in the sets without additives (NA) and sets with 1% boron compound additive (BA) with regard to the synthesis time and temperature.

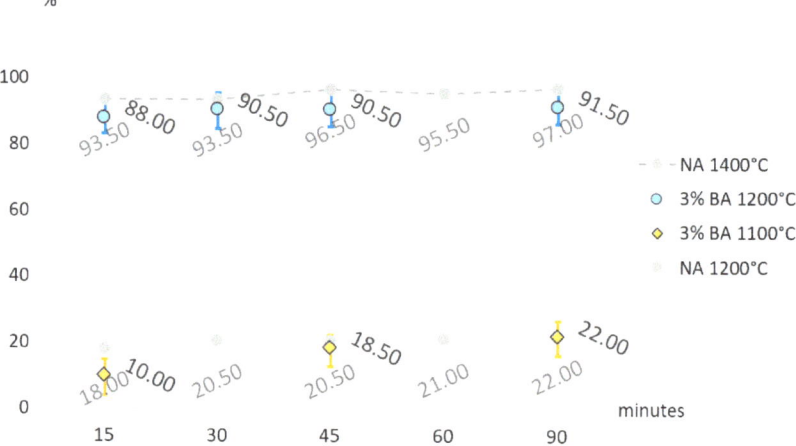

Figure 6. Comparison of the amount of CA$_2$ in the sets without additives (NA) and sets with 3% boron compound additive (BA) with regard to the synthesis time and temperature.

For samples not doped with mineralizers, at synthesis temperatures of 1200 °C and 1300 °C, all mentioned phases are present in the analyzed specimens, except for calcium borate. However, at higher temperatures the phases CaO, C$_3$A, and C$_{12}$A$_7$ gradually disappear, followed by CA and Al$_2$O$_3$ phases. It is worth noting that in the case of samples doped with boron additives, only the specimen synthesized at 1100 °C with a 3% boron additive contained, in addition to other minerals, the phase calcium borate Ca$_3$(BO$_3$)$_2$, detectable by X-ray diffraction. In specimens doped with 3% boron and subjected to higher temperatures, the presence of this compound was not observed. Specimens doped with 1% boron synthesized at 1200 °C and 1300 °C consist of phases CA, CA$_2$, and αAl$_2$O$_3$. Similarly, in the case of synthesis with a 3% boron addition, the same phases are present, but at lower thermal treatment temperatures, i.e., 1100 °C and 1200 °C.

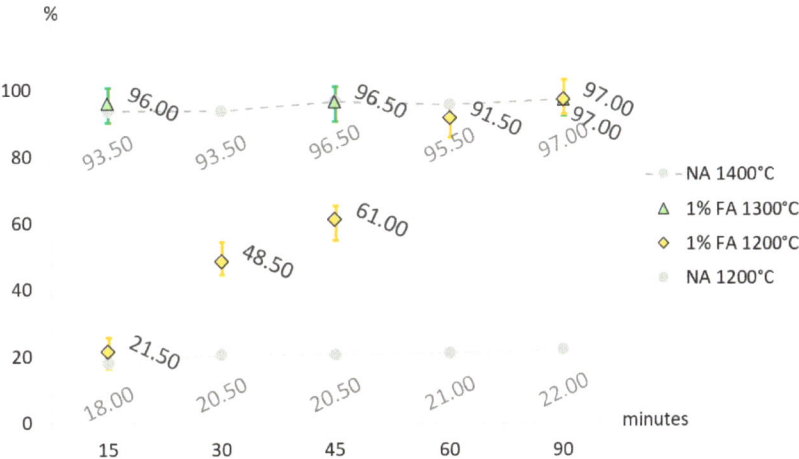

Figure 7. Comparison of the amount of CA$_2$ in the sets without additives (NA) and sets with 1% fluorine compound additive (FA) with regard to synthesis time and temperature.

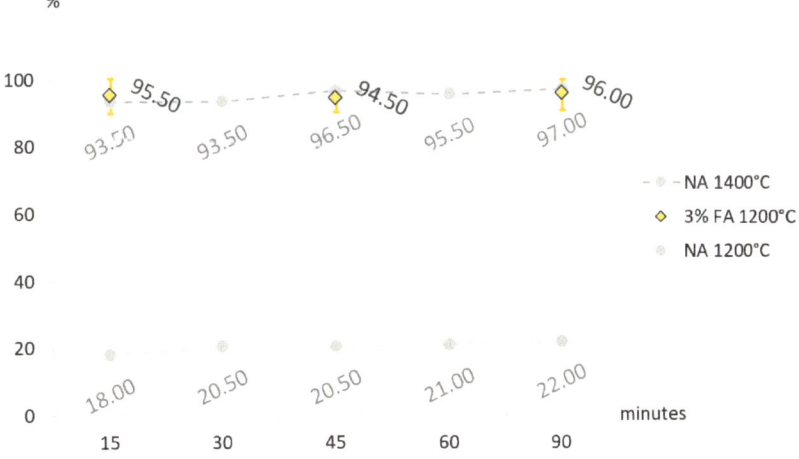

Figure 8. Comparison of the amount of CA$_2$ in the sets without additives (NA) and sets with 3% fluorine compound additive (FA) with regard to synthesis time and temperature.

An interesting observation is that in the case of specimens doped with a fluorine compound, throughout the entire range of applied temperatures and thermal treatment times, no presence of fluorine compounds with CaO and Al$_2$O$_3$ was detected. Only during the synthesis of specimens at 1200 °C with a 1% fluorine addition, the examined samples consisted of CaO, C$_{12}$A$_7$, CA, CA$_2$, and αAl$_2$O$_3$. In other cases at temperatures of 1200 °C and 1300 °C, complete synthesis of calcium aluminate CA$_2$ occurred.

The analysis of the results presented in the tabular form (Tables 9–12) for specimens without additives and with mineralizing additives indicates that in most cases, complete reaction of the samples was achieved after approximately 60 min of thermal treatment.

The presented and discussed results of the analysis of Figures 5–7 allow us to conclude that doping the primary raw material set with boron and fluorine compounds reduces and accelerates the synthesis temperature of calcium aluminate CA$_2$. Taking the temperature of 1500 °C as a reference point for the full synthesis temperature of CA$_2$ for the set without mineralizers, it can be demonstrated that the addition of fluorine has a stronger effect on reducing the temperature and speeding up the synthesis. For a 3% fluorine admixture, the temperature of full CA_2 synthesis was 1100 °C, and for a 1% addition, it was 1200 °C. Doping with boron in an amount of 3% resulted in reducing the synthesis temperature of calcium aluminate to 1200 °C, and for a 1% addition, it was 1300 °C. The discussed results apply to cases of thermal treatment lasting 60 and 90 min. It is also important to remember that the achieved results of temperature reduction in synthesis apply only to the case of the raw material set prepared according to the description in this study.

The obtained results of the statistical analysis indicate that for all considered cases, determining the percentage content of calcium aluminate using XRD techniques is associated with fluctuations in the standard deviation values ranging from 0.75 to 4.30, while the standard error values range from 0.3 to 2.4.

3.2. Research on the Unit Cell

Concurrently with the qualitative–quantitative XRD studies, an initial attempt was made to determine the parameters of the unit cell of the obtained samples to demonstrate potential differences in the structure and dimensions of the CA_2 phase unit cell, particularly in the case of doping the raw material set with mineralizers. The parameters of the unit cells were determined based on two specialized computer programs: "Powder X" and "Unitcell". Each of these programs required a separate file from the obtained diffractometric data. The results of the preliminary studies of the unit cells of the "pure" calcium aluminate phase and the phase doped with mineralizing additives are presented in table below.

The analysis of the results presented in Table 14 leads to the conclusion that the determined parameters of the unit cells obtained in the experiment for the CA_2 phases do not deviate from the parameters of the unit cells of phases with calcium aluminate as reported in the cited literature [14]. This suggests that both the quality and quantity of the introduced mineralizing additives do not cause changes in the parameters of the unit cells of the obtained samples.

Table 14. The unit cell parameters of CA_2 according to the datasheets for the synthesized preparations calculated using the "Unitcell" computer program [24].

Phase	Unit Cell Parameters [Å]			
	a	b	c	β
According to the literature [6]	12.82	8.84	5.42	107.50
No additives	12.86	8.88	5.43	106.80
1% B	12.86	8.87	5.43	106.87
1% F	12.87	8.87	5.43	106.87

3.3. Hydration Studies

The qualitative X-ray analysis showed that the samples obtained through thermal treatment of both "pure" and mineralized sets consist solely of the phase $CaO \cdot 2Al_2O_3$ (marked with X in Table 15). Since no other phases were found in the examined samples besides calcium aluminate, it was assumed that the synthesis process of this phase was complete. The temperature of full synthesis of calcium aluminate (CA_2) for the "pure" set without mineralizers was determined to be 1500 °C with a thermal treatment duration of 60 min (see Table 15). A detailed qualitative XRD analysis of the results indicates that no other phases are present in the examined samples besides the calcium aluminate phase. This observation suggests that in the case of "pure" samples and those mineralized with additives, phases resulting from reactions between calcium oxide and boron compounds or calcium oxide and fluorine compounds do not form, nor do phases formed from aluminum oxide compounds with boron compounds or aluminum oxide compounds with fluorine compounds [13].

Table 15. The selected synthesis temperatures for samples subjected to hydration. Samples marked with X consist solely of the $CaO \cdot 2Al_2O_3$ phase. [1] No additive—phase CA_2 synthesized without the addition of boron and fluorine, 1%B—phase CA_2 synthesized with the addition of 1% boron, 1%F—phase CA_2 synthesized with the addition of 1% fluorine.

Phase [1]	Synthesis Temperature and Time			
	1300 [°C]		1500 [°C]	
	60 [min]	90 [min]	60 [min]	90 [min]
No additive	-	-	X	-
1% B	-	X	-	-
1% F	-	X	-	-

The above discussion should also be applied to the formation of ternary compounds consisting of oxides B_2O_3, Al_2O_3, and CaO, and also compounds resulting from the reaction of a fluoride compound with aluminum oxide and calcium oxide [12,13,25]. Doping the raw material set with a 1% (by weight) fluoride compound resulted in the completed synthesis of the CA_2 phase at 1300 °C, while in the raw material set with a 1% boron compound, complete synthesis of calcium aluminate occurred at 1200 °C. The completed reaction of samples for the "pure" CA_2 phase and for the CA_2 phase with fluoride occurred after 60 min of thermal treatment at the mentioned temperatures, whereas for the CA_2 phase with boron, it was achieved after 90 min. Preliminary studies of the unit cell parameters of the "pure" CA_2 compound and phases doped with selected mineralizers demonstrate full compliance with data in the scientific literature [13,26]. Interestingly, the addition of boron or fluorine compounds in an amount not exceeding 1% does not induce changes in the unit cell parameters corresponding to the dimensions of the "pure" CA_2 phase. These findings are supported by the analysis of the results presented in Table 14. The unit cell parameters of the "pure" CA_2 phase align with those of phases doped with mineralizers, and the differences in dimensions between literature data and those obtained in this study range from 0.02 to 0.04%, falling within the measurement error range. Examining the hydration process of the obtained calcium aluminate preparations, as shown in Figure 8 and Tables 16–19, it can be observed that for the "pure" CA_2 phase, after 1 day of reaction with water, the degree of reaction is approximately 15%. By the third day of hardening, this degree reaches about 35%, gradually increasing over time and reaching a maximum value of about 62% after 28 days of the hydration process [27]. For the CA_2 phase with a 1% boron addition, initially, slight progress in the hydration process is observed, resulting in a degree of reaction of about 10% after the first day. In the time interval between the first and third day of the hydration process, the increase in the hydration degree "α" for this phase is minimal, amounting to only about 5%. After 3 hydration days, there was a significant acceleration of this process, with the determined value of "α" after 7 days being about 47% [28,29]. Between 7 and 28 days of reaction of this phase with water, there is a gradual slowing down of the pace of this process, reaching only a 50% degree of reaction after 28 days. Observing the hydration process of the CA_2 phase doped with a 1% fluoride addition, it can be noted that after one day of reaction with water, only a 7% degree of reaction is achieved, and then in the interval between the first and seventh day of the hydration process, the level reached is 53% [27]. Between the 7th and 28th day of the water reaction process, a stabilization of the pace of this process is observed, which reaches approximately 70% reaction completion at the final stage, after 28 days [16,30,31].

Analyzing the relationship between the hydration time of the samples and the degree of reaction "α" of the CA_2 phases leads to the conclusion that for the "pure" CA_2 preparation without mineralizers, the hydration process is faster in the period from 1 to 3 days after water mixing compared to preparations doped with boron and fluoride compounds (Figure 9). For phases doped with fluoride and boron compounds, a faster increase in the hydration process rate is observed between 3 and 7 days. After 7 days of hydration, both the "pure" preparation and the preparations doped with mineralizers have similar hydration degrees "α", ranging from 45% to about 51%. The addition of fluoride compounds induces a more intense reaction with water for the CA_2 phase doped with fluoride than for the "pure" phase. Conversely, in the case of doping the CA_2 phase with boron compounds, the reaction process with water is slower compared to the "pure" calcium aluminate phase. Regarding the dynamics of the hydration process of the "pure" CA_2 phase, it can be stated that doping this phase with fluoride compounds accelerates hydration, which consequently may lead to higher mechanical properties of the cement paste. Conversely, doping the "pure" CA_2 phase with boron compounds may lead to the opposite effect. Slower hydration processes of this phase may result in slower hydration processes overall. In summary, it can be concluded that doping the basic raw material set with boron and fluoride compounds reduces and accelerates the synthesis temperature of calcium aluminate. If we take the temperature of 1500 °C as the reference point for the full synthesis temperature of the CA_2 phase for the set without mineralizers, we can demonstrate that boron addition has a stronger effect on reducing the synthesis temperature. Doping with 1% boron compound resulted in lowering the synthesis temperature of calcium aluminate to 1200 °C.

Table 16. Results of the statistical analysis for quantitative studies of the CA_2 phase, no additives.

Hydration Time [Days]	Average	Median	Standard Deviation	Standard Error	Number of Samples
1	14.95	14.13	2.72	0.90	9
3	34.09	33.41	1.90	0.67	8
7	44.23	44.46	2.51	0.84	9
14	52.90	52.59	2.37	0.80	9
28	61.00	60.96	3.31	1.17	8

Table 17. Results of the statistical analysis for quantitative studies of the CA_2 phase, 1%B additive.

Hydration Time [Days]	Average	Median	Standard Deviation	Standard Error	Number of Samples
1	9.10	8.94	1.65	0.58	8
3	15.08	13.35	2.86	1.08	7
7	45.96	45.35	1.63	0.54	9
14	49.14	49.27	1.58	0.53	9
28	51.15	50.88	1.96	0.74	7

Table 18. Results of the statistical analysis for quantitative studies of the CA_2 phase, 1%F additive.

Hydration Time [Days]	Average	Median	Standard Deviation	Standard Error	Number of Samples
1	6.87	14.13	2.72	0.90	9
3	20.69	33.41	1.90	0.67	8
7	51.19	44.46	2.51	0.84	9
14	58.29	52.59	2.37	0.80	9
28	67.76	60.96	3.31	1.17	8

Table 19. Results of the hydration analysis.

Phase	Degree of Hydration α [%]				
	1	2	7	14	28
No additives	14.95	34.09	44.23	52.90	60.00
1%F	6.87	20.69	51.19	58.29	67.76
1% B	9.10	15.08	45.96	49.14	51.15

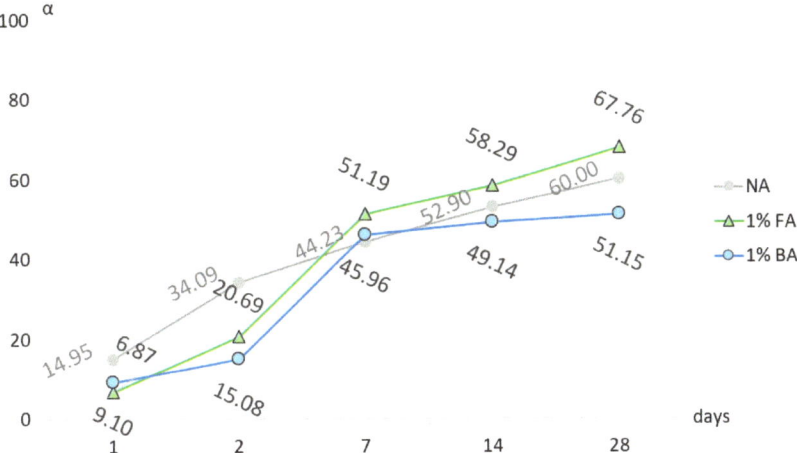

Figure 9. The time dependence of the hydration level (α in %) of CA$_2$ without additives (NA) as well as CA$_2$ with boron (BA) and fluorine (FA) compounds.

The quantitative analysis of the hydrated samples was complemented by thermogravimetric analysis of the sets containing calcium bicarbonate, both pure and doped with ions modifying the synthesis temperature. A detailed analysis is included in Table 20. The results of the quantitative analysis, carried out using the X-ray method, in comparison with the thermogravimetric analysis, show a reasonably high agreement. Higher percentages of mass loss are recorded for the TG method. The information obtained indicates that, in addition to the crystalline phases determined by powder methods, there are amorphous or poorly crystallized phases, which can certainly include aluminum hydroxides.

Table 20. Results of thermogravimetric analysis of calcium bicarbonate samples sintered at 1000 °C [wt%].

Hydration Time [Days]	CA$_2$ Phase, No Additives	CA$_2$ Phase, 1%B Additive	CA$_2$ Phase, 1%F Additive
1	15.24	9.45	7.08
3	35.59	15.83	21.72
7	47.90	46.58	55.79
14	54.25	50.19	62.62
28	64.44	52.42	71.86

Samples of calcium aluminates were subjected to hydration processes in order to demonstrate the influence of the addition of modifying ions on the synthesis temperature of the preparations as well as on the morphology of the hydration products. Microstructure analysis was carried out after 1, 3, 7 and 14 days of the hydration process. For the pure phase without modifying additives, the analysis was carried out after 1 and 14 days of the hydration process. In the case of samples doped with 1% boron compound and 1% fluorine compound, the analysis of the microstructure image of the sample was carried out after 3 and 7 days according to the order used previously. In addition to the standard images showing the sample microstructure, X-ray surface elemental distribution (samples 1 and 3) and quantitative elemental analysis of the micro-areas (samples 2 and 4) were performed. Microstructure studies were carried out according to the quantitative and qualitative analysis of the hydrated samples, as described in Section 3.3.

Figure 10 shows the morphology of the non-hydrated CA_2 phase (panels a–c). Quantitative sample analysis based on the EDS method in the mapped areas showed visible changes in the elemental concentration of calcium (highlighted in red in panel c), particularly at the left edge of the image.

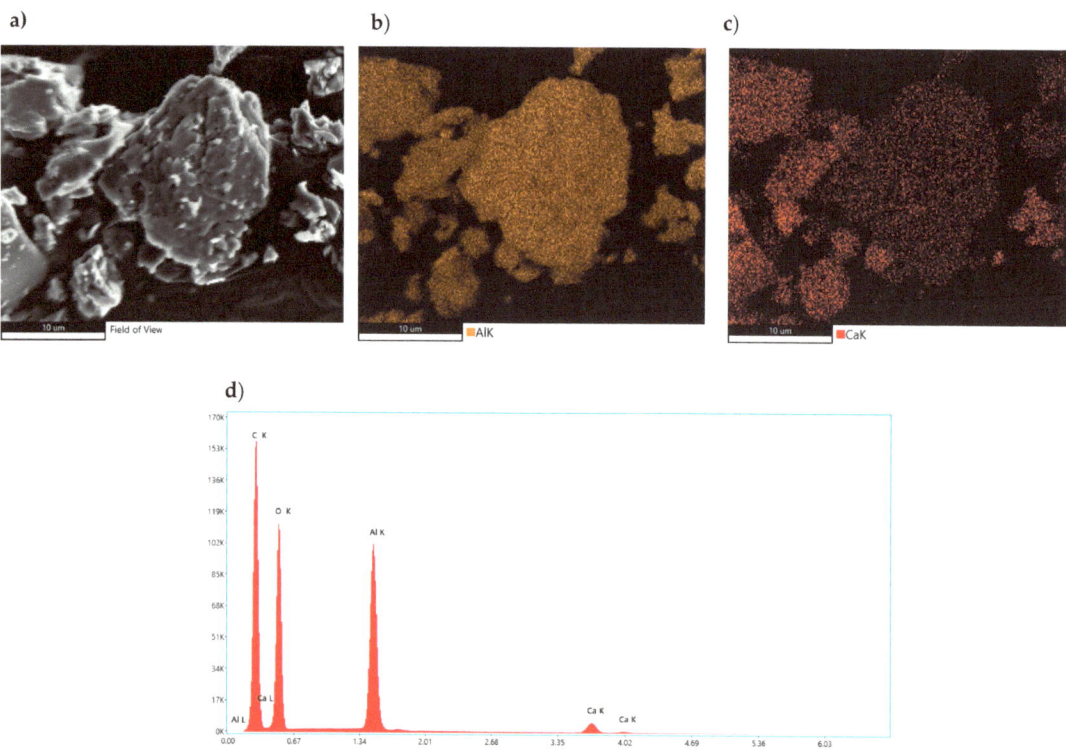

Figure 10. Morphology of CA_2 sample with analysis of the distribution of elements on its surface and average quantitative analysis of the studied micro-area. (**a**) SEM photo of the tested sample, (**b**) Al concentration, (**c**) Ca concentration, (**d**) summary EDS analysis of elements in the examined area.

These observations correlate with the image showing the location of the aluminum element, clearly lowering its concentration in areas with higher calcium elemental content. These results are consistent with the qualitative analysis performed by XRD (panel d), as the sample, in addition to the predominant CA_2 phase content, also contains aluminum-poor and calcium-rich phases, such as C_3A and $C_{12}A_7$. Quantitative analysis of the average concentration values of the abovementioned elements from the entire sample surface confirms the domination of the aluminum-rich phase.

Figure 11 shows the morphology of CA_2 without the addition of modifying compounds (panel e) after 1 day of the hydration process. An analysis of the concentration of elements present in the examined micro-areas on the samples' surface (panel f) and an average elemental composition analysis were carried out (panel g). The microphotograph of the sample shows a grain, on the surface of which two areas with a microstructure showing fundamental differences can be identified. The area in the red box is characterized by the presence of overlapping lamellar conglomerate plates growing from the grain surface (most likely products of hexagonal calcium aluminates), resembling crumpled films.

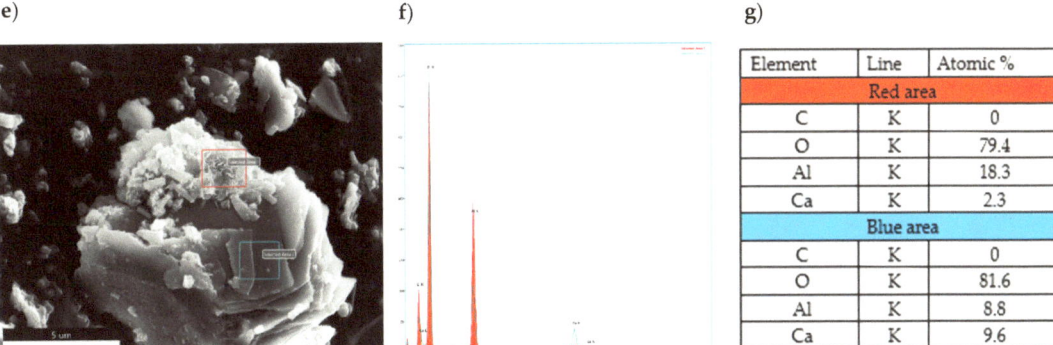

Figure 11. Morphology of CA_2 sample subjected to hydration process for 1 day with analysis of the distribution of elements on its surface and average quantitative analysis of the studied micro-area. (**e**) SEM photo of the tested sample, (**f**) EDS analysis of elements in the examined (red and blue) areas, (**g**) quantitative elemental composition of the areas examined in the SEM image.

The blue-framed area represents a flat surface not assigned to any particular crystalline phase. The analysis of the micro-areas combined with the quantitative EDS plot showed a fundamental difference in the aluminum content between the examined areas, with 52% less aluminum present in the area marked with the blue frame compared to the area marked with the red frame. This fact may indicate that hydrated calcium aluminates with different Al/Ca atomic ratios are present in the sample in addition to the non-hydrated hydration products. The varying Al/Ca ratios may also be related to the fact that the phase composition of the sample before hydration was varied, with compounds such as C_3A, $C_{12}A_7$, and CA present in addition to the predominant CA_2 phase content.

Further microscopic investigations focused on the analysis of a sample containing 1% fluoride compound, in which the hydration process was stopped on its seventh day (Figure 12h–j). Elemental concentrations on the samples' surface were mapped and an average elemental composition analysis was presented (panel k). Photographs of the surface showed the presence of a very compact, non-porous microstructure with varying crystal shapes. Cascading lamellar crystals exceeding the size of 1 μm are noticeable on the sample, with much smaller clusters of lamellar crystals in between, whose agglomerations resemble the image of a bent film. In the case of the sample in question, there was a high homogeneity of areas showing concentrations of calcium and aluminum. In the case of the aluminum element, there are local areas where its quantity is higher. XRD analysis confirmed the above findings, showing that the composition is dominated by the CA_2 phase, with CA and Al_2O_3 also present. The resulting mean micro-area X-ray dispersion analysis showed a significantly higher ratio of aluminum to calcium, thus confirming the presence of aluminum-rich compounds. The analyzed sample fragment did not show the presence of fluorine compounds.

In the photographs of the sample CA_2 doped with 1% of boron compound subjected to hydration process for 3 days (Figure 13), one can observe the occurrence of a varied microstructure in which small areas are distinguished, forming visible clusters of crystallites with diameters much smaller than 1 micrometer. In some places, large lamellar sheets located perpendicular to the flat surface of the grain can be observed. The conglomerates of small-sized crystallites resemble a spherical shape giving rise to the assumption that these are regular hydrated calcium aluminates formed by local conversion of hexagonal crystals. Spot atomic analysis of the chemical composition of selected areas in the sample was carried out. It can be concluded that for both the red and blue areas there are no significant differences in elemental composition although a higher aluminum content was found in the red rectangle.

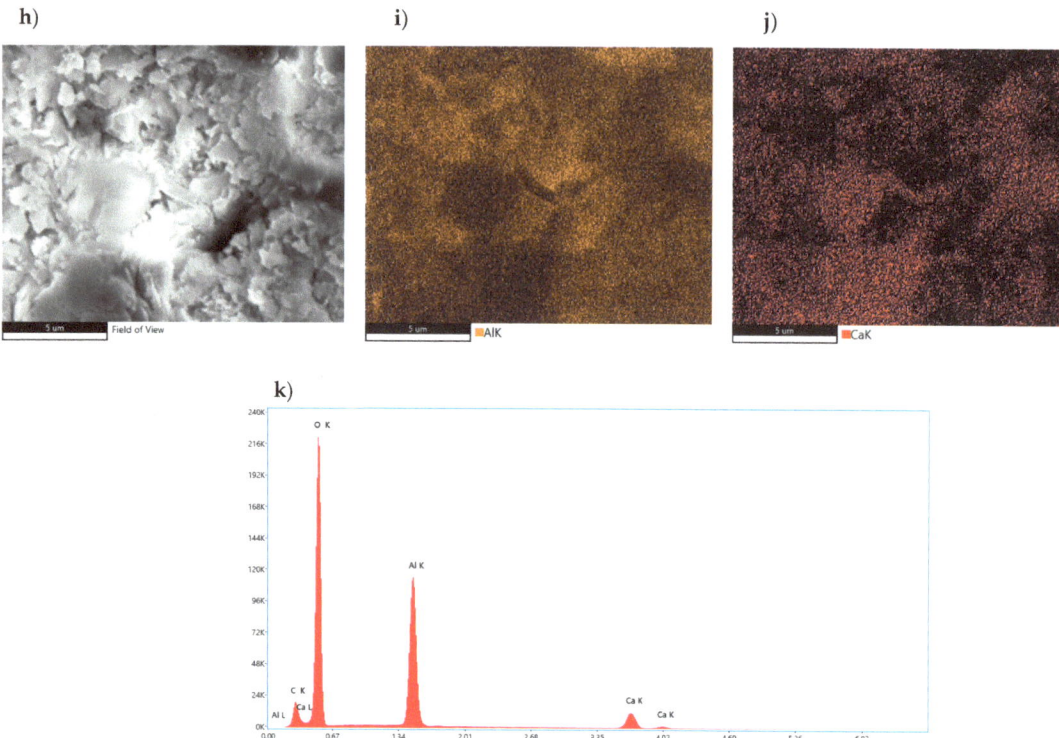

Figure 12. Morphology of CA$_2$ sample doped with 1% of fluoride compound subjected to hydration process for 7 days with analysis of the distribution of elements on its surface and average quantitative analysis of the studied micro-area. (**h**) SEM photo of the tested sample, (**i**) Al concentration, (**j**) Ca concentration, (**k**) summary EDS analysis of elements in the examined area.

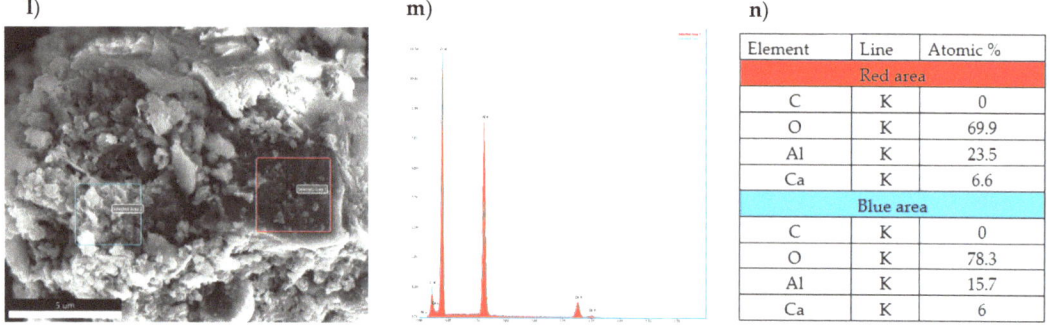

Figure 13. Morphology of CA$_2$ sample doped with 1% of boron compound subjected to hydration process for 3 days with analysis of the distribution of elements on its surface and average quantitative analysis of the studied micro-area. (**l**) SEM photo of the tested sample, (**m**) EDS analysis of elements in the examined (red and blue) areas, (**n**) quantitative elemental composition of the areas examined in the SEM image.

The same relationship is apparent in the EDS spectra. Due to the chemical composition of the analyzed sample, it is reasonable to find higher aluminum contents in relation to calcium. X-ray dispersion analysis shows that there are no boron-derived elements in the studied area.

Analysis of the microstructure of the CA_2 sample after 14 days of the hydration process (panel o) revealed the presence of a rather compact microstructure (Figure 14). In it, hexagonal layers of lamellae and clusters of cubic phases can be observed, which are accumulated in the compact surface of the test sample. In order to show the variation and the concentration of atoms, the entire study area was mapped. The "p" panel shows the concentration of aluminum atoms, while the "q" panel presents the arrangement of calcium atoms on the sample surface. Analysis of the morphology of the sample showed heterogeneity in the distribution of atoms in the different areas. In the image showing the areas of calcium atoms, areas of higher concentration of calcium atoms can be clearly seen, while this relationship is not so apparent for aluminum atoms. This relationship suggests the presence of different hydration products, which vary in calcium ion content. The average content of the individual atoms in the sample clearly indicates the dominance of aluminum atoms over calcium atoms.

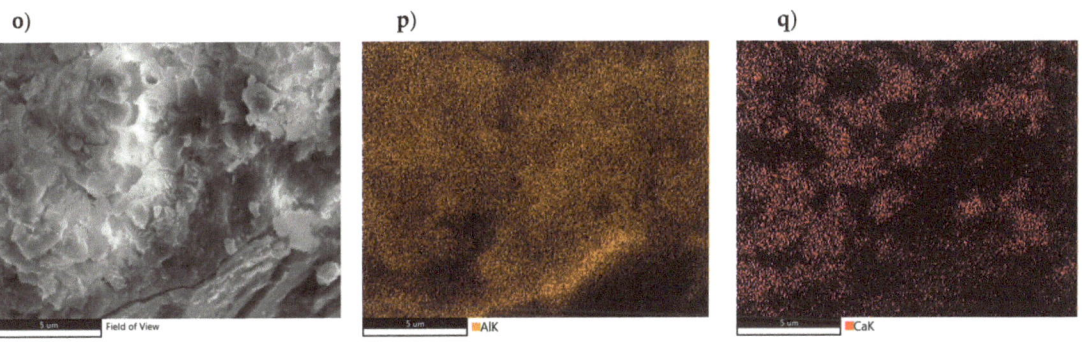

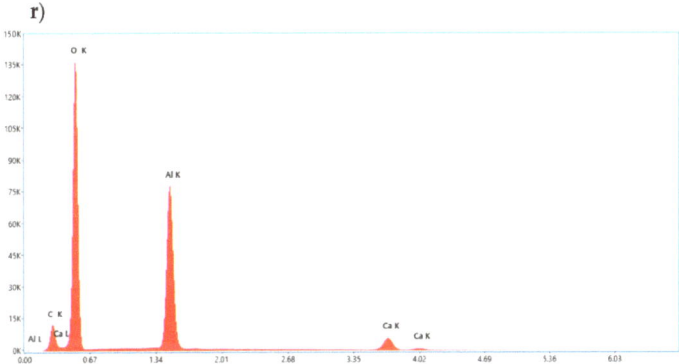

Figure 14. Morphology of CA_2 sample subjected to hydration process for 14 days with analysis of the distribution of elements on its surface and average quantitative analysis of the studied micro-area. (**o**) SEM photo of the tested sample, (**p**) Al concentration, (**q**) Ca concentration, (**r**) summary EDS analysis of elements in the examined area.

The discussed research results pertain to cases of thermal treatment lasting 60 and 90 min. Analyzing the hydration process and evaluating the potential consequences caused by introducing mineralizing additives, it can be inferred that using fluoride compounds is a more favorable option than doping CA_2 with boron compounds. Preferring the use of a fluorine-based mineralizer may lead to an increase in the mechanical properties typically achieved by the hardening paste.

Considering the above, it can be assumed that the research objective of accurately selecting the type and amount of additives has resulted in a significant reduction in the synthesis temperature of CA_2 phases, as well as provided various insights into the speed and dynamics of the hydration process between water and the "pure" and boron- and fluoride-doped calcium aluminate phases ($CaO \cdot 2Al_2O_3$).

4. Conclusions

Based on the conducted research and analysis of the obtained results, the following scientific conclusions can be drawn:

- The full synthesis temperature of calcium aluminate (CA_2) for the raw material set without mineralizing additives was 1500 °C.
- Adding a fluoride compound to the raw material set, with a 3% admixture, resulted in full synthesis of CA_2 already at a temperature of 1100 °C, while with a 1% addition, this temperature was 1200 °C.
- Adding a boron compound to the raw material set, with a 3% admixture, led to full synthesis of CA_2 already at a temperature of 1200 °C, while with a 1% addition, this temperature was 1300 °C.
- Complete reaction of the samples for all examined cases was achieved after approximately 60 min of thermal treatment at the aforementioned temperatures.
- Adding a fluoride compound to the raw material set, over the entire range of temperatures and thermal treatment times, did not lead to the formation of fluoride compounds with CaO and Al_2O_3. Only the preparation synthesized at 1100 °C with a 3% admixture contained, among other minerals, calcium borate phase ($Ca_3(BO_3)_2$).
- The use of boron and fluoride compounds in amounts up to 1% as mineralizing additives in the synthesis processes of the CA_2 phase does not cause changes in the unit cell parameters of the obtained calcium aluminate phases.
- The degree of hydration of calcium aluminate phases without mineralizing additives ranges from 15 to 60% of the degree of reaction between 1 and 28 days of the process.
- The degree of hydration of calcium aluminate with a 1% fluoride addition ranges from 7 to 70% between 1 and 28 days of the process.
- The degree of hydration of calcium aluminate phases doped with boron compounds at 1% ranges from 9 to 51% between 1 and 28 days of the hydration process.
- Adding fluoride compounds to the CA_2 phase accelerates the hydration process, which consequently may lead to better mechanical properties of the paste, while the addition of boron compounds results in a slowing down of the reaction process with water.

Limitations of the research are as follows:

- Although SEM studies were conducted, along with a detailed analysis of element mapping it is challenging to determine the precise quantities of fluorine or boron compounds that alter the elementary cell of calcium aluminate and their impact on altering the dynamics of the hydration process, as both compounds vaporize at temperatures around 800 °C.
- In the current state of the research it is impossible to determine the exact correlation between the temperature reduction during synthesis depending on the addition of temperature-lowering ions in calcium aluminates. That is due to the uncontrolled evaporation process of boron and fluorine compounds.
- As both, boron and fluoride compounds undergo evaporation processes, at this stage it is not possible to specifically indicate the percentage contribution or determine the mechanism of their reaction regarding the reduction of eutectic points in the studied systems.
- The authors intend to address the above issues in future studies.

The presented conclusions confirm the influence of fluoride and boron compound additives on the synthesis and hydration process of calcium aluminate. They also suggest further research to better understand and utilize these properties in practice. Further studies should focus on determining the levels of mineralizing additives where potential changes in the unit cell parameters of the forming phases could be observed. Additionally, efforts

should be directed towards determining other physicochemical properties of hydrated calcium aluminate doped with fluoride and boron ions.

Author Contributions: Conceptualization, M.P.; methodology, M.P.; software, M.P. and T.B.; validation, M.P.; formal analysis, M.P.; investigation, M.P.; resources, M.P.; data curation, M.P. and A.S.; writing—original draft preparation, M.P.; writing—review and editing, M.P., A.S., K.D., M.J., M.S. and T.B.; visualization, M.P., A.S., K.D., M.J., M.S. and T.B.; results discussion, M.P., A.S. and M.J.; final conclusions, M.P., A.S. and M.J.; supervision, M.P., A.S., K.D. and T.B.; project administration, K.D., M.P., A.S. and M.J.; funding acquisition, K.D., M.J. and A.S. All authors have read and agreed to the published version of the manuscript.

Funding: This work was supported by the subsidy of the Ministry of Education and Science for the AGH University of Science and Technology in Kraków (Project No 16.16.160.557). The publication was co-financed within the framework of the Polish Ministry of Science and Higher Education program: Regional Initiative Excellence, for the years 2019–2022 (No. 005/RID/2018/19), with a financing amount of PLN 12,000,000.

Institutional Review Board Statement: Not applicable.

Informed Consent Statement: Not applicable.

Data Availability Statement: All measurement data are included in this publication. For enquiries, please contact michal.pyzalski@agh.edu.pl.

Conflicts of Interest: The authors declare no conflicts of interest.

References

1. Kurdowski, W. *Chemia Cementu*; PWN: Warszawa, Poland, 1991.
2. Durczak, K.; Pyzalski, M.; Brylewski, T.; Sujak, A. Effect of Variable Synthesis Conditions on the Formation of Ye'elimite-Aluminate-Calcium (YAC) Cement and Its Hydration in the Presence of Portland Cement (OPC) and Several Accessory Additives. *Materials* **2023**, *16*, 6052. [CrossRef] [PubMed]
3. Pyzalski, M.; Brylewski, T.; Sujak, A.; Durczak, K. Changes in the Phase Composition of Calcium Aluminoferrites Based on the Synthesis Condition and Al_2O_3/Fe_2O_3 Molar Ratio. *Materials* **2023**, *16*, 4234. [CrossRef] [PubMed]
4. Pyzalski, M.; Dąbek, J.; Adamczyk, A.; Brylewski, T. Physicochemical Study of the Self-Disintegration of Calcium Orthosilicate ($\beta \rightarrow \gamma$) in the Presence of the $C_{12}A_7$ Aluminate Phase. *Materials* **2021**, *14*, 6459. [CrossRef] [PubMed]
5. Krawczenko, I.W. *Glinoziemistyj Cement*; Gosstroizdat: Moscow, Russia, 1961.
6. Kuzniecowa, T.W. *Aliuminatnyje i Sulfoaliuminatnyje Cementy*; Stroiizdat: Moscow, Russia, 1986.
7. Sawków, J.; Sulikowski, J. Wpływ składu fazowego ogniotrwałych cementów glinowych na ich własności użytkowe. *CWG* **1976**, *10*, 281–293.
8. Chatterjee, A.K. An Update on the Binary Calcium Aluminates Appearing in Aluminous Cements. In Proceedings of the International Conference of Calcium Aluminate Cements, Edinburgh, UK, 16–19 July 2001.
9. Goetz-Neuhoeffer, F.; Klaus, S.R.; Neubauer, J. Kinetics of CA and CA2 dissolution determined QXRD and corresponding enthalpies of reactions. In Proceedings of the Conference of Calcium Aluminate Cements, Avignon, France, 18–21 May 2014.
10. Pöllmann, H.; Kaden, R.; Stöber, S. Crystal structures and XRD data of new calcium aluminate cement hydrates. In Proceedings of the Conference of Calcium Aluminate Cements, Avignon, France, 18–21 May 2014.
11. Haden, R.; Pöllmann, H. Hydraulic phases in the system $BaO-Al_2O_3$. In Proceedings of the Conference of Calcium Aluminate Cements, Avignon, France, 18–21 May 2014.
12. Pragnąca, M. Studies on the Hydration of Calcium Bicarbonate. Diploma Thesis, AGH University of Cracow, Cracow, Poland, 2002. (In Polish).
13. Pöllmann, H. Mineralogy and crystal chemistry of calcium aluminate cement. In Proceedings of the International Conference of Calcium Aluminate Cements, Edinburgh, UK, 16–19 July 2001.
14. Midgley, H.G. Quantitative Determination of Phases in High alumina Cement Clinkers by X-ray diffraction. *Cem. Concr. Res.* **1976**, *6*, 217–223. [CrossRef]
15. Brylicki, W.; Derdacka, A.; Gawlicki, M.; Małolepszy, J. *Technologia Budowlanych Materiałów Wiążących 2—Cement*; WsiP: Warszawa, Poland, 1983.
16. Uliasz-Bocheńczy, K.A.; Pawluk, A.; Pyzalski, M. Characteristics of ash from the combustion of biomass in fluidized bed boilers. *Gospodarka surowcami mineralnymi. Miner. Resour. Manag.* **2016**, *32*, 149–162. [CrossRef]
17. Sulikowski, J. *Cement—Produkcja i Zastosowanie*; Arkady: Warszawa, Poland, 1981.
18. Cullity, B.D. *Podstawy Dyfrakcji Promieni Rentgenowskich*; PWN: Warszawa, Poland, 1964.
19. Rayflex Inc. *Computer Software "Analyze"*; SEIFERT; Rayflex Inc.: Bridgeport, CT, USA, 1996.
20. Redler, L. Quantitative X-Ray Diffraction Analysis of high Alumina Cements. *Cem. Concr. Res.* **1991**, *21*, 873–884. [CrossRef]

21. Snellings, R.; Salze, A.; Scrivener, K.L. Use of X-ray diffraction to quantify amorphous supplementary cementitious materials in anhydrous and hydrated blended cements. *Cem. Concr. Res.* **2014**, *64*, 89–98. [CrossRef]
22. Le Saout, G.; Kocaba, V.; Scrivener, K. Application of the Rietveld method to the analysis of anhydrous cement. *Cem. Concr. Res.* **2011**, *41*, 133–148. [CrossRef]
23. Taylor, H.F.W. *The Chemistry of Cement*; Thomas Telford: London, UK; New York, NY, USA, 1964; Volume 2.
24. Holland, T.J.B.; Redfern, S.A.T. *UNITCELL*: A nonlinear least-squares program for cell-parameter refinement and implementing regression and deletion diagnostics. *J. Appl. Cryst.* **1997**, *30*, 84. [CrossRef]
25. Majumdar, S.D.; Sarkar, R.; Vajifdar, P.P.; Narayanan, S.; Cursetji, R.M.; Chatterjee, A.K. User Friendly High Refractory Calcium Aluminate Cement. In Proceedings of the CAC: Calcium Aluminate Cements, Edinburgh, UK, 16–19 July 2001.
26. Khaliq, W.; Khan, H.A. High temperature material properties of calcium aluminate cement concrete. *Constr. Build. Mater.* **2015**, *94*, 475–487. [CrossRef]
27. Zapataa, J.F.; Gomezc, M.; Coloradoa, H.A. Structure-property relation and Weibull analysis of calcium aluminate cement pastes. *Mater. Charact.* **2017**, *134*, 9–17. [CrossRef]
28. Son, H.M.; Park, S.M.; Jang, J.G.; Lee, H.K. Effect of nano-silica on hydration and conversion of calcium aluminate cement. *Constr. Build. Mater.* **2018**, *169*, 816–825. [CrossRef]
29. Bizzozero, J.; Scrivener, K.L. Limestone reaction in calcium aluminate cement–calcium sulfate systems. *Cem. Concr. Res.* **2015**, *76*, 159–169. [CrossRef]
30. Ivanov, R.C.; Angulski da Luz, C.; Zorel, H.E., Jr.; Pereira Filho, J.I. Behavior of calcium aluminate cement (CAC) in the presence of hexavalent chromium. *Cem. Concr. Compos.* **2016**, *73*, 114–122. [CrossRef]
31. Shang, X.; Ye, G.; Zhang, Y.; Li, H.; Hou, D. Effect of micro-sized alumina powder on the hydration products of calcium aluminate cement at 40 °C. *Ceram. Int.* **2016**, *42*, 14391–14394. [CrossRef]

Disclaimer/Publisher's Note: The statements, opinions and data contained in all publications are solely those of the individual author(s) and contributor(s) and not of MDPI and/or the editor(s). MDPI and/or the editor(s) disclaim responsibility for any injury to people or property resulting from any ideas, methods, instructions or products referred to in the content.

Article

Study on the Effect of Citric Acid-Modified Chitosan on the Mechanical Properties, Shrinkage Properties, and Durability of Concrete

Zhibin Qin [1], Jiandong Wu [2], Zhenhao Hei [3], Liguo Wang [1,*], Dongyi Lei [4], Kai Liu [5] and Ying Li [4]

1. School of Materials Science and Engineering, Southeast University, Nanjing 211189, China; qinzhibinseu@163.com
2. Shandong Provincial Communications Planning and Design Institute Group Co., Ltd., Jinan 250101, China; jdwu_mail@163.com
3. Qingdao Haifa Real Estate Co., Ltd., Qingdao 266033, China; zhenhaohei@126.com
4. School of Civil Engineering, Qingdao University of Technology, Qingdao 266033, China; leidongyi@qut.edu.cn (D.L.); liying@qut.edut.cn (Y.L.)
5. Jiangsu China Construction Ready Mixed Concrete Co., Ltd., Nanjing 210033, China; lk230131@126.com
* Correspondence: wlg_seu@sina.com

Abstract: As an environmentally friendly natural polymer, citric acid-modified chitosan (CAMC) can effectively regulate the hydration and exothermic processes of cement-based materials. However, the influence of CAMC on the macroscopic properties of concrete and the optimal dosage are still unclear. This work systematically investigates the effects of CAMC on the mixing performance, mechanical properties, shrinkage performance, and durability of concrete. The results indicated that CAMC has a thickening effect and prolongs the setting time of concrete. CAMC has a negative impact on the early strength of concrete, but it is beneficial for the development of the subsequent strength of concrete. With the increase in CAMC content, the self-shrinkage rate of concrete samples decreased from 86.82 to 14.52 $\mu\varepsilon$. However, the CAMC-0.6% sample eventually expanded, with an expansion value of 78.49 $\mu\varepsilon$. Moreover, the long-term drying shrinkage rate was decreased from 551.46 to 401.94 $\mu\varepsilon$. Furthermore, low-dose CAMC can significantly reduce the diffusion coefficient of chloride ions, improve the impermeability and density of concrete, and thereby enhance the freeze–thaw cycle resistance of concrete.

Keywords: citric acid-modified chitosan; fresh mix performance; mechanical properties; shrinkage performance; durability performance

1. Introduction

Mass concrete structures frequently experience a cracking issue brought on by temperature stress. Temperature fractures have a number of detrimental impacts on high volume concrete structures, including alkali aggregate reaction, steel corrosion, concrete carbonation, and even a major impact on the durability of concrete, which can result in early building retirement and significant losses [1–3]. Therefore, timely control of the shrinkage amplitude of concrete is of positive significance for controlling cracks.

Chitosan, as a natural biopolymer, has good biocompatibility and biodegradability [4,5]. Its degradation products are non-toxic, highly adsorptive, and non-carcinogenic, and are widely used in the biomedical, wastewater treatment, flocculation, and civil engineering fields [6–9]. The previous literature [10,11] has shown that chitosan, as a polymer structure added to cement-based materials, has great application prospects in improving the early fracture toughness of cement-based materials, improving the rheological properties of cement-based materials, and regulating the heat release process of cement hydration due to the interaction between its molecular chains and the surface of cement particles.

However, chitosan's effectiveness as a polymer additive in cement-based structures is limited by its insolubility in alkaline pH conditions. Therefore, many scholars have modified chitosan to improve the properties of cement-based materials. The impact of three distinct modified chitosans on cement mortar's fresh mixing performance was examined by M. Lasheras-Zubiate [12]. The findings demonstrated that the ionic derivative carboxymethyl chitosan (CMCH) postponed the hydration of cement particles and cut the slump of cement mortar by fifty percent. Nevertheless, the slump was only marginally alleviated by the viscosity promoters with a larger molecular weight, these being hydroxypropyl methyl cellulose (HPMC) and hydroxypropyl citrulline (HPC). Bezerra [13] investigated the effects of chitosan and latex on the mechanical properties and durability of concrete. When 2% chitosan and latex were added, the strength increased and polymer fibers were found on the fracture surface of the composite material. Ustinova's [14] research results showed that the addition of chitosan increased the strength of cement components. The introduction of modified chitosan is beneficial to reducing the overall pore volume in cement components and increasing their frost resistance and bacterial resistance. It was found in our survey of the literature that the application research of chitosan and its modified products in the construction industry is still quite limited. Additionally, they are mostly used as additives and coatings in cement slurry or slurry mixtures. It is worth noting that most studies have focused on using chitosan-modified products as high-efficiency water-reducing agents to investigate their workability, setting properties, etc., [15–17] without conducting systematic research on their fresh properties, mechanical properties, or durability. In order to obtain more accurate results to design and control the impact of chitosan and its modified products on the performance of concrete, it is necessary to choose more effective modified products to study the basic properties of concrete, especially to conduct a comprehensive investigation of the fresh-mix performance, mechanical properties, and durability of concrete.

Based on previous research, citric acid-modified chitosan (CAMC) was prepared through acylation reaction and applied to cement-based materials, effectively regulating the hydration process of the cement and, as a result, the rise in temperature and the temperature cracks inside mass concrete [18–20]. Therefore, this study systematically investigated the effects of CAMC on the fresh-mix performance, mechanical properties, shrinkage performance, and durability of concrete, providing data support for the further promotion and application of CMAC in concrete.

2. Materials and Methods

2.1. Materials

P II 52.5 Portland cement (OPC) (CEM PII 52.5 supplied by Xiaoyetian Co., Ltd., Nanjing, China) and a fly ash (FA) (Shenzhen Daote Technology Co., Ltd., Shenzhen, China), as cementitious material, are adopted in this work. The oxide composition of OPC is obtained through XRF testing analysis and the mineral composition of cement is calculated based on the oxide content Bogue model [21], and its basic physical and mechanical properties and composition are shown in Tables 1 and 2. The auxiliary cementitious material is 2 μm special grade FA. The fineness modulus of river sand is 2.86, the apparent density is 2650 kg/m^3, and the bulk density is 1650 kg/m^3. The basalt aggregate has an apparent density of 2720 kg/m^3, consisting of continuously graded small stones (5 mm~10 mm) and large stones (10 mm~20 mm). The oxide compositions of basalt aggregate and river sand are shown in Table 3. The CAMC used in this experiment was citric acid-modified chitosan prepared in the laboratory. The modification preparation method and the characteristics of the polymer-modified products can be found in references [18–20].

Table 1. Physical properties of OPC.

Density (g/cm³)	Specific Surface Area (m²/kg)	Water Demand (wt.%)	Initial Setting Time/min	Final Setting Time/min	Flexural Strength/MPa		Compressive Strength/MPa	
					3 d	28 d	3 d	28 d
3.12	372	30	186	275	5.10	8.15	30.75	54.04

Table 2. Chemical and mineral compositions of portland cement and fly ash (wt.%).

Material	CaO	SiO$_2$	Al$_2$O$_3$	Fe$_2$O$_3$	SO$_3$	MgO	K$_2$O	Na$_2$O	TiO$_2$	Loss
OPC	63.62	19.70	4.45	2.93	2.93	1.28	0.68	0.12	0.27	3.92
FA	17.60	65.67	6.84	0.06	-	0.08	0.04	0.035	0.015	9.639
OPC wt.%	C$_3$S 63.6	C$_2$S 15.1	C$_3$A 7.2	C$_4$AF 6.4	Gypsum (CaSO$_4$·2H$_2$O) 2.9	CaCO$_3$ 4.8				

Table 3. Oxide composition of basalt aggregate and river sand (wt.%).

Material	SiO$_2$	Al$_2$O$_3$	Fe$_2$O$_3$	MgO	CaO	Na$_2$O	K$_2$O	TiO$_2$	MnO	P$_2$O$_5$	Loss
Basalt aggregate	48.77	13.95	13.09	4.86	8.08	2.32	1.40	4.18	0.18	0.38	2.35
River sand	82.47	9.26	2.02	0.557	3.66	0.94	0.83	0.11	-	-	0.16

2.2. Methods

In order to compare the effects of different amounts of CAMC on the fresh mixing performance, mechanical properties, and durability of concrete, 0%, 0.2%, 0.4%, and 0.6% of the cement mass were added to the concrete, with the mixing proportions shown in Table 4.

Table 4. Mix proportions of concrete.

Component	C0	CAMC-0.2%	CAMC-0.4%	CAMC-0.6%
OPC (kg/m³)	336	336	336	336
FA (kg/m³)	95	95	95	95
Natural sand (kg/m³)	789	789	789	789
Basalt (kg/m³)	1079	1079	1079	1079
Polycarboxylate water reducer (wt.%)	0.2	0.2	0.2	0.2
CAMC (wt.%)	0	0.2	0.4	0.6
Water/binder ratio	0.4	0.4	0.4	0.4

2.2.1. Fresh Mixing Performance

The slump and spread of fresh concrete are measured using a concrete slump meter, and the test method is based on the standard GB/T50080-2002 [22]. The setting time of fresh concrete slurry with different types and contents of admixtures is measured using a penetration resistance meter, and the test method is JTGE30-2005.

2.2.2. Mechanical Properties

The quasi-static compressive performance test of concrete is conducted on an electric servo hydraulic testing machine with a range of 3000 kN. The compressive strength and tensile splitting strength tests are conducted using cubic specimens (each 100 × 100 × 100 mm³). The compressive strength of concrete is determined at a constant loading rate of 0.8–1.0 MPa/s, and the tensile splitting strength of concrete is determined at a constant loading rate of 0.05–0.08 MPa/s.

2.2.3. Shrinkage Performance

The early autogenous shrinkage test of fresh concrete is conducted using the SBT-AS 200 autogenous shrinkage tester. Two parallel samples are formed for each mixing ratio, and the test samples are placed in a corrugated PE tube with a length of 420 ± 5 mm and a diameter of 50 ± 0.2 mm to test the early autogenous shrinkage of the concrete. Then, the sample is placed directly in a standard curing room for curing and testing (temperature of 23 ± 2 °C and relative humidity of 60 ± 2%). The zero point of autogenous shrinkage is based on the initial setting time of the system, which is determined using the penetration resistance method. The calculation method for early autogenous shrinkage strain is shown in the formula:

$$\varepsilon_a(t) = \xi(t) - \xi(0) \tag{1}$$

$\varepsilon_a(t)$ is the self-shrinking strain at time t, $\xi(t)$ is the measured linear strain at time t, $\xi(0)$ is the measured linear strain at time 0.

The dry shrinkage performance of the samples was tested according to the GB/T50080-2002 [22]. Three groups of $100 \times 100 \times 515$ mm^3-prism specimens were made for each mixing ratio sample, and repeated experiments were conducted. After the specimens were formed, they were sent to a standard curing room, and after 48 h of mold curing, the molds were removed. Then, the sample was placed in a standard curing room for curing. The change in the vertical length of the cylindrical specimens was recorded with a microcaliper, and the long-term drying shrinkage rate was calculated.

$$\varepsilon_d(t) = \frac{x(t) - x(0)}{l} \tag{2}$$

where $\varepsilon_d(t)$ is the drying shrinkage strain at time t, $x(t)$ is the microcaliper measurement at time t (mm), $x(0)$ is the initial reading of the test piece placed in the environmental chamber (mm), and l is the initial vertical length.

2.2.4. Durability Performance

(1) Rapid chloride ion migration coefficient method (RCM)

Perform a concrete RCM test according to GB/T 50082-2019 [23], and place the test sample on a concrete chloride ion electrotransportation coefficient tester after vacuum water retention (Beijing RCM-NTB). After power-on, use a 0.1 molAgNO$_3$ color indicator solution to test the chloride ion migration depth. After about 15 min, a white silver nitrate precipitate can be seen at the sample interface. Use a marker to draw the color boundary and divide it into 10 equal parts along the interface. Measure the distance from the boundary to the bottom of the sample and take the average as the penetration depth, accurate to 0.1 mm. Calculate the RCM coefficient of the mortar according to Equation (3).

$$D_{RCM} = \frac{0.0239(273+T)L}{(U-2)t}\left(X_d - 0.0238\sqrt{\frac{(273+T)LX_d}{U-2}}\right) \tag{3}$$

where D is the non-steady state ion migration velocity of soil, in 10^{-12} m^2/s; T is the average of the initial and end temperatures of the anode solution, in °C; U is the absolute value of the applied voltage in the experiment, in V; L is the thickness of the concrete specimen, in mm; X_d is the penetration depth of chloride ions, in mm; t is the energization time, in h.

(2) Concrete Freezing-Thawing Cycle Test

Refer to GB/T 50082-2019 [23] for the freezing–thawing cycle test of concrete. Before the freezing–thawing cycle, the test block was first kept in the standard curing room for 24 days, then immersed in water for 4 days to reach saturation. After removing the sample, the surface moisture was wiped clean with a cloth, and the weight was measured and recorded. Then, the test block was placed in a concrete rapid freezing–thawing test machine for the freezing–thawing cycle test. After the test reached the desired number of cycles, the test block was weighed and ultrasonic testing was performed.

(1) Quality loss

The quality loss rate is calculated according to formula (4):

$$\Delta m = \frac{m_0 - m_t}{m_0} \times 100\% \qquad (4)$$

where Δm is the mass loss rate of the test block at erosion age t; m_0 is the initial mass of the test block; m_t is the mass of the test block at erosion age t.

(2) Relative dynamic elastic modulus

The dynamic elastic modulus of concrete is measured using a non-metallic ultrasonic tester. Ultrasonic waves with a frequency of 54 Hz are used to test the dynamic elastic modulus of concrete specimens with different erosion ages, and the relative change in the dynamic elastic modulus is calculated. The calculation method can be obtained by using Equations (5) and (6).

$$E = \frac{(1+v)(1-2v)\rho V^2}{1-v} \qquad (5)$$

$$E_{rd} = \frac{E_n}{E_0} = \frac{V_n^2}{V_0^2} = \frac{t_0^2}{t_n^2} \qquad (6)$$

where E represents the dynamic elastic modulus, in MPa; E_0 and E_n represent the dynamic elastic modulus before and at the erosion age n, respectively, in MPa; E_{rd} represents the relative dynamic elastic modulus, 1; v represents the Poisson's ratio of cement-based materials, 1; ρ represents the density of cement-based materials, in kg/m^3; V represents the ultrasonic velocity, in m/s; t_0 represents the ultrasonic time before erosion, in μs; t_n represents the ultrasonic time at the erosion age n, in μs.

2.2.5. Mercury Injection Pressure Test (MIP)

The pore structure distribution of the hardened paste was tested and analyzed using the AutoPore IV 9500 mercury porosimeter from the American company Microtek. Prior to testing, the test sample is soaked in anhydrous ethanol for termination of hydration, and dried to constant weight in a 50 °C vacuum drying oven. The pore size analysis range was from approximately 4 nm to approximately 300 μm, and high-purity mercury (99.99%) was used for the test at room temperature.

3. Results and Discussion

3.1. Working Performance

The change patterns of the slump and the spread of fresh concrete with different CAMC contents are shown in Figure 1. With the change in CAMC content, the slump of the blank group is 335 mm, while the slump of the concrete with 0.2% CAMC added decreases to 315 mm. When CAMC was further added to 0.4% and 0.6%, the slump decreased to 300 mm and 290 mm, respectively. The corresponding spread is 210 mm, 202 mm, 195 mm, and 190 mm. This indicates that, with the increase in CAMC content, the fresh mixing performance of concrete is affected. CAMC has a slight thickening effect. Although CAMC increases the water solubility of chitosan (CTS) and changes its molecular structure, the conclusion obtained from this experiment is the same as that of adding natural chitosan, which also has a thickening effect.

CTS is mainly extracted from chitin and consists of glucosamine and acetyl glucosamine units. When CTS is added, it increases the viscosity of concrete, resulting in a decrease in the fluidity of the mixture [12]. Moreover, its effect is almost independent of the dose. The larger the molecular weight of the CTS, the greater the detected thickening effect, which is caused by the increase in entanglement and crosslinking between chains in calcium-rich systems. The setting time at low doses is mainly affected by the molecular weight of the polymer, while the degree of deacetylation at high doses is the main controlling factor. Due to the interaction between the polymer and the cement particles, CTS in cement mortar also has a retarding effect. However, as a modified derivative of CTS, CAMC has a more significant effect on workability and plays a greater role in thickening.

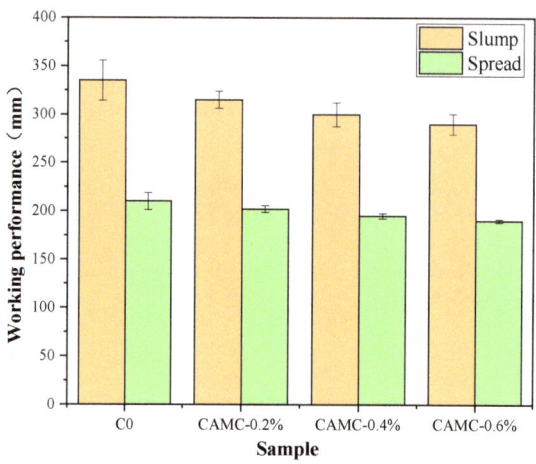

Figure 1. Working performance of premixed concrete.

3.2. Setting Time

Figure 2 shows the variation in the setting time of fresh concrete under different CAMC content conditions. It can be observed from Figure 2 that the initial setting time of the blank group concrete is approximately 4 h. After the addition of CAMC, the initial setting times increase to 5.5 h, 15 h, and 26 h, respectively, and the final setting times also increase from 8 h to 9.5 h, 21 h, and 33 h, respectively. The hardening process of cement paste is mainly the formation of a calcium silicate hydrate C-S-H phase. The addition of CAMC leads to the interaction between the CAMC and cement minerals, which delays the diffusion of water to the unhydrated phase, resulting in the delay of the formation of C-S-H and prolonging the setting time. The increase in setting time induced by the presence of CTS and its derivatives has been reported in previous studies [24].

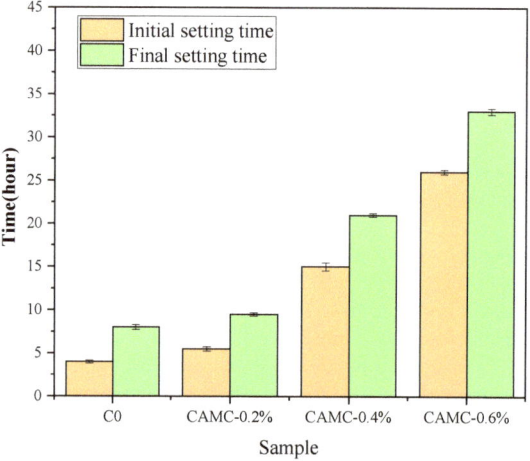

Figure 2. Setting time of fresh concrete.

3.3. Mechanical Properties

3.3.1. Compressive Strength

The compressive strength of concrete with various CAMC contents at different ages can be found in [18]. Figure 3 shows a strength development index of concrete with different CAMC contents and ages. In the early stage of hardening, CAMC had a negative impact on the initial strength of the concrete. On the first day, the compressive strengths of the C0 and CAMC-0.2% samples reached 21.2 MPa and 22.1 MPa, respectively. As the CAMC content increased to 0.4%, the compressive strength of the CAMC-0.4% sample decreased to 15.2 MPa. After 3 days, the compressive strength values of all samples were above 30 MPa, and the early strength loss caused by the CAMC basically disappeared. It is worth noting that, when the compressive strength was measured at 28 days, the strength of the concrete increased with the increase in CAMC content. The compressive strength of C0 was 49.6 MPa, and the compressive strengths of concrete with added CAMC increased by 0.6%, 17.8%, and 21.4%, respectively. From the strength development index (Figure 3, it can be seen that after 3 days of age, with the increase in CAMC content, the strength of the sample increases more rapidly. This indicated that, although CAMC as an additive added to concrete had a negative impact on early strength, it was beneficial for subsequent strength development.

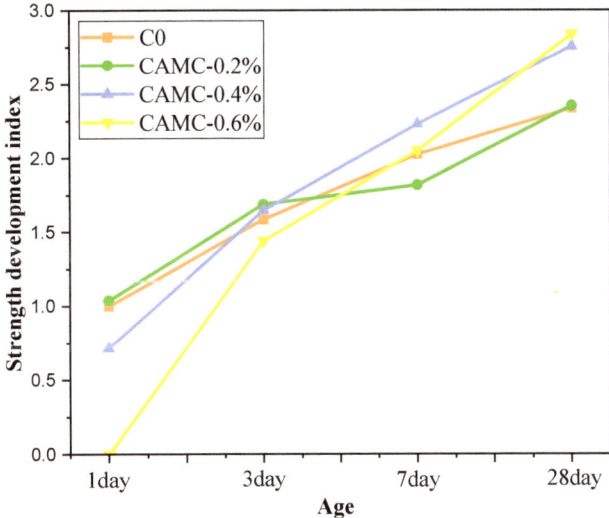

Figure 3. Compressive strength development index of concrete with various CAMC contents at different ages.

3.3.2. Splitting Tensile Strength

The splitting tensile strength of concrete under different CAMC content conditions is shown in Figure 4. The figure mainly shows that the splitting tensile strength decreases with the increase in CAMC content in the early stage of hardening, while after the hardening is stable, the splitting tensile strength increases slightly except for CAMC-0.6%. In the early stage of hydration, the splitting tensile strength decreases significantly. At 3 days of age, the splitting tensile strength of the blank group is 2.88 MPa, and after adding different amounts of CAMC, the splitting tensile strengths decrease to 2.74 MPa, 2.51 MPa, and 1.08 MPa, respectively. When hydration progresses to 7 days of age, the splitting tensile strength of the blank group reaches 3.25 MPa, and after adding CAMC, the splitting tensile strengths increase to 3.25 MPa, 3.43 MPa, and 3.11 MPa, respectively. Among the samples, the splitting tensile strengths of higher CAMC content samples increase faster, with CAMC-0.4% increasing by 36.65% compared to 3 days of age, while CAMC-0.6% increases by 188%

compared to 3 days of age. This is mainly because, before 3 days of age, CAMC causes a significant delay in the hydration of cement paste, and as the hydration time increases, the hydration rate accelerates and the strength increases significantly. When the curing age reaches 28 days, the splitting tensile strength still has a significant increase compared to 7 days, with the blank group and CAMC-0.2% reaching 4.09 MPa, while CAMC-0.4% increases to 4.32 MPa, and CAMC-0.6% shows a slight decrease compared to the blank group, with a splitting tensile strength of 3.94 MPa. CTS bonds cement particles together through its viscosity, rather than accelerating the production rate of crystallization products. Concrete slump and expansion experiments also show that CAMC increases the viscosity of cement paste while also improving its mechanical strength to some extent. However, when the CAMC content is too high, it can affect the hydration process of cement, which in turn affects the development of later splitting tensile strength.

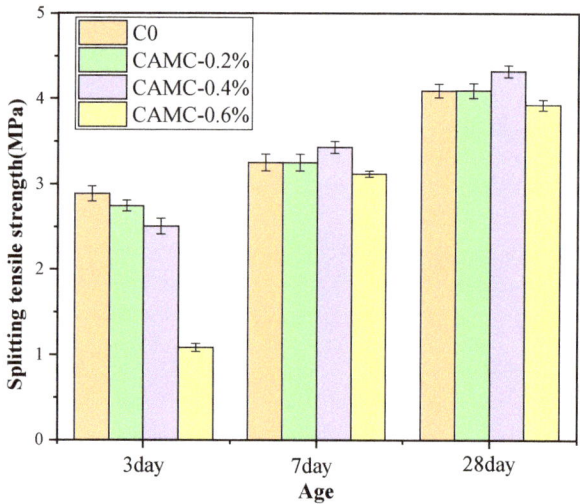

Figure 4. Splitting tensile strength of concrete with different dosages of CAMC.

3.4. Shrinkage

3.4.1. Autogenous Shrinkage

The non-contact mortar shrinkage tester was used to detect the autogenous shrinkage development of low-temperature-rise concrete from the initial setting stage to the early hydration stage within 7 days. The test results are shown in Figure 5. From the early autogenous shrinkage variation trend of the four groups of proportions, it can be seen that the first three groups all experienced significant shrinkage in the early stage, while the CAMC-0.6% group experienced a slight expansion phenomenon. Among them, the blank group and the CAMC-0.2% group both experienced a relatively rapid shrinkage at the earliest stage, followed by an expansion behavior. After the expansion reached its peak, the sample experienced a second shrinkage. Based on its early sample shrinkage behavior, the entire change stage of the sample can be divided into three stages: shrinkage, expansion, and then shrinkage [25]. The four characteristic values of ΔP, ΔE, ΔH, and ΔS represent the changes in shrinkage behavior in the three stages, where ΔP is the first-stage shrinkage change value, ΔE is the absolute value of the early expansion phenomenon in the second stage, ΔH is the change value after the expansion reaches its maximum value and then shrinks up to day 7 in the third stage, and ΔS is the final shrinkage change value of the system during the entire early autogenous shrinkage process from the initial setting time to 7 days.

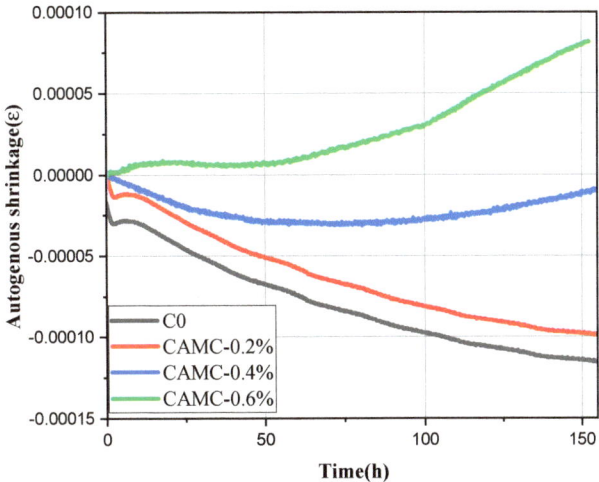

Figure 5. Effect of CAMC on early autogenous shrinkage of cement paste.

As shown in Table 5, the blank group and CAMC-0.2% group had a shrinkage value of 12.91 and 12.59 με for ΔP, respectively, while the CAMC-0.4% and CAMC-0.6% groups did not exhibit the initial shrinkage phenomenon. When developing into the second stage of early expansion, the blank group and CAMC-0.2% sample group had an expansion value of 15.8 and 18.9 με for ΔE, respectively, with the CAMC-0.2% sample group having a slightly increased expansion value compared to the blank group in the second stage. Similarly, the CAMC-0.4% sample did not exhibit any expansion phenomenon, while the CAMC-0.6% sample had an expansion value of 9.29 με for ΔE. When developing into the third stage of re-shrinkage, it was found that the first three groups of samples had shrinkage values of 87.44, 88.07, and 14.85 με for ΔH, respectively, while the CAMC-0.6% sample continued to expand with an expansion value of 69.2 με. When analyzing the final shrinkage value of the entire system, it was found that the shrinkage of the sample decreased with an increasing CAMC content, with the blank group, CAMC-0.2%, and CAMC-0.4% samples having shrinkage values of 86.82 με, 70.89 με, and 14.52 με, respectively. However, the CAMC-0.6% sample ultimately expanded, with an expansion value of 78.49 με.

Table 5. The characteristic value of early autogenous shrinkage change in cement paste.

Number	$\Delta P / \times 10^{-6}$ m	$\Delta E / \times 10^{-6}$ m	$\Delta H / \times 10^{-6}$ m	$\Delta S / \times 10^{-6}$ m
C0	−12.91	15.8	−87.44	−86.82
CAMC-0.2%	−12.59	18.9	−88.07	−70.89
CAMC-0.4%	-	-	−14.85	−14.52
CAMC-0.6%	-	9.29	69.2	78.49

There are currently three main theories that can explain the mechanism of autogenous shrinkage: surface tension theory, disjoining pressure theory, and capillary tension theory. The surface tension theory suggests that a decrease in humidity leads to a reduction in the adsorbed water layer between particles within the gelling material, resulting in an increase in surface tension and macroscopic shrinkage of the material [26]. The disjoining pressure theory suggests that the separation pressure between solid particles is a combination of complex forces such as van der Waals forces and layer repulsion. When the humidity decreases, the separation pressure between solid particles decreases and shrinkage occurs [27]. The capillary tension theory suggests that a decrease in humidity leads to the formation of a meniscus at the interface between the gas phase and the liquid phase, resulting in macroscopic shrinkage [28–30]. The hydration of cement particles proceeds with the consumption of water and the formation of a porous structure, as shown in Figure 6a. In capillary theory, due to the existence of surface tension, a curved liquid surface forms in unsaturated pores. This meniscus causes capillary stress in the pores, resulting in a decrease in volume [31]. The lower the water–cement ratio, the higher the autogenous shrinkage rate, the finer the pore structure, and the lower the porosity. The consumption of free water gradually causes it to enter small pores from large pores to achieve thermodynamic equilibrium. Due to the significant reduction in pore volume, the consumption of water in low-water–cement-ratio mixtures results in faster changes in pore saturation. This helps to reduce the relative humidity faster in low-water–cement-ratio slurries. In capillary pores with a low relative humidity, the radius of curvature of the meniscus is smaller, so the capillary stress is larger and the autogenous shrinkage Is also larger [32–34].

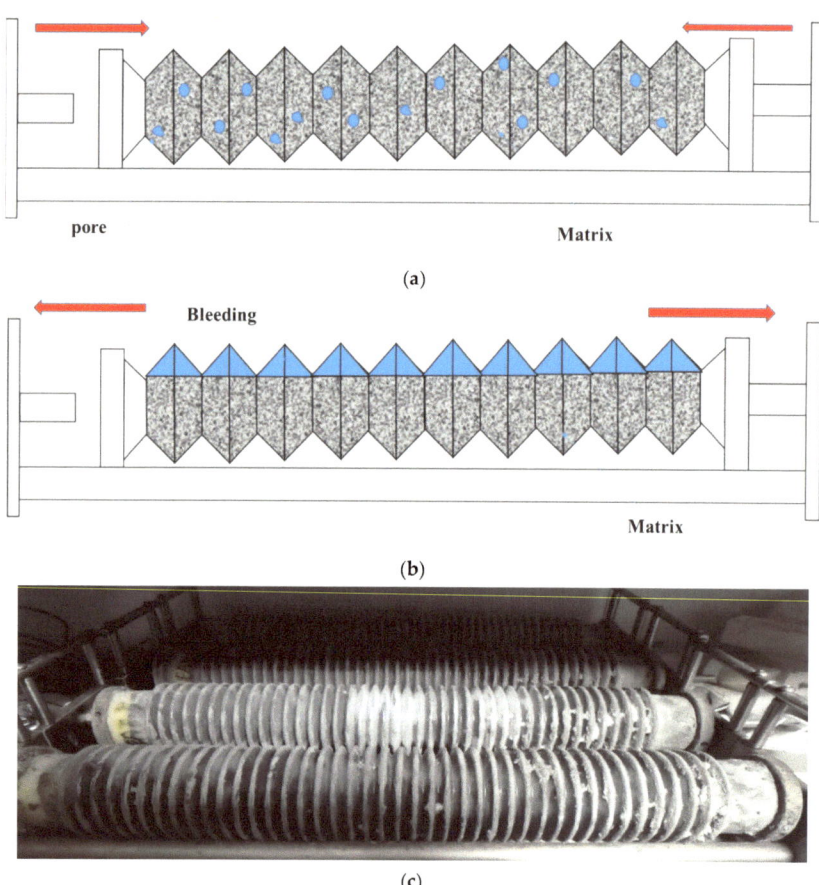

Figure 6. The effect of moisture distribution on the self-shrinking properties of concrete; (**a**) dense slurry, (**b**) bleeding, (**c**) test sample.

Through the analysis of the above data, it is found that, during the early autogenous shrinkage process, some samples undergo a process of micro-expansion. The expansion source in this stage mainly includes three reasons: first, the expansion pressure caused by the formation of hydration products (CH, AFt), followed by the thermal expansion phenomenon caused by the heat release of hydration [35,36], and then the influence of slurry bleeding and its reabsorption [29,30]. When the sample contains a high content of CAMC, the hydration process is significantly delayed, and there is a significant amount of excess free water in the cement paste, resulting in a "bleeding-like" phenomenon (as shown in Figure 6b,c). Previous ^{1}H-NMR research results have also confirmed that CAMC can cause a "bleeding-like" phenomenon [18]. In most cases, the reabsorption of secreted water is the most important reason for this. The free water content and relative humidity inside the high-water–cement-ratio sample are still high, and the capillary stress is relatively small, so the autogenous shrinkage is very small, and the expansion phenomenon occurs [37,38].

3.4.2. Shrinkage Analysis at Different Stages of the Entire Process

The drying shrinkage is shown in Figure 7. The long-term drying shrinkage rate of low-temperature-rise concrete decreases with time, and the long-term drying shrinkage gradually decreases with the increase in CAMC content. The drying shrinkage value of the blank group at 90 days is about 551.46 με. After adding 0.2% CAMC, the long-term drying shrinkage value decreases by 42 με. As the CAMC content increases, the long-term drying shrinkage decreases to 500.97 and 401.94 με. The long-term drying shrinkage of the four groups of samples is basically consistent with the change law of autogenous shrinkage described in the previous section.

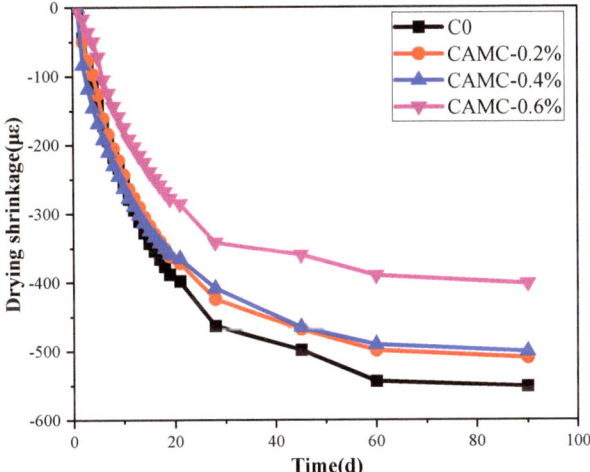

Figure 7. Effect of CAMC on the drying shrinkage of low-temperature-rise concrete.

The overall shrinkage change in low-temperature-rise concrete during the whole process is mainly the sum of the early autogenous shrinkage before demolding and the long-term drying shrinkage after demolding, and was determined in a standard environment. The early autogenous shrinkage and long-term drying shrinkage of low-temperature-rise concrete have been tested and measured in our previous article. Table 6 analyzes the early autogenous shrinkage and long-term drying shrinkage data of the four groups of samples. The specific division method can be found in the literature [25]. A_3 in Table 6 is the early autogenous shrinkage values of the four groups of samples in Figure 5, A_{14} is the stable autogenous shrinkage values at 14 days, T_{14} is the long-term total shrinkage values of the four groups of low-temperature-rise concrete after demolding at 14 days, and E_{14} is the total shrinkage values of the low-temperature-rise concrete after initial setting. E_{14} can be calculated by the following formula:

$$E_{14} = A_3 + T_{14} \tag{7}$$

Table 6. Shrinkage value of the whole process of low-temperature-rise concrete mixed with CAMC.

Number	A_3 (µε)	A_{14} (µε)	T_{14} (µε)	E_{14} (µε)	D_{14} (µε)
C0	−146.19	−86.82	−327.18	−473.37	−386.55
CAMC-0.2%	−166.78	−70.89	−304.85	−471.63	−400.74
CAMC-0.4%	−170.15	−14.52	−313.59	−483.74	−469.22
CAMC-0.6%	−145.67	78.49	−225.24	−370.91	−449.4

D_{14} is the drying shrinkage value of concrete at the age of 14 days, and D_{14} can be calculated by the following formula:

$$D_{14} = E_{14} - A_{14m} \quad (8)$$

The analysis of Table 6 reveals that no expansion was observed in any of the low-temperature-rise concrete samples, indicating that the shrinkage caused by drying can balance (or mask) the expansion observed in the early autogenous shrinkage phase. The overall trend is that the total shrinkage value decreases with the increase in CAMC content, where the total shrinkage value of the blank group is 473.37 µε, and after the addition of 0.6% CAMC, the total shrinkage value decreases to 370.91 µε. The decrease in the total shrinkage of the CAMC-0.6% sample is due to the expansion phenomenon of autogenous shrinkage and the reduction in long-term drying shrinkage. The total shrinkage rate mainly depends on the drying shrinkage rate, as the observed trends are the same in both cases. This phenomenon is due to the larger magnitude of drying shrinkage compared to autogenous shrinkage.

3.5. Durability

3.5.1. Chloride Ion Diffusion Coefficient

Figure 8 shows the results of the chloride diffusion coefficient of low-temperature-rise concrete samples with different CAMC content. According to Figure 8, the average D_{RCM} of the blank concrete sample is 7.24×10^{-12} m^2/s. With the increase in CAMC content, the chloride diffusion coefficient of low-temperature-rise concrete samples shows a trend of first decreasing and then increasing. The chloride penetration resistance coefficients of low-temperature-rise concrete are in the order of CAMC-0.6% > blank group > CAMC-0.4% > CAMC-0.2%, with average DRCMs of 6.28×10^{-12} m^2/s, 6.64×10^{-12} m^2/s, and 9.63×10^{-12} m^2/s, respectively. Several parameters, such as the water–cement ratio, type of cement and admixtures, curing conditions, the existence of chemical erosion, and characteristics of micro-cracks, can affect chloride penetration. Chloride ions can penetrate into concrete through diffusion caused by concentration gradients and capillary forces, which are related to the volume and size of pores and micro-cracks, as well as their interconnections. Under low-CAMC-content conditions, the chloride diffusion coefficient can be significantly reduced, which plays an important role in improving the durability of low-temperature-rise concrete. By comparing the mechanical properties and shrinkage performance, it can also be found that low-content CAMC can improve the mechanical properties and durability of concrete. Although 0.6% CAMC can still improve the mechanical properties and shrinkage performance of low-temperature-rise concrete in the later stage of hydration, an excessive CAMC content can lead to an increase in the porosity of the concrete structure in the later stage of hydration. The increase in porosity allows salt solutions to pass through the porous regions and cause precipitation in internal pores and voids, resulting in higher permeability of the concrete, leading to an increase in the chloride diffusion coefficient of low-temperature-rise concrete.

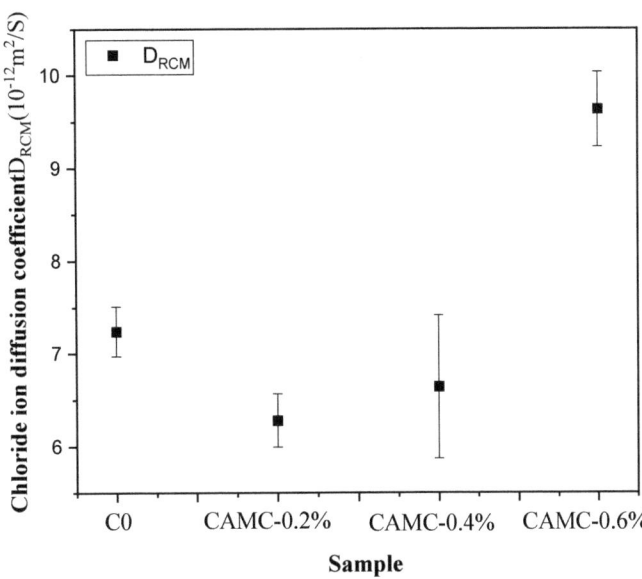

Figure 8. Chloride diffusion coefficients of different samples.

3.5.2. Freeze–Thaw Cycle

The quality changes in low-temperature-rise concrete under different freeze–thaw cycles with different amounts of CAMC are shown in Table 7. When the low-temperature-rise concrete is exposed to freeze–thaw conditions, the quality loss in the low-temperature-rise concrete increases with the increase in the number of freeze–thaw cycles. As shown in Figure 9a, the quality loss rate of the low-temperature-rise concrete decreases almost linearly with the number of freeze–thaw cycles. At the same time, the addition of CAMC can reduce the quality loss in the low-temperature-rise concrete. When the number of freeze–thaw cycles reaches 150, the surface of the blank group sample exhibits significant peeling, and the coarse aggregate is exposed significantly, with a quality loss rate exceeding 5% (Figure 9c). When 0.2% and 0.4% of CAMC are added, the appearance of the sample remains basically intact, and no peeling of the sample surface occurs. After 200 freeze–thaw cycles, the quality loss rates are finally 2.39% and 3.62%, respectively. When the amount of CAMC is increased to 0.6%, the surface of the sample begins to exhibit surface erosion, and although the quality loss rate is only 1.79%, it can be observed from the relative dynamic modulus (Figure 9b) that the relative dynamic modulus of the CAMC-0.6% sample after 50 freeze–thaw cycles is lower than that of other samples with different amounts of CAMC. As the amount of CAMC increases, the relative dynamic modulus of the low-temperature-rise concrete shows a trend of first increasing and then decreasing. However, after 125 freeze–thaw cycles, the blank group cannot obtain an ultrasonic wave velocity, indicating that the internal structure of the sample has been damaged.

Freeze–thaw cycles can significantly weaken the quality, elastic modulus, and other properties of concrete. The causes of these destructive phenomena can be attributed to water pressure, osmotic pressure, ice crystallization pressure, microscopic ice lenses, and thermal effects that result from the mismatch between ice and solid phases. As the frozen pore solution expands within the pores [39,40] and micro-cracks expand and merge, the connectivity of the pores can be enhanced, further increasing the permeability and accelerating damage. The pore structure determines the degree of freeze–thaw damage, and the pore size distribution of concrete is wide, ranging from 0.5 nm to several centimeters. When air is trapped in the pores, it forms a bubble with a diameter from 10 μm to 1 cm [41]. During hydration, the water in the pores is consumed, and C-S-H gel can be filled into the pores. Therefore, the capillary pore volume decreases while the gel pore volume increases. As shown in Figure 10, the pore structures of the concrete samples after 150 freeze–thaw cycles showed a significant change. The pore diameter of the control group mainly ranged from 10 to 100 μm. With an increase in CAMC, the porosity in concrete decreases, and the pore size distribution mainly ranges from 10 to 100 nm. Compared with the control group, the total porosity decreases from a high of 30% to around 15%, indicating that, after the addition of CAMC, harmful pores (>200 nm) in the concrete matrix gradually transform into less harmful pores (20–50 nm) and harmless pores (<20 nm) [42,43]. The addition of CAMC can increase the impermeability of the concrete matrix and improve the density of the concrete.

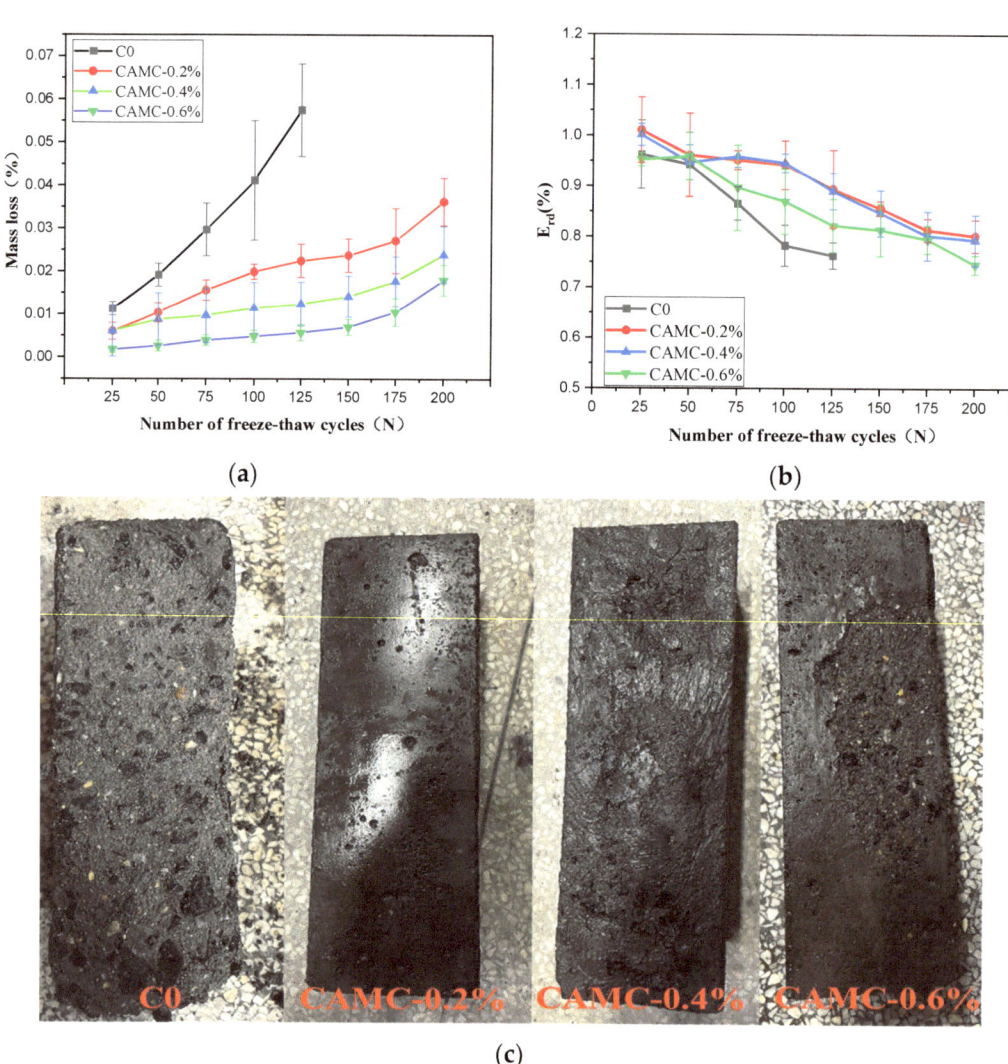

Figure 9. Freeze–thaw cycle test results of samples with different CAMC contents; (**a**) mass loss rate; (**b**) relative dynamic elastic modulus; (**c**) appearance of samples after 150 freeze–thaw cycles.

Table 7. Mass changes in samples with different CAMC contents.

Sample	Initial Value	Number of Cycles							
		25	50	75	100	125	150	175	200
C0-1	7.98	7.90	7.84	7.80	7.77	7.62	damage	damage	damage
C0-2	7.66	7.56	7.49	7.41	7.33	7.17	damage	damage	damage
C0-3	7.42	7.34	7.29	7.17	7.02	6.95	damage	damage	damage
CAMC-0.2%-1	7.65	7.62	7.58	7.54	7.51	7.50	7.49	7.48	7.41
CAMC-0.2%-2	7.82	7.77	7.75	7.71	7.65	7.61	7.60	7.54	7.49
CAMC-0.2%-3	7.68	7.62	7.58	7.54	7.53	7.52	7.51	7.50	7.41
CAMC-0.4%-1	7.66	7.56	7.54	7.53	7.52	7.52	7.51	7.48	7.42
CAMC-0.4%-2	7.44	7.42	7.39	7.39	7.38	7.37	7.35	7.31	7.28
CAMC-0.4%-3	7.44	7.42	7.41	7.40	7.38	7.37	7.36	7.35	7.30
CAMC-0.6%-1	7.52	7.51	7.50	7.50	7.49	7.49	7.48	7.46	7.39
CAMC-0.6%-2	7.42	7.41	7.41	7.39	7.39	7.38	7.37	7.35	7.31
CAMC-0.6%-3	7.78	7.76	7.75	7.74	7.73	7.72	7.71	7.67	7.61

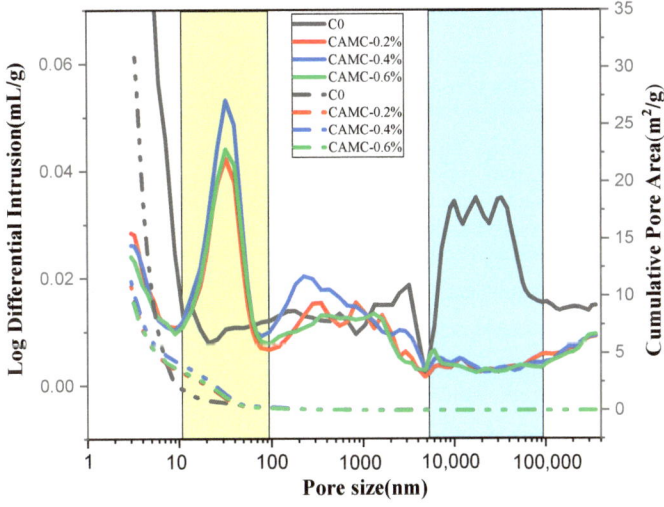

Figure 10. Effect of freeze–thaw cycles on concrete pore structure.

4. Conclusions

This work systematically studies the influence of CAMC on the fresh concrete properties, macroscopic mechanical properties, shrinkage behavior, and durability of concrete. The main conclusions are as follows:

(1) CAMC enhances the water solubility of CTS, and the interaction between CAMC and cement particles increases the consistency of cement paste. As the amount of CAMC increases, the slump and spreading of concrete decrease. Meanwhile, CAMC prolongs the setting time of concrete. When the content of CAMC is too high, it will affect the hydration process of cement, delay the formation of early C-S-H and hydration products, and have a negative impact on the early compressive strength and splitting tensile strength. However, with the acceleration of the hydration rate in the later stage, it is beneficial to the development of subsequent strength.

(2) The early shrinkage stages of the studied specimens can be divided into three phases: shrinkage, expansion, and re-shrinkage. With the increase in CAMC content, the early shrinkage values of the specimens were 86.82 με, 70.89 με, and 14.52 με, respectively. When the CAMC content was 0.6%, the hydration process was significantly delayed, and a large amount of excess free water in the cement paste led to a "bleeding-like"

phenomenon, resulting in expansion behavior with an expansion value of 78.49 με. The total shrinkage value and drying shrinkage value exhibited the same trend. As the CAMC content increased, the shrinkage value decreased. The total shrinkage value of the blank group was 473.37 με, while the total shrinkage value decreased to 370.91 με after adding 0.6% CAMC. The total shrinkage rate was mainly determined by the drying shrinkage rate.

(3) At low CAMC dosages, the chloride ion diffusion coefficient can be significantly reduced, which plays a certain role in improving the durability of low-temperature-rise concrete. The chloride penetration resistance coefficient of low-temperature-rise concrete follows the order: CAMC-0.6% > blank group > CAMC-0.4% > CAMC-0.2%. After adding CAMC, harmful pores in the concrete matrix gradually transform into less harmful and non-harmful pores. However, a high dosage of CAMC can increase the porosity of the concrete structure in the later stages of hydration, leading to an increase in the chloride ion diffusion coefficient of low-temperature-rise concrete. Adding CAMC can reduce the mass loss of low-temperature-rise concrete. With the increase in CAMC content, the relative dynamic modulus of low-temperature-rise concrete exhibits a trend of first increasing and then decreasing.

Author Contributions: Writing—original draft, Z.Q.; Formal analysis, J.W. and Y.L.; methodology, Z.H.; Supervision, Data curation, L.W.; Project administration, Formal analysis, D.L. and K.L. All authors have read and agreed to the published version of the manuscript.

Funding: The authors greatly acknowledge the National Natural Science Foundation of China Joint Fund for regional innovation and development(U23A20658). General Program of the National Natural Science Foundation of China (52350004). State Key Laboratory of High Performance Civil Engineering Materials (2020CEM001). Jiangsu Province Excellent Post-Doctoral Program (2023ZB030). National Natural Science Foundation of China (U23A20673, U22A20244).

Institutional Review Board Statement: Not applicable.

Informed Consent Statement: Not applicable.

Data Availability Statement: Data presented in this study are available on request from the corresponding authors due to restrictions privacy.

Conflicts of Interest: Author Jiandong Wu was employed by Shandong Provincial Communications Planning and Design Institute Group Co., Ltd. Author Zhenhao Hei was employed by Qingdao Haifa Real Estate Co., Ltd. Author kai Liu was employed by Jiang Su China Construction Ready Mixed Concrete Co., Ltd. The remaining authors declare that the research was conducted in the absence of any commercial or financial relationships that could be construed as a potential conflict of interest.

References

1. Wang, L.; Ju, S.; Chu, H.; Liu, Z.; Yang, Z.; Wang, F.; Jiang, J. Hydration process and microstructure evolution of low exothermic concrete produced with urea. *Constr. Build. Mater.* **2020**, *248*, 118640. [CrossRef]
2. Liu, Z.; Jiang, J.; Jin, X.; Wang, Y.; Zhang, Y. Experimental and numerical investigations on the inhibition of freeze–thaw damage of cement-based materials by a methyl laurate/diatomite microcapsule phase change material. *J. Energy Storage* **2023**, *68*, 107665. [CrossRef]
3. Liu, Z.; Zang, C.; Hu, D.; Zhang, Y.; Lv, H.; Liu, C.; She, W. Thermal conductivity and mechanical properties of a shape-stabilized paraffin/recycled cement paste phase change energy storage composite incorporated into inorganic cementitious materials. *Cem. Concr. Compos.* **2019**, *99*, 165–174. [CrossRef]
4. Xie, D.; Yue, J.; Yankun, L.U. Effect of Sodium Selenite-Chitosan Compound Preservative on Storability of Kumquats. *Asian Agric. Res.* **2021**, *13*, 5.
5. Siddiqui, M.T.H.; Baloch, H.A.; Nizamuddin, S.; Mubarak, N.M.; Hossain, N.; Zavabeti, A.; Srinivasan, M. Synthesis and optimization of chitosan supported magnetic carbon bio-nanocomposites and bio-oil production by solvothermal carbonization co-precipitation for advanced energy applications. *Renew. Energy* **2021**, *178*, 587–599. [CrossRef]
6. Vo, T.S.; Vo TT, B.C.; Tran, T.T.; Pham, N.D. Enhancement of water absorption capacity and compressibility of hydrogel sponges prepared from gelatin/chitosan matrix with different polyols. *Prog. Nat. Sci. Mater. Int.* **2022**, *32*, 54–62. [CrossRef]
7. Chen, S.; Aladejana, J.T.; Li, X.; Bai, M.; Shi, S.Q.; Kang, H.; Li, J. A strong, antimildew, and fully bio-based adhesive fabricated by soybean meal and dialdehyde chitosan. *Ind. Crops Prod.* **2023**, *194*, 116277. [CrossRef]

8. Nistico, P.B.P. Chitosan and its char as fillers in cement-base composites: A case study. *Bol. La Soc. Esp. Ceram. Y Vidr.* **2020**, *59*, 186–192. [CrossRef]
9. Choi, H.Y.; Bae, S.H.; Choi, S.J. Synthesis of catechol-conjugated chitosan and its application as an additive for cement mortar. *Bull. Korean Chem. Soc.* **2022**, *1*, 43. [CrossRef]
10. Pan, Z.H.; Cai, H.P.; Jiang, P.P.; Fan, Q.Y. Properties of a calcium phosphate cement synergistically reinforced by chitosan fiber and gelatin. *J. Polym. Res.* **2006**, *13*, 323–327. [CrossRef]
11. Alkhraisat, M.H.; Rueda, C.; Jerez, L.B.; MariO, F.T.; Torres, J.; Gbureck, U.; Cabarcos, E.L. Effect of silica gel on the cohesion, properties and biological performance of brushite cement. *Acta Biomater.* **2010**, *6*, 257–265. [CrossRef] [PubMed]
12. Lasheras-Zubiate, M.; Navarro-Blasco, I.; Fernández, J.M. Studies on chitosan as an admixture for cement-based materials: Assessment of its viscosity enhancing effect and complexing ability for heavy metals. *J. Appl. Polym. Sci.* **2010**, *120*, 242–252. [CrossRef]
13. Bezerra, U.T.; Ferreira, R.M.; Castro-Gomes, J.P. The Effect of Latex and Chitosan Biopolymer on Concrete Properties and Performance, Key Engineering Materials. *Key Eng. Mater.* **2011**, *466*, 37–46. [CrossRef]
14. Ustinova, Y.V.; Nikiforova, T.P. Cement Compositions with the Chitosan Additive. *Procedia Eng.* **2016**, *153*, 810–815. [CrossRef]
15. Lv, S.; Cao, Q.; Zhou, Q.; Lai, S.; Gao, F. Structure and Characterization of Sulfated Chitosan Superplasticizer. *J. Am. Ceram. Soc.* **2013**, *96*, 1923–1929. [CrossRef]
16. Lv, S.; Liu, J.; Zhou, Q.; Huang, L.; Sun, T. Investigation of removal of Pb(II) and Hg(II) by a novel cross-linked chitosan-poly(aspartic acid) chelating resin containing disulfide bond. *Colloid Polym. Sci.* **2014**, *292*, 2157–2172.
17. Lv, S.H. *High-performance Superplasticizer Based on Chitosan, Biopolymers and Biotech Admixtures for Eco-Efficient Construction Materials Book*; Elsevier Inc.: Amsterdam, The Netherlands, 2016; pp. 131–150.
18. Wang, L.; Ju, S.; Wang, L.; Wang, F.; Sui, S.; Yang, Z.; Liu, Z.; Chu, H.; Jiang, J. Effect of citric acid-modified chitosan on the hydration and microstructure of Portland cement paste. *J. Sustain. Cem. -Based Mater.* **2023**, *12*, 83–96. [CrossRef]
19. Wang, L.; Wang, F.; Sui, S.; Ju, S.; Qin, Z.; Su, W.; Jiang, J. Adsorption capacity and mechanism of citric acid-modified chitosan on the cement particle surface. *J. Sustain. Cem. -Based Mater.* **2023**, *12*, 893–906. [CrossRef]
20. Wang, L.; Zhang, Y.; Guo, L.; Wang, F.; Ju, S.; Sui, S.; Liu, Z.; Chu, H.; Jiang, J. Effect of citric-acid-modified chitosan (CAMC) on hydration kinetics of tricalcium silicate (C3S). *J. Mater. Res. Technol.* **2022**, *21*, 3604–3616. [CrossRef]
21. Vanguri, S.; Palla, S.; Chaturvedi, S.K.; Mohapatra, B.N. Phase quantification of Indian industrial clinkers containing minor oxides. *ZKG Cem. Lime Gypsum* **2022**, *75*, 42–45.
22. *Standard GB/T50080-2002*; Standard Test Method for Performance of Ordinary Concrete Mixtures. Ministry of Construction of the PRC: Beijing, China, 2003.
23. *Standard GB/T50082-2009*; Standard for Testing Methods for Long-Term Performance and Durability of Ordinary Concrete. Ministry of Construction of the PRC: Beijing, China, 2010.
24. Lasheras-Zubiate, M.; Navarro-Blasco, I.; Alvarez, J.I.; Fernández, J.M. Interaction of carboxymethylchitosan and heavy metals in cement media. *J. Hazard. Mater.* **2011**, *194*, 223–231. [CrossRef] [PubMed]
25. Zhou, J. *Design, Preparation, and Shrinkage Reduction Mechanism of Low Shrinkage Ultra-High Performance Concrete*; Southeast University: Nanjing, China, 2019. (In Chinese)
26. Odler, I.; Hagymassy, J., Jr.; Yudenfreund, M.; Hanna, K.M.; Brunauer, S. Pore structure analysis by water vapor adsorption. IV. Analysis of hydrated portland cement pastes of low porosity. *J. Colloid Interface Sci.* **1972**, *38*, 265–276.
27. Ferraris, C.; Wittmann, F.H. Shrinkage mechanisms of hardened cement paste. *Research* **1987**, *17*, 453–464. [CrossRef]
28. Hua, J.; Huang, L.; Luo, Q.; Chen, Z.; Xu, Y.; Zhou, F. Prediction on the shrinkage of concrete under the restraints of steel plates and studs based on the capillary tension theory. *Materials* **2020**, *258*, 119499. [CrossRef]
29. Huang, L.; Hua, J.; Kang, M.; Zhou, F.; Luo, Q. Capillary tension theory for predicting shrinkage of concrete restrained by reinforcement bar in early age. *Materials* **2019**, *210*, 63–70. [CrossRef]
30. Mehta, P.K. Concrete. Structure, properties and materials. *Cem. Concr. Res.* **1986**, *16*, 790–800.
31. Tam, V.W.; Wang, K.; Tam, C.M. Assessing relationships among properties of demolished concrete, recycled aggregate and recycled aggregate concrete using regression analysis. *J. Hazard. Mater.* **2008**, *152*, 703–714. [CrossRef] [PubMed]
32. Henkensiefken, R.; Bentz, D.; Nantung, T.; Weiss, J.J.C. Volume change and cracking in internally cured mixtures made with saturated lightweight aggregate under sealed and unsealed conditions. *Cem. Concr. Compos.* **2009**, *31*, 427–437. [CrossRef]
33. Huang, C.-C.; Chen, L.; Gu, X.; Zhao, M.; Nguyen, T. The effects of humidity and surface free energy on adhesion force between atomic force microscopy tip and a silane self-assembled monolayer film. *J. Mater. Res.* **2010**, *25*, 556–562. [CrossRef]
34. Nguyen, S.T. Generalized Kelvin model for micro-cracked viscoelastic materials. *Eng. Fract. Mech.* **2014**, *127*, 226–234. [CrossRef]
35. Holt, E.J.C. Contribution of mixture design to chemical and autogenous shrinkage of concrete at early ages. *Cem. Concr. Res.* **2005**, *35*, 464–472. [CrossRef]
36. Maruyama, I.; Teramoto, A.J.C. Temperature dependence of autogenous shrinkage of silica fume cement pastes with a very low water–binder ratio. *Cem. Concr. Res.* **2013**, *50*, 41–50. [CrossRef]
37. Mohr, B.; Hood, K.J.C. Influence of bleed water reabsorption on cement paste autogenous deformation. *Cem. Concr. Res.* **2010**, *40*, 220–225. [CrossRef]
38. Bjøntegaard, Ø. *Thermal Dilation and Autogenous Deformation as Driving Forces to Self-Induced Stresses in High Performance Concrete*; Norges Teknisk-Naturvitenskapelige Universitet: Trondheim, Norway, 1999.

39. Guthrie, W.S.; Lay, R.D.; Birdsall, A.J. Effect of reduced cement contents on frost heave of silty soil: Laboratory testing and numerical modeling. In Proceedings of the Transportation Research Board 86th Annual Meeting Compendium of Papers, Washington, DC, USA, 21–25 January 2007.
40. Li, L.; Shao, W.; Li, Y.; Cetin, B.J.G. Effects of climatic factors on mechanical properties of cement and fiber reinforced clays. *Engineering* **2015**, *33*, 537–548. [CrossRef]
41. Cho, T.J.C. Prediction of cyclic freeze–thaw damage in concrete structures based on response surface method. *Constr. Build. Mater.* **2007**, *21*, 2031–2040. [CrossRef]
42. Renhe, Y.; Baoyuan, L.; Zhongwei, W. Study on the pore structure of hardened cement paste by saxs. *Cem. Concr. Res.* **1990**, *20*, 385–393. [CrossRef]
43. Liguo, W.; Dapeng, Z.; Shupeng, Z.; Hongzhi, C.; Dongxu, L. Effect of Nano-SiO$_2$ on the Hydration and Microstructure of Portland Cement. *Nanomaterials* **2016**, *6*, 241. [CrossRef]

Disclaimer/Publisher's Note: The statements, opinions and data contained in all publications are solely those of the individual author(s) and contributor(s) and not of MDPI and/or the editor(s). MDPI and/or the editor(s) disclaim responsibility for any injury to people or property resulting from any ideas, methods, instructions or products referred to in the content.

Article

Investigating the Impact of Superabsorbent Polymer Sizes on Absorption and Cement Paste Rheology

Nilam Adsul [1], Jun-Woo Lee [1] and Su-Tae Kang [2,*]

[1] Department of Civil Engineering, Daegu University, Gyeongsan 38453, Republic of Korea; adsulnil@daegu.ac.kr (N.A.); dlwnsdn518@naver.com (J.-W.L.)
[2] Department of Architecture Engineering, Daegu University, Gyeongsan 38453, Republic of Korea
* Correspondence: stkang@daegu.ac.kr

Abstract: This study aims to understand the water retention capabilities of Superabsorbent Polymers (SAPs) in different alkaline environments for internal curing and to assess their impact on the rheological properties of cement paste. Therefore, the focus of this paper is on the absorption capacities of two different sizes of polyacrylic-based Superabsorbent Polymers : SAP A, with an average size of 28 µm, and SAP B, with an average size of 80 µm, in various solutions, such as pH 7, pH 11, pH 13, and cement filtrate solution (pH 13.73). Additionally, the study investigates the rheological properties of SAP-modified cement pastes, considering three different water-to-cement (w/c) ratios (0.4, 0.5, and 0.6) and four different dosages of SAPs (0.2%, 0.3%, 0.4%, and 0.5% by weight of cement). The results showed that the absorption capacity of SAP A was higher in all solutions compared to SAP B. However, both SAPs exhibited lower absorption capacity and early desorption in the cement filtrate solution. In contrast to the absorption results in pH 13 and cement filtrate solutions, the rheological properties, including plastic viscosity and yield stress, of the cement paste with a w/c ratio of 0.4 and 0.5, as well as both dry and wet (presoaked) SAPs, were higher than those of the cement paste without SAP, indicating continuous absorption by SAP. The viscosity and yield stress increased over time with increasing SAP dosage. However, in the mixes with a w/c ratio of 0.6, the values of plastic viscosity and yield stress were initially lower for the mixes with dry SAPs compared to the reference mix. Additionally, cement pastes containing wet SAP showed higher viscosity and yield stress compared to the pastes containing dry SAP.

Keywords: cement paste; water-to-cement ratio; superabsorbent polymer; absorption capacity; plastic viscosity; yield stress

1. Introduction

Superabsorbent polymers (SAPs) are highly versatile materials renowned for their remarkable liquid absorption capabilities, allowing them to absorb liquid volumes many times their weight. These properties make them widely applicable in diverse industrial sectors, including agriculture, healthcare, effluent treatment, and civil construction [1]. The integration of SAP into cementitious materials has been studied extensively, such as admixtures, internal curing agents, shrinkage reducers, self-healing agents, and freeze and thaw resistance enhancers [2,3]. Many researchers have investigated the incorporation of SAPs in different types of concrete, such as high-performance concrete, high-strength concrete, self-compacting concrete, ultra-high-performance concrete, and alkali-activated slag mortar, in combination with various supplementary cementitious materials [4–8].

Water absorption into SAPs occurs due to osmotic pressure, which expands the space between polymer chains and cross-links. This osmotic pressure is driven by the concentration gradient of mobile ions between the gel and the solution. Therefore, the presence of dissolved ions such as K^+, Na^+, Mg^{2+}, and Ca^{2+} in the surrounding solution affects the osmotic pressure and, thus, influences the swelling capacity of SAPs [2,9,10].

Citation: Adsul, N.; Lee, J.-W.; Kang, S.-T. Investigating the Impact of Superabsorbent Polymer Sizes on Absorption and Cement Paste Rheology. *Materials* **2024**, *17*, 3115. https://doi.org/10.3390/ma17133115

Academic Editors: Yuan Gao, Junlin Lin and Xupei Yao

Received: 29 May 2024
Revised: 17 June 2024
Accepted: 21 June 2024
Published: 25 June 2024

Copyright: © 2024 by the authors. Licensee MDPI, Basel, Switzerland. This article is an open access article distributed under the terms and conditions of the Creative Commons Attribution (CC BY) license (https://creativecommons.org/licenses/by/4.0/).

The absorption capacity and rate of SAPs can be modified based on chemical compositions, physical attributes, and the properties of the surrounding fluid, including composition, ionic concentration, pH, temperature, and pressure [9,11]. Additionally, the particle size, quantity, and chemical composition of SAPs play a crucial role in their performance in concrete, impacting mechanical stability during mixing and rheological properties such as yield stress, plastic viscosity, and thixotropy over time [12–15]. Likewise, the type and properties of SAPs influence workability and setting times [16,17].

Different types and sizes of SAPs have varying effects on fresh concrete properties because of their different absorption characteristics. For instance, incorporating acrylamide/acrylic sodium copolymer SAP (with mean sizes of 471.3 and 95.1 μm) with additional water initially reduces yield stress, followed by an increase over time, accompanied by a simultaneous reduction in plastic viscosity. In contrast, incorporating acrylamide-based SAP (with a mean size of 63 μm) with extra water leads to a continuous rise in yield stress and plastic viscosity over time as the dosage of SAP increases, owing to its lack of desorption properties [12]. In a study by Ma et al. (2019), acrylamide/acrylic acid copolymers with average diameters of 125 μm and 840 μm were utilized. The mortar with the larger SAP particle size exhibited higher maximum shear stress, plastic viscosity, and relative thixotropy compared to the mortar with the smaller SAP size [18]. On the other hand, Lee et al. (2018) studied two types of SAPs, where polyacrylate-based SAP exhibited a lower swelling ratio and slower swelling recovery compared to polyacrylate-co-acrylamide-based SAP, as it is more prone to the calcium ions present in the solution [19].

The absorption properties of SAPs are influenced by the w/c ratio, superplasticizer, and supplementary cementitious materials, subsequently affecting the rheological properties of cementitious materials with SAP [13]. Higher w/c ratios lead to more excellent SAP absorption, causing more significant changes in Bingham parameters for mixes with SAP [10,13,20]. As SAP content increases, more dosage of water-reducing agent is needed to maintain workability. However, excess SAP content without additional water or superplasticizer reduces the workability of the cementitious material [20,21]. The method of SAP addition, whether presoaked or dry, significantly affects slump more than particle size. Concrete with presoaked SAP with additional water showed an increased slump, while concrete with dry SAP exhibited a decreased slump with an increased SAP dosage [22,23].

However, certain aspects of the performance of different sizes of SAP, especially very fine particle sizes, in varying dosages without additional water within a cementitious environment, particularly regarding rheological properties, require further investigation. This will help determine the effect of SAP alone in cement paste and identify the suitable dosage to enhance the properties of cement paste.

This study investigates the absorption and desorption characteristics of SAPs in different pH and cement filtrate solutions. The aim is to understand how long SAP can absorb solutions and when it will start the desorption process in various solution mediums. Additionally, this study explores the impact of two different sizes of SAPs as internal curing agents on the plastic viscosity and yield stress of cement paste. The cement paste was prepared with varying dosages of SAPs (0.2% to 0.5%) and three different w/c ratios (0.4 to 0.6). Furthermore, the study examines the influence of both dry and wet SAP on the rheological properties. The results of the absorption and desorption by SAP in different solutions will contribute to understanding the changes in viscosity and yield stress of cement paste with SAP over time.

2. Materials and Methods

2.1. Material and Sample Preparation

In this study, Ordinary Portland Cement (Type I, Hanil cement, Seoul, Republic of Korea) and Polyacrylic-based SAP from TPY Co. Ltd., Hwaseong-si, Republic of Korea, in two different sizes: SAP A (avg. size 28 μm-TPY 900) and SAP B (avg. size 80 μm-TPY 502) were utilized. The scanning Electron Microscope (SEM) (S-4300, Hitachi Ltd., Tokyo, Japan) images of the particle size of both SAPs are depicted in Figure 1. The cement paste was prepared using three different water-to-cement ratios (0.4, 0.5 and 0.6) and varying dosages of SAP (0.2%, 0.3%, 0.4% and 0.5% by weight of cement) as outlined in Table 1 and tested for rheological properties. The addition of SAP was carried out in two different ways: dry (D) and wet (W), with SAP A denoted as A and SAP B as B. During experimental work, it was noticed that a w/c ratio of less than 0.4 resulted in a highly viscous mix, which was not suitable for rheology testing. Therefore, the w/c ratios were selected above 0.4.

(a) SAP A (avg. size 28 μm) (b) SAP B (avg. size 80 μm)

Figure 1. SEM images of the particle sizes of SAP A and SAP B.

Table 1. Mix proportions of cement paste with different w/c ratios and dosages of SAPs.

Sample	w/c	SAP/c (%)	Sample	w/c	SAP/c (%)	Sample	w/c	SAP/c (%)
D-1A	0.4	0.2	D-17B	0.5	0.2	W-33A	0.6	0.2
D-2A	0.4	0.3	D-18B	0.5	0.3	W-34A	0.6	0.3
D-3A	0.4	0.4	D-19B	0.5	0.4	W-35A	0.6	0.4
D-4A	0.4	0.5	D-20B	0.5	0.5	W-36A	0.6	0.5
D-5A	0.5	0.2	D-21B	0.6	0.2	W-37B	0.4	0.2
D-6A	0.5	0.3	D-22B	0.6	0.3	W-38B	0.4	0.3
D-7A	0.5	0.4	D-23B	0.6	0.4	W-39B	0.4	0.4
D-8A	0.5	0.5	D-24B	0.6	0.5	W-40B	0.4	0.5
D-9A	0.6	0.2	W-25A	0.4	0.2	W-41B	0.5	0.2
D-10A	0.6	0.3	W-26A	0.4	0.3	W-42B	0.5	0.3
D-11A	0.6	0.4	W-27A	0.4	0.4	W-43B	0.5	0.4
D-12A	0.6	0.5	W-28A	0.4	0.5	W-44B	0.5	0.5
D-13B	0.4	0.2	W-29A	0.5	0.2	W-45B	0.6	0.2
D-14B	0.4	0.3	W-30A	0.5	0.3	W-46B	0.6	0.3
D-15B	0.4	0.4	W-31A	0.5	0.4	W-47B	0.6	0.4
D-16B	0.4	0.5	W-32A	0.5	0.5	W-48B	0.6	0.5

Throughout mix preparation and rheology testing, room conditions were maintained at 20 ± 2 °C and 30 ± 5% Relative Humidity (RH). The water used for cement paste preparation was consistently kept at 20 °C using a water bath (Joanlab Equipment Co., Ltd., Huzhou, China).

The mixing procedure for all cement pastes used in the present study is as follows:

(a) Dry SAP mixing in cement paste:
Initially, all dry materials, such as cement and dry SAP, were mixed at speed 2 (95 rpm) using a Hobart mixer for 60 s. Afterward, water was added to the dry materials and mixed at speed 2 (95 rpm) for 30 s. Then, the mixer was stopped, and the cement paste was scraped from the walls and bottom of the mixing bowl (50 s). After that, the cement paste was mixed at speed 2 (95 rpm), and the bottom of the mixing bowl was scraped again (40 s). Finally, the paste was mixed at speed 4 (135 rpm) for 60 s.

(b) Wet SAP mixing in cement paste:
The dry SAP and water were mixed and left to presoak for 30 min. The dry cement alone was mixed at speed 2 (95 rpm) to remove lumps (if any) for 60 s. After pre-soaking the SAP, the SAP and water mixture was added to the cement and mixed at speed 2 (95 rpm) for 60 s. Then, the mixer was stopped, and the cement paste was scraped from the walls and bottom of the mixing bowl for 30 s. Furthermore, the paste was mixed at speed 2 (95 rpm) and again the bottom of the bowl was scraped (60 s). Finally, the paste was mixed at speed 4 (135 rpm) for 60 s.

2.2. Testing Methods

2.2.1. Absorption Capacity of SAP

Initially, SAP absorption was studied in three different buffer solutions—pH 7, pH 11, and pH 13—to investigate the impact of pH on its absorption. The pH 7 represented water, pH 11 represented a mildly alkaline concrete admixture environment, and pH 13 represented a highly alkaline cementitious environment.

Additionally, absorption capacity was tested in a cement filtrate solution. The cement filtrate solution was prepared using cement and deionized water with a w/c ratio of 5. The cement and water paste were continuously stirred using a mixer for 24 h at a constant speed, and after that, the liquid was filtered.

Furthermore, following RILEM guidelines [24,25], two testing methods—the 'tea-bag method' and the 'filtration Method'—were used to investigate the absorption capacity of SAP, as shown in Figures 2 and 3, respectively. In the tea-bag method, a dry tea bag was initially weighed as mass m_1, and a dry tea bag containing 0.2 g of SAP was weighed as mass m_2. The tea bag was then sealed to prevent any dry or swollen SAP from leaking out. Further, the SAP-filled tea bag was soaked in a beaker filled with the test solution (about 200 mL). During this process, the beaker was covered with a thin plastic film to prevent carbonation and evaporation. The soaked SAP-filled tea bag was removed at specific time intervals: 1, 5, 10, 30, 60 min, 3 h and 24 h after the SAP/liquid contact time. The tea bag was wrapped with a dry tissue, and a weight of 1 kg was placed on it for 60 s to remove surface water before measuring the weight. This was performed to avoid errors caused by manual pressure applied by hand while removing surface water from the tea bag [26]. Afterwards, the weight m_3 was recorded. Furthermore, m_0 was calculated by taking into consideration the average dry and wet weight of the tea bag as shown in Equation (1), where n is the number of tea bags used (total of 10 tea bags), m_{Ai} and m_{Bi} are the individual masses of the dry and wet tea bags, respectively. Consequently, the absorption capacity was calculated according to Equation (2).

$$m_0 = \frac{1}{n}\sum_{i=1}^{n}(m_{Bi} - m_{Ai}) \qquad (1)$$

$$\text{Absorption capacity by tea-bag method} = \frac{m_3 - m_2 - m_0}{m_2 - m_1} \qquad (2)$$

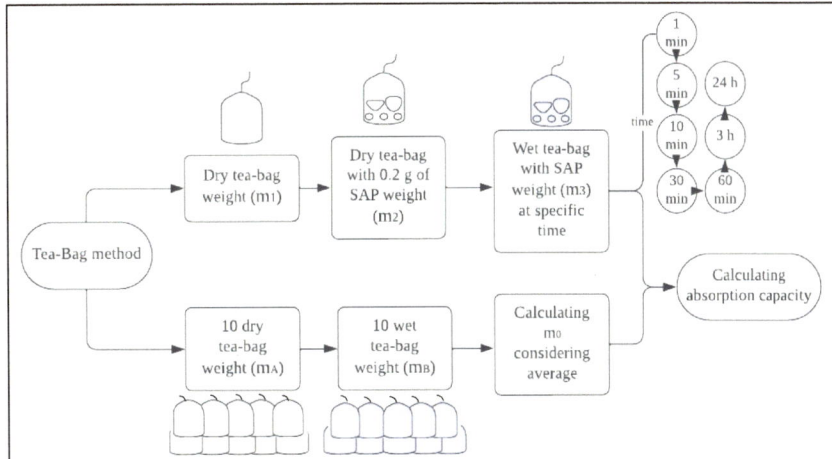

Figure 2. Tea-bag method.

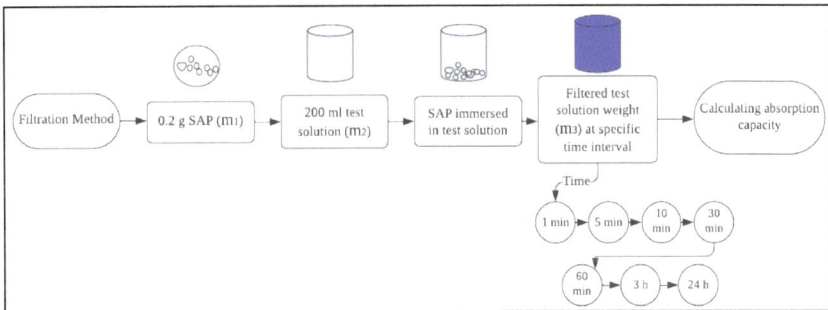

Figure 3. Filtration method.

In the filtration method, the dry amount of 0.2 g of SAP was measured as mass m_1, and the test solution of 200 mL was measured as mass m_2. The test fluid was added to a beaker, and dry SAP was immersed. Immediately, the beaker was tightly sealed using a thin plastic film to prevent evaporation and carbonation. The entire solution was filtered after time intervals of 1, 5, 10, 30, 60 min, 3 h and 24 h from the SAP/liquid contact time, and the amount of filtered fluid was determined as mass m_3. The filter paper (12–15 μm mesh size) used for filtering was saturated in the test solution before use, and the filtration was carried out using a funnel with filter paper. Different containers with new solutions and SAP were prepared for each testing interval. Equation (3) was used to calculate absorption capacity by filtration method.

$$\text{Absorption capacity by filtration method} = \frac{m_2 - m_3}{m_1} \qquad (3)$$

2.2.2. Rheological Testing

Cement pastes, immediately after mixing, were poured into a cylinder, and rheological tests using a Brookfield DV2T rheometer (Brookfield AMETEK, Middleborough, MA, USA) were conducted. The rheometer was connected to a PC, and data were recorded in RheocalcT software (https://www.brookfieldengineering.com/products/software/rheocalct, accessed from 14 January 2024). During testing, the cylinder was covered with thin plastic to prevent evaporation. In this study, the Bingham model was adopted, which includes yield stress and viscosity, as shown in Equation (4),

$$\tau = \tau_0 + \mu\gamma \qquad (4)$$

where τ is the shear stress, τ_0 is the yield stress, μ is the plastic viscosity of the cementitious suspensions, and γ is the shear strain rate.

The torque required to rotate the spindle and the shear rate were determined. The samples were pre-sheared for 30 s at 33.15 s^{-1}, followed by a 10 s rest. Then, the shear rate was ramped up from 0 to 33.15 s^{-1} and down from 33.15 to 0 s^{-1}, which was repeated throughout the test duration of about 35 min, as shown in Figure 4. The shear rate at each step was maintained for 10 s to measure a stable shear stress. The curves obtained during the decreasing rate were more consistent. Furthermore, a linear regression was performed on the experimental data, plotting shear stress against shear rate. The slope and intercept of the plotted regression line were used to determine the plastic viscosity and yield stress [27]. The evaluations were conducted at 5, 10, 15, 20, 25, 30, and 35 min.

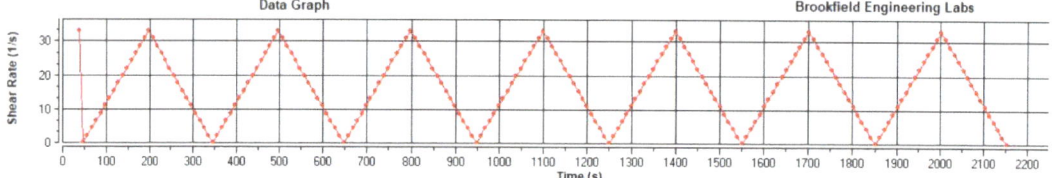

Figure 4. Shear rate applied during the test duration.

3. Results and Discussion

3.1. Absorption Capacities

The absorption capacities of two sizes of SAPs, namely SAP A, with an avg. size of 28 μm, and SAP B, with an avg. size of 80 μm, were studied in different solutions (pH 7, pH 11, pH 13, and cement filtrate solution with a pH of 13.73) using both the tea-bag and filtration methods. These tests were conducted at intervals of 1, 5, 10, 30, 60 min, 3 h, and 24 h.

From the results of the tea-bag method presented in Figure 5, it is evident that SAP A exhibited a higher absorption capacity than SAP B despite having a smaller particle size. Likewise, a study by Kazemian and Shafei (2024) revealed that the fine-sized SAP exhibited a 10% higher absorption capacity when compared with the large SAP particles [28]. The smaller size of SAP A led to a delayed initial absorption, attributed to gel-blocking, as illustrated in Figure 6. Gel blocking is a characteristic of very fine SAP particles smaller than 100 μm. In this scenario, minimal absorption occurs on the surface when SAP encounters liquid, causing the slightly swollen particles to adhere to each other, forming clusters that contain a significant amount of unswollen SAP, hindering further disaggregation and absorption [2,29,30].

SAP's initial absorption capacity was lower in all pH solutions; however, significantly lower initial absorption was noted in the pH 11 solution, as shown in Figure 5b. The absorption test was conducted twice to confirm the changes in SAP absorption in the pH 11 solution; however, it showed a similar trend each time. This may be due to small particles aggregating and swelling, leading to gel blocking. Over time, SAP A's absorption capacity increased from 23.59 g/g at 1 min to 49.35 g/g at 3 h in pH 11. Furthermore, SAP A demonstrated an increase in absorptivity in all three pH solutions, with the highest absorption in pH 11 (46.65 g/g at 24 h) compared to pH 7 (36.74 g/g at 24 h) and pH 13 (33.20 g/g at 24 h).

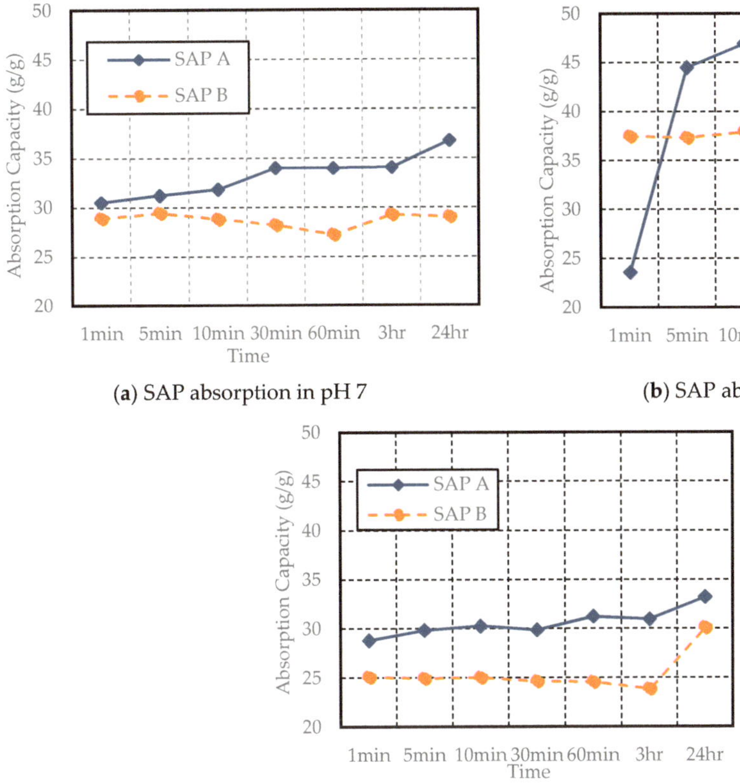

(a) SAP absorption in pH 7

(b) SAP absorption in pH 11

(c) SAP absorption in pH 13

Figure 5. SAP absorption testing using the tea-bag method in pH 7, pH 11, and pH 13 solution.

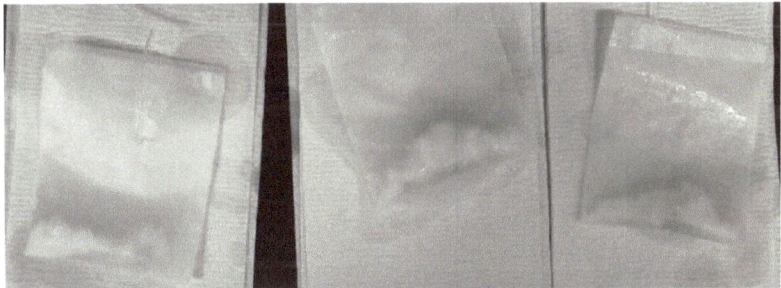

Figure 6. Gel blocking observed in the tea-bag method by SAP A in the first few minutes.

In contrast, SAP B exhibited signs of water release in all cases after 30–60 min. However, in pH 13, absorption capacity was significantly increased from 23.89 g/g at 3 h to 30.10 g/g at 24 h. The increase in pH from 11 to 13 resulted in a decrease in absorption capacity for both SAPs, consistent with studies conducted by [11,26].

Both SAPs showed higher absorptivity in the filtration method than the tea-bag method [25,31]. Like the tea-bag method, SAP A's absorption capacity was higher in all three pH solutions than SAP B's, as shown in Figure 7. In the pH 7 solution, both SAPs initially exhibited increased absorption, followed by a significant decreasing trend after 60 min. In pH 11, SAP A showed lower early absorption (66.27 g/g at 1 min of contact with solution) than SAP B (70.84 g/g at 1 min). Over time, a reduction in absorption capacity in pH 11 solution was observed after 30 min by SAP A and after 10 min by SAP B. Furthermore, in the filtration method, the absorption by both SAPs was lower in pH 13 compared to pH 7 and pH 11, consistent with a study by [32]. In pH 13, SAP A demonstrated a fluctuating trend, reaching its peak at 54.93 g/g at 60 min, while SAP B exhibited varying capacities, with a minimum of 31.09 g/g at the 3 h mark and generally lower values compared to SAP A across the testing periods, including a notable increase for both SAPs at 24 h.

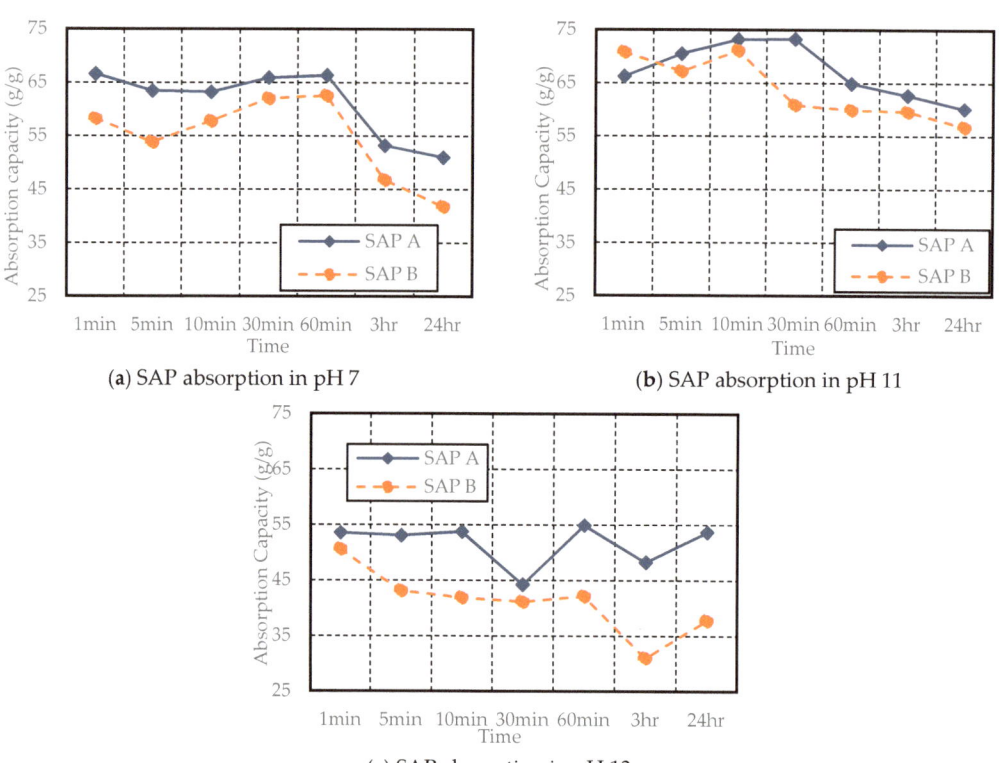

Figure 7. SAP absorption testing using filtration method in pH 7, pH 11, and pH 13 solution.

Both SAPs demonstrated lower absorption in the cement filtrate solution with a higher pH of 13.73, as shown in Figure 8, similar to the findings in [33]. This may be due to the dissolved ions (K^+, Na^+, Mg^{2+}, and Ca^{2+}) in the cement filtrate solution affecting the osmotic pressure and hence influencing the swelling capacity of SAP [2,10]. SAP A and SAP B also formed a white egg-shell-like crust in the cement filtrate solution during both tests. This phenomenon may be attributed to the exposure of SAP to a highly alkaline fluid with an elevated ion concentration, particularly calcium ions [25,34], as shown in Figure 9. The decreasing trend in the absorptivity of polyacrylic-based SAP, as shown in Figure 8, closely resembles the absorptivity observed by Zhong et al. (2020) [31] using acrylic acid and acrylamide-based SAP in the cement filtrate solution in both the test methods. Similarly, in a study by Yun et al. (2017), sodium-polyacrylate-based SAP [35] exhibited a trend similar to that observed in this study. The absorption by SAP A using the tea-bag and filtration methods at 1 min was 22.68 g/g and 48.06 g/g, respectively, which then decreased to 2.19 g/g and 26.54 g/g after 24 h. Similarly, for SAP B, the absorption capacity using the tea-bag and filtration methods at 1 min was 29.74 g/g and 46.39 g/g, respectively, which then decreased to 2.31 g/g and 23.55 g/g after 24 h, respectively.

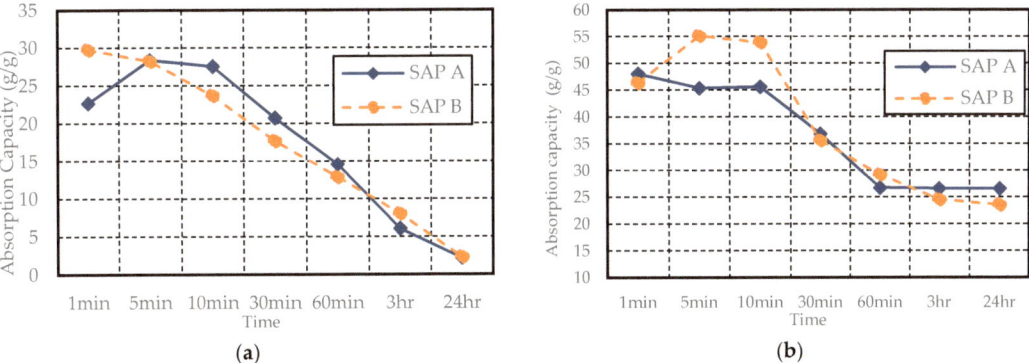

Figure 8. SAP absorption testing in cement filtrate solution (pH 13.73) (**a**) tea-bag method (**b**) filtration method.

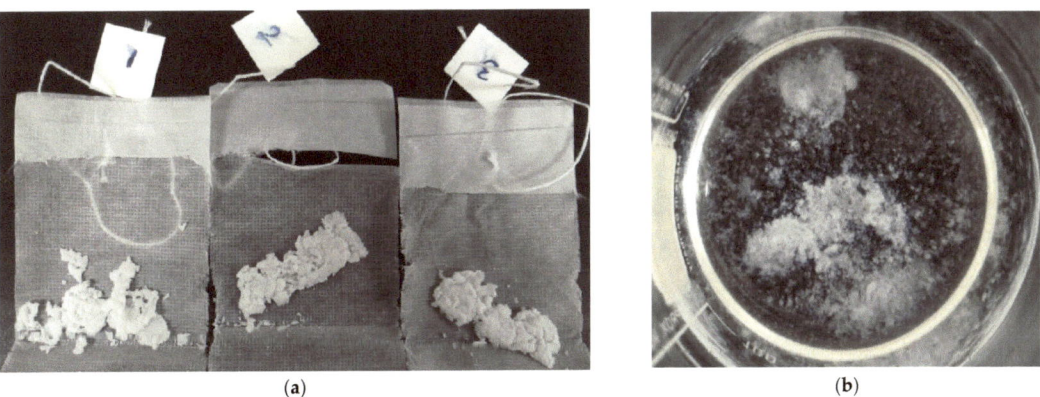

Figure 9. Formation of white egg shell-like crust during (**a**) tea-bag and (**b**) filtration methods in Cement Filtrate.

3.2. Rheological Properties

Figures 10–15 illustrate the variations over time in plastic viscosity and yield stress for cement paste with varying dry SAP A and SAP B content (ranging from 0.2 to 0.5% weight of cement) and three different w/c ratios (0.4, 0.5, and 0.6). Despite the observed desorption behavior of SAP in the cement filtrate solution, the introduction of SAP led to elevated plastic viscosity and yield stress [12,36]. Both SAPs exhibited higher values of plastic viscosity and yield stress in mixes with w/c ratios of 0.4, 0.5, and 0.6, containing different dosages of dry and wet SAP compared to reference mixes.

Examining Figure 10a,b for cement paste with a w/c ratio of 0.4, it is evident that the plastic viscosity increased with the dosage of SAP A and SAP B from 0.2% to 0.5%, compared to the reference mixes. Similar behavior was observed in a study carried out by [37]. However, after 35 min, a reduction in viscosity was noted for D-4A (1.73 Pa·s) compared to D-1A (1.75 Pa·s) and D-3A (1.86 Pa·s) mixes. Whereas, cement paste with a lower content of SAP B (D-13B) showed higher viscosity (1.76 Pa·s) at 30 min than mixes with a higher SAP dosage, i.e., D-14B (1.67 Pa·s) and D-15B (1.60 Pa·s).

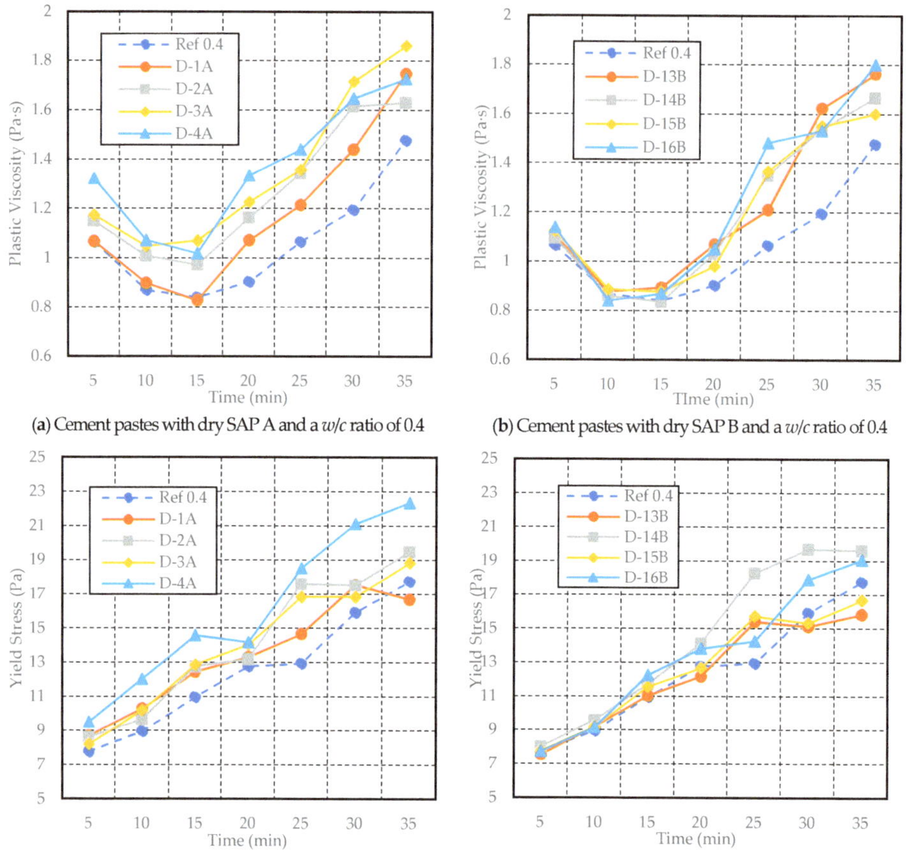

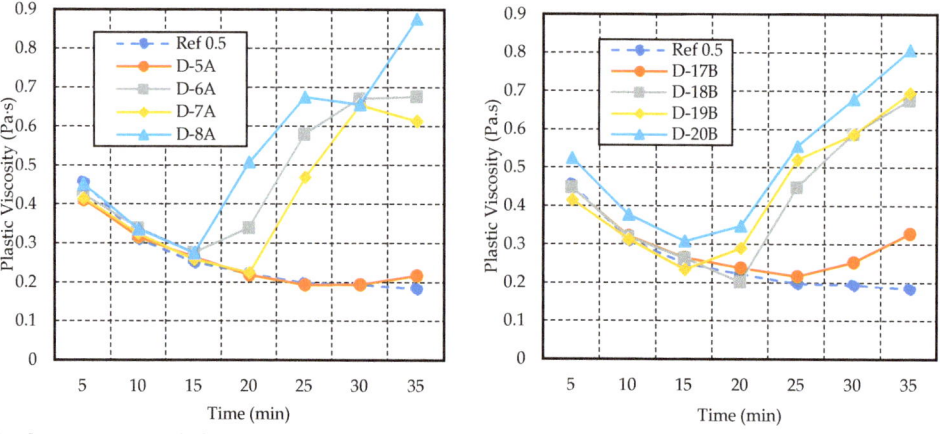

(a) Cement pastes with dry SAP A and a *w/c* ratio of 0.4   (b) Cement pastes with dry SAP B and a *w/c* ratio of 0.4

(c) Cement pastes with dry SAP A and a *w/c* ratio of 0.4   (d) Cement pastes with dry SAP B and a *w/c* ratio of 0.4

Figure 10. Variations in yield stress and plastic viscosity with time for cement paste with a w/c ratio of 0.4 and different dosages of dry SAP A and B.

(a) Cement pastes with dry SAP A and a *w/c* ratio of 0.5   (b) Cement pastes with dry SAP B and a *w/c* ratio of 0.5

Figure 11. *Cont.*

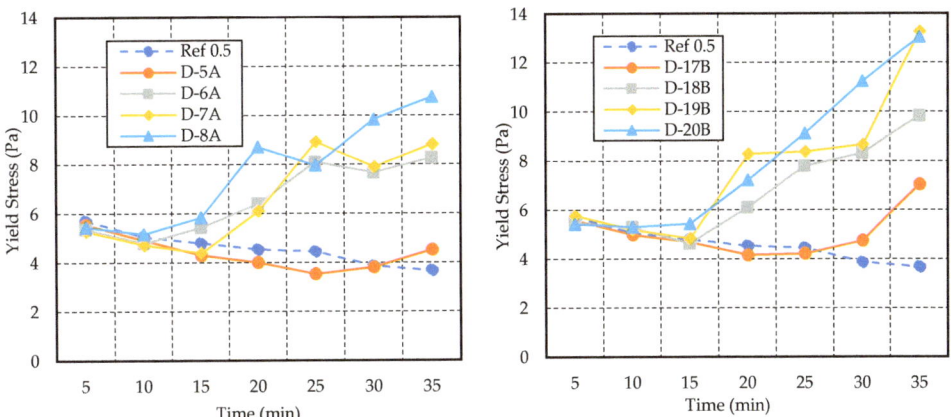

(c) Cement pastes with dry SAP A and a *w/c* ratio of 0.5 (d) Cement pastes with dry SAP B and a *w/c* ratio of 0.5

Figure 11. Variations in yield stress and plastic viscosity with time for cement paste with a *w/c* ratio of 0.5 and different dosages of dry SAP A and B.

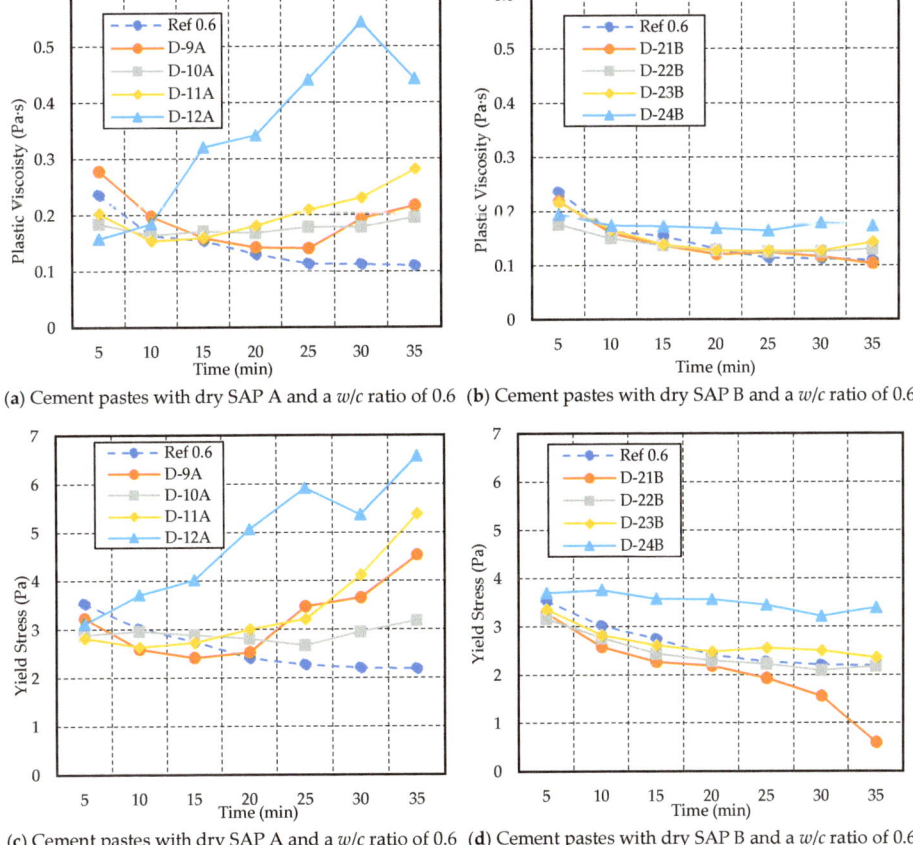

(a) Cement pastes with dry SAP A and a *w/c* ratio of 0.6 (b) Cement pastes with dry SAP B and a *w/c* ratio of 0.6

(c) Cement pastes with dry SAP A and a *w/c* ratio of 0.6 (d) Cement pastes with dry SAP B and a *w/c* ratio of 0.6

Figure 12. Variations in yield stress and plastic viscosity with time for cement paste with a *w/c* ratio of 0.6 and different dosages of SAP A and B.

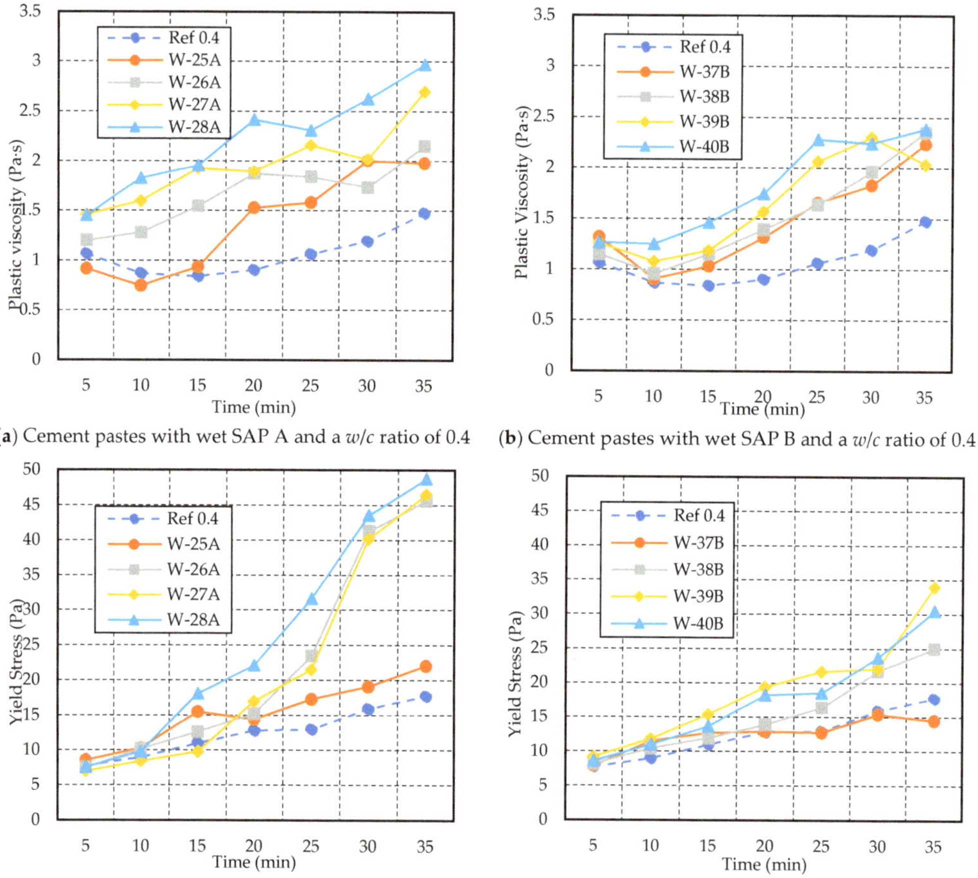

(a) Cement pastes with wet SAP A and a *w/c* ratio of 0.4   (b) Cement pastes with wet SAP B and a *w/c* ratio of 0.4

(c) Cement pastes with wet SAP A and a *w/c* ratio of 0.4   (d) Cement pastes with wet SAP B and a *w/c* ratio of 0.4

Figure 13. Variations in yield stress and plastic viscosity with time for cement paste with a *w/c* ratio of 0.4 and different dosages of wet SAP A and B.

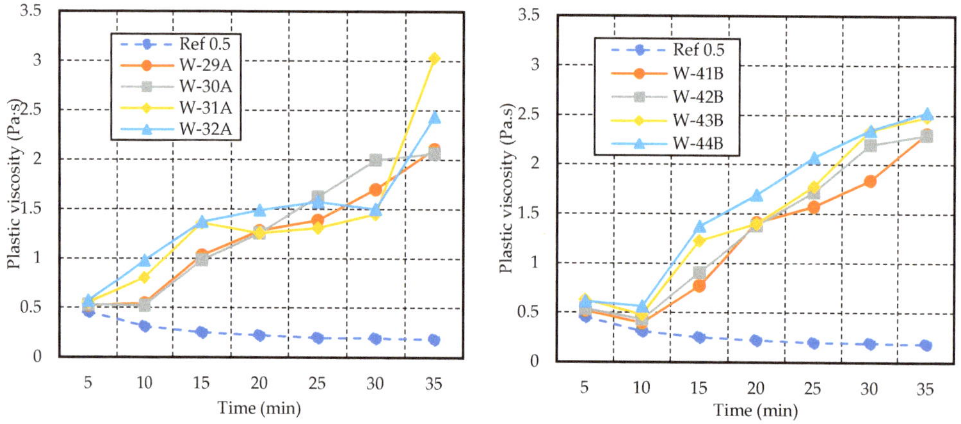

(a) Cement pastes with wet SAP A and a *w/c* ratio of 0.5   (b) Cement pastes with wet SAP B and a *w/c* ratio of 0.5

Figure 14. *Cont.*

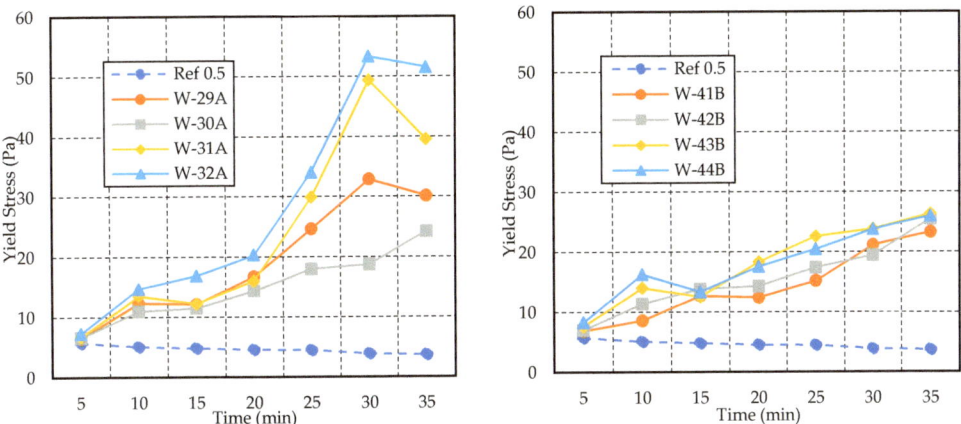

(c) Cement pastes with wet SAP A and a *w/c* ratio of 0.5   (d) Cement pastes with wet SAP B and a *w/c* ratio of 0.5

Figure 14. Variations in yield stress and plastic viscosity with time for cement paste with a *w/c* ratio of 0.5 and different dosages of wet SAP A and B.

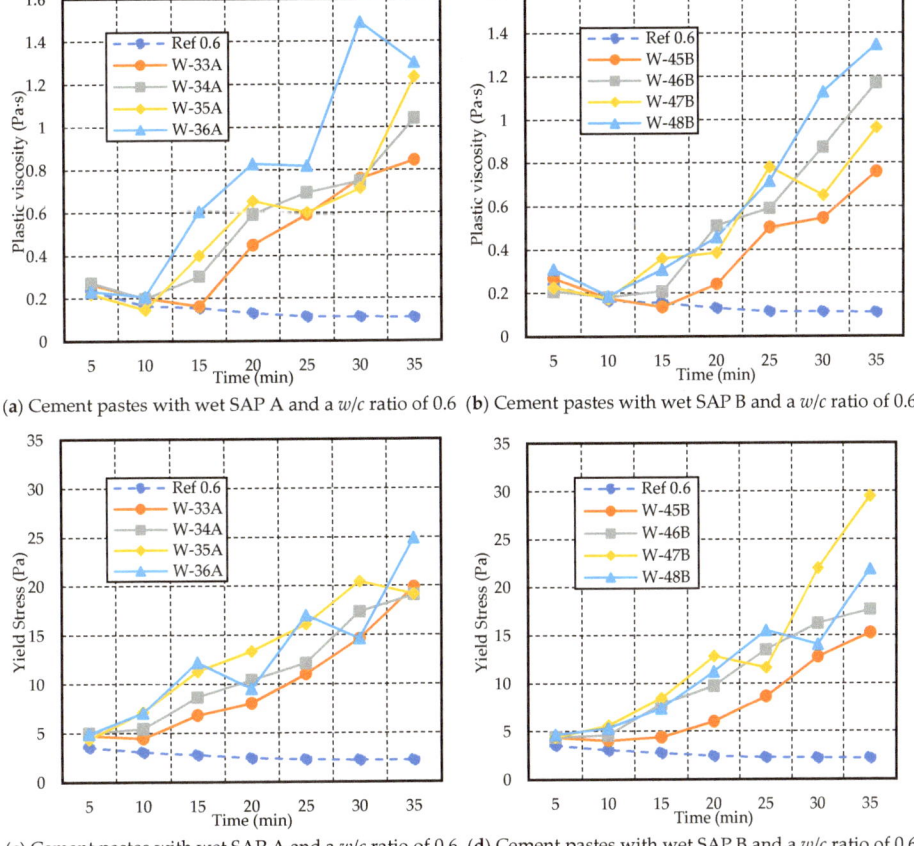

(a) Cement pastes with wet SAP A and a *w/c* ratio of 0.6   (b) Cement pastes with wet SAP B and a *w/c* ratio of 0.6

(c) Cement pastes with wet SAP A and a *w/c* ratio of 0.6   (d) Cement pastes with wet SAP B and a *w/c* ratio of 0.6

Figure 15. Variations in yield stress and plastic viscosity with time for cement paste with a *w/c* ratio of 0.6 and different dosages of wet SAP A and B.

The yield stress of 0.3% and 0.4% SAP A (D-2A and D-3A) showed nearly similar trends and mixes with 0.2% SAP B (D-13B) and 0.4% SAP B (D-15B) also exhibited almost similar behavior throughout the test duration as shown in Figure 10c,d. Reduction in yield stress was observed for D-1A from 17.55 Pa to 16.69 Pa (30 to 35 min) compared to the reference mix from 15.93 to 17.74 Pa. Similarly, the yield stress reduced for D-13B from 15.43 Pa (at 25 min) to 15.14 Pa (at 30 min), with a sudden increase by 15.85 Pa (at 35 min), which was lower than the reference mix. The D-15B mix rose from 15.35 Pa at 30 min to 16.70 Pa at 35 min, which was also lower than the reference mix. Meanwhile, D-14B showed a drastic increase in yield stress from 14.09 Pa at 20 min to 19.62 Pa at 35 min.

For cement paste with a w/c ratio of 0.5 (Figure 11), mixes with 0.3% to 0.5% of dry SAP A and B exhibited higher plastic viscosity and yield stress compared to the reference and 0.2% SAP mix. This aligns with the findings of Oh and Choi (2023) [38]. The viscosity of paste with 0.2% of dry SAP A (D-5A) and the reference showed minimal variation, except for a slight increase observed after 30 min, from 0.19 Pa·s to 0.22 Pa·s. On the other hand, the paste with 0.2% SAP B (D-17B) showed a higher increase in plastic viscosity, reaching 0.45 Pa·s at 5 min, followed by a decreasing trend, with a slight increase observed between 20 min and 35 min, ranging from 0.24 Pa·s to 0.33 Pa·s, compared to the reference mix, which ranged from 0.22 Pa·s to 0.18 Pa·s. The mix containing 0.3% SAP A (D-6A) showed higher viscosity than the 0.4% SAP A (D-7A).

The yield stress of the paste with 0.2% SAP A and B was lower than the reference mix from 5 min to 25 min, after which an increase was noted, as shown in Figure 11c,d. A higher dosage of SAP A (D-8A) and B (D-20B) led to an increase in yield stress. At 35 min, it was the highest, i.e., 10.74 Pa and 13.04 Pa, respectively, compared to the reference mix, which was 3.68 Pa.

Figure 12 illustrates the plastic viscosity and yield stress for cement paste with a w/c ratio of 0.6. Increasing the dosage of SAP A and SAP B resulted in an increase in plastic viscosity and yield stress over time, closely aligning with the findings of Senff et al. (2015) [39]. However, the mix containing a higher dosage of SAP A (D-12A) showed a drastic increase in plastic viscosity and yield stress over time compared to the other mixes. D-9A, containing 0.2% SAP, initially showed an increase in plastic viscosity, followed by a slight decrease. On the other hand, D-21B exhibited the opposite behavior, experiencing a decrease in yield stress throughout the test duration. When compared, the rheological properties of cement paste with a w/c ratio of 0.6 are significantly influenced by SAP A rather than SAP B.

At 30 min, the yield stress of the paste with 0.2% SAP A (D-9A) showed an increase to 4.53 Pa, compared to 3.18 Pa for D-10A and 2.19 Pa for Ref-0.6. Furthermore, higher dosages of both SAP A and B showed higher yield stress compared to the rest of the mixes, which is consistent with the findings in [39].

Based on the above results, a common observation was noted, which was lower initial plastic viscosity and yield stress in the mixes containing SAP. This can be attributed to SAPs absorbing Ca^{2+} and releasing Na^+ and K^+ ions into the pore solution. Ca^{2+} initially binds with SAP, thereby reducing the initial swelling. However, over time, the bound Ca^{2+} is displaced, allowing the swelling to gradually recover. Higher alkalinity tends to increase swelling, whereas it decreases with increased calcium concentration. The greater the degree of ion exchange, the lower the swelling of SAP [19].

The cement pastes with a w/c of 0.4 and wet SAPs (presoaked for 30 min) showed higher plastic viscosity and yield stress with an increase in the dosage of SAP compared to the reference mixes and cement pastes with dry SAP, as shown in Figure 13. However, adding SAP A resulted in greater viscosity when compared to mixes with SAP B. The higher dosages of SAP A (W-28A) and SAP B (W-40B) showed an increase in viscosity, specifically 2.97 Pa·s and 2.39 Pa·s, respectively, in comparison to the viscosity of the Ref-0.4 mix, which measured 1.48 Pa·s at the end of the test duration. Furthermore, W-37B exhibited an increase in plastic viscosity from 1.32 Pa·s (at 20 min) to 2.24 Pa·s (at 35 min), which was quite close to W-38B, where it increased from 1.39 Pa·s (at 20 min) to 2.36 Pa·s (at 35 min).

Regarding yield stress, W-25A increased with time, while W-37B showed a decrease, as shown in Figure 13c,d, compared to Ref 0.4. The higher dosage of SAP A and SAP B mixes resulted in higher yield stress compared to the rest of the mixes. W-27A exhibited an increase in yield stress from 16.98 Pa (at 20 min) to 46.54 Pa (at 35 min), which is quite similar to W-26A, ranging from 15.19 Pa (at 20 min) to 45.61 Pa (at 35 min). On the other hand, W-39B depicted more yield stress, ranging from 9.19 Pa (at 5 min) to 33.99 Pa (at 35 min), than W-40B, which ranged from 8.60 Pa (at 5 min) to 30.51 Pa (at 35 min).

From Figures 14 and 15, a drastic increase in plastic viscosity and yield stress was observed as the dosage of wet SAP increased in the cement paste with w/c ratios of 0.5 and 0.6. The behavior of both wet SAPs with varying dosages followed almost a similar trend for cement paste with w/c ratios of 0.5 and 0.6. However, the plastic viscosity and yield stress increased when the w/c ratio was lower, such as for the cement paste with a w/c ratio of 0.5, compared to the cement paste with a w/c ratio of 0.6.

Similarly, SAP A had a more significant impact on the plastic viscosity and yield stress of the cement paste with a w/c ratio of 0.5 than SAP B. The plastic viscosity of W-31A ranged from 1.45 Pa·s to 3.03 Pa·s (30 to 35 min), surpassing that of W-32A, which ranged from 1.50 Pa·s to 2.44 Pa·s. As the dosage of SAP B increased, the plastic viscosity also increased, as shown in Figure 14b. The yield stress of W-32A was higher, ranging from 7.27 Pa to 51.55 Pa over the test duration, whereas for W-44B, the yield stress ranged from 8.29 Pa to 25.96 Pa, as illustrated in Figure 14c,d.

From Figure 15a,b, both SAP A and B showed a similar trend of increasing plastic viscosity with an increase in dosage over the testing duration. The W-36A mix depicted higher plastic viscosity, ranging from 0.23 Pa·s to 1.30 Pa·s (from 5 min to 35 min), whereas W-48B ranged from 0.31 Pa·s to 1.35 Pa·s. The plastic viscosity for W-35A and W-47B (0.4% SAP) exhibited fluctuations during the test duration. However, there was an overall increase over the test period compared to the reference mix. Specifically, for W-35A, there was a continuous increase in plastic viscosity from 0.22 Pa·s at 5 min to 1.23 Pa·s at 35 min. Similarly, for W-47B, the plastic viscosity increased from 0.23 Pa·s at 5 min to 0.96 Pa·s at 35 min.

From Figure 15c,d, cement paste with 0.5% SAP, i.e., W-36A and W-48B, demonstrated varying yield stress trends throughout the test duration. Specifically, for W-36A, the yield stress increased from 4.89 Pa at 5 min to a peak of 24.91 Pa at 35 min. In contrast, for W-48B, the yield stress also increased over time, starting at 4.64 Pa at 5 min and reaching 21.86 Pa at 35 min. Both W-36A and W-48B exhibited an upward trend in yield stress over the 35 min test period, with W-36A showing a more pronounced increase and finishing with a higher peak value compared to W-48B. The W-47B mix exhibited the highest yield stress compared to the rest of the mixes, ranging from 4.42 Pa to 29.48 Pa throughout the duration.

The w/c ratio and SAP dosage significantly impact plastic viscosity and yield stress. Plastic viscosity and yield stress were higher in the mixes with a w/c ratio of 0.4, decreasing as the w/c ratio increased to 0.5 and 0.6. Mixes with a w/c ratio of 0.6 showed very low plastic viscosity and yield stress, as they were highly flowy during the test, whereas mixes with a w/c ratio of 0.4 became very thick over the test duration.

In a study by Dang et al. (2017), the addition of wet SAP in cement-based mixes increased the slump largely compared to the slump of mixes with dry SAP when additional internal curing water was added. However, without the additional internal curing water, the slump decreased severely, and the slump was less affected by the change in particle size of SAP [22]. Similarly, in this study, the addition of wet SAP resulted in a more significant increase in plastic viscosity and yield stress compared to the mixes with dry SAP. This may be due to the addition of presoaked SAP reducing the effective water content, equivalent to reducing water. Without providing additional water, it may lead to an increase in plastic viscosity and yield stress. Additionally, when the dry SAP is mixed with cement, the intricate ion composition in the cement mix prevents the SAP from fully absorbing the water. Consequently, fine cement particles cling to the SAP particles, elevating the friction [23], which could also be the reason for the lower plastic viscosity and yield stress of cement paste with dry SAP compared to cement paste with wet SAP.

During wet SAP mixing, SAP A and SAP B were presoaked in water for 30 min; however, the addition of SAP A in water resulted in gel blocking, which was similar to what was observed during the absorption testing. Even after 30 min of presoaking SAP A in water, the particles were not fully swollen, as shown in Figure 16. This phenomenon does not occur in dry SAP mixing because the cement and dry SAP were mixed before adding water. On the other hand, SAP B, which had a larger particle size, was not affected by gel blocking and was fully absorbed during the 30 min. The gel blocking by SAP A was very severe when a higher dosage of SAP and a w/c ratio of 0.4 were used. In cement paste with wet SAP A mixes, plastic viscosity and yield stress were also affected by gel blocking caused by SAP A.

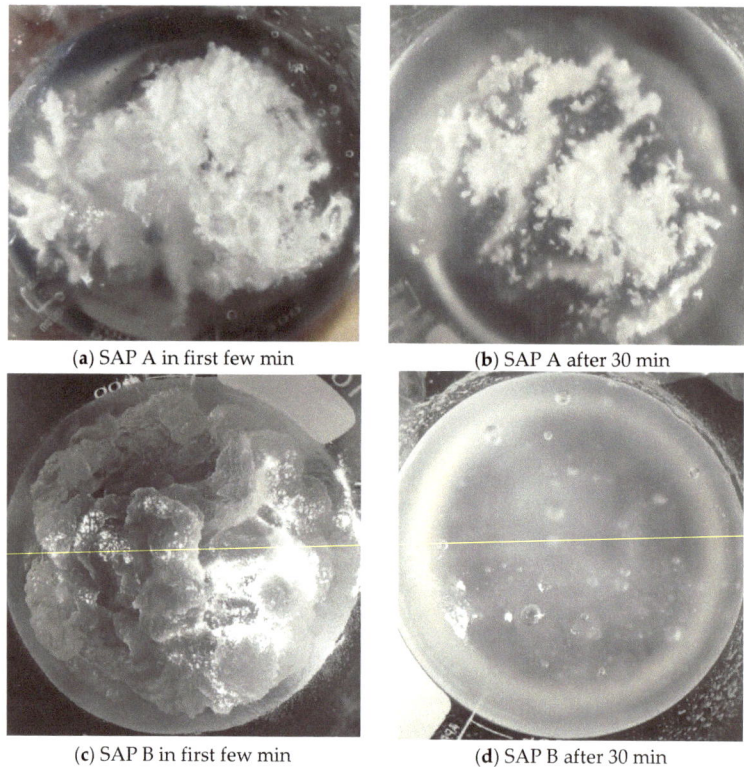

(a) SAP A in first few min (b) SAP A after 30 min

(c) SAP B in first few min (d) SAP B after 30 min

Figure 16. SAP A and SAP B mixed with water for 30 min presoaking.

In order to prevent gel-blocking, it is suggested that dry SAP be used and mixed with cement before adding water. Alternatively, if presoaked SAP is used in the cement paste, using SAP with a particle size above 100 μm is recommended to avoid gel blocking [2,29].

3.3. Comparative Analysis and Applications

The SAP used in this study showed higher absorption in pH 7 and pH 11 solutions, which correspond to neutral water and mildly alkaline concrete admixture environments. This indicates that the SAP has good water retention properties despite the alkaline environment.

Based on the results of the absorption capacity in pH 13 and cement filtrate solution, it was observed that the absorption capacity of both SAPs gradually decreased. However, in cement paste, both SAPs exhibited a continuous increase in plastic viscosity and yield stress throughout the test. This suggests that SAPs continuously absorb water in cement paste, demonstrating their ability to retain water in a cementitious environment.

Based on these findings, the polyacrylic-based SAPs used in this study can be effectively employed for internal curing purposes. They can effectively maintain moisture and contribute to promoting hydration in various construction materials and applications where low workability is necessary, such as plastering and grouting. For highly workable cementitious composites, the addition of a superplasticizer can be considered in SAP-modified cementitious materials.

4. Conclusions

This paper assesses the absorption capacity of two different sizes of SAP in various pH environments, examining their impact on the rheological properties of cement paste prepared using three different w/c ratios: 0.4, 0.5, and 0.6, with varying SAP dosages ranging from 0.2 to 0.5% weight of cement. The findings lead to the following conclusions:

1. The absorption capacity of SAP A (avg. size of 28 μm) surpasses that of SAP B (avg. size of 80 μm) across all three pH solutions in both methods, such as the tea-bag and filtration methods. The filtration method depicted a higher absorption capacity than the tea-bag method. Notably, SAP A exhibits lower initial absorption in pH 11 due to its finer size, attributed to the gel-blocking effect.

2. In the tea-bag method, SAP A absorbs more solution in pH 11 (46.65 g/g at 24 h) compared to pH 7 (36.74 g/g at 24 h) and pH 13 (33.20 g/g at 24 h). SAP B displays signs of water release in all instances after 30–60 min. However, in pH 13, a significant increase in absorption capacity is observed, rising from 23.89 g/g at 3 h to 30.10 g/g at 24 h. The absorption capacity of both SAPs decreases as the pH of the solution increases from 11 to 13.

3. Both SAPs demonstrated lower absorption in the cement filtrate solution (pH of 13.73). The absorption by SAP A, using the tea-bag and filtration methods at 1 min, was 22.68 g/g and 48.06 g/g, respectively, then decreased to 2.19 g/g and 26.54 g/g after 24 h. Similarly, for SAP B, the absorption capacity using the tea-bag and filtration methods at 1 min was 29.74 g/g and 46.39 g/g, respectively, which then decreased to 2.31 g/g and 23.55 g/g after 24 h, respectively.

4. The introduction of SAP leads to higher plastic viscosity and yield stress despite the inherent desorption behavior observed in the cement filtrate solution. Both SAPs exhibit increased plastic viscosity and yield stress in mixes with w/c ratios of 0.4, 0.5, and 0.6, containing varying dosages of dry and wet SAP compared to reference mixes. Furthermore, adding wet SAP resulted in a more significant increase in plastic viscosity and yield stress compared to mixes with dry SAP.

5. Cement pastes with SAP and a w/c ratio of 0.4 exhibited higher plastic viscosity and yield stress. As the w/c ratio increased to 0.5 and 0.6, the viscosity and yield stress decreased, but they remained higher than the reference mixes without SAP. Mixes containing SAP and a w/c ratio of 0.6 demonstrated significantly lower plastic viscosity and yield stress compared to those with a w/c ratio of 0.4. This indicates that mixes with a w/c ratio of 0.6 were highly fluid, while those with a w/c ratio of 0.4 became very thick due to the addition of SAP.

6. Both SAPs in dry form increased the plastic viscosity and yield stress of cement paste with w/c ratios of 0.4, 0.5, and 0.6. However, when the mix contained 0.2% SAP, the plastic viscosity and yield stress were almost similar to the reference mixes at w/c ratios of 0.5 and 0.6. This suggests that incorporating 0.2% SAP has a significantly lesser impact on the rheological properties, making it a viable option for practical applications.

The results obtained in this study for cement paste with SAP will further be considered in evaluating the impact of SAP in concrete and mortar, specifically for understanding changes in fresh concrete properties, microstructure analysis, and the impact on mechanical properties.

Author Contributions: Data curation, N.A.; Investigation, N.A.; Validation, N.A.; Writing—original draft, N.A.; Data curation, J.-W.L.; Investigation, J.-W.L.; Validation, J.-W.L.; Conceptualization, S.-T.K.; Writing—review and editing, S.-T.K.; Supervision, S.-T.K.; Funding acquisition, S.-T.K. All authors have read and agreed to the published version of the manuscript.

Funding: This research was supported by the National Research Foundation of Korea (NRF) grant funded by the Korean Government (Ministry of Science and ICT) (No. RS-2023-00251506).

Institutional Review Board Statement: Not applicable.

Informed Consent Statement: Not applicable.

Data Availability Statement: The original contributions presented in the study are included in the article, further inquiries can be directed to the corresponding author.

Conflicts of Interest: The authors declare no conflicts of interest.

References

1. Venkatachalam, D.; Kaliappa, S. Superabsorbent polymers: A state-of-art review on their classification, synthesis, physicochemical properties, and applications. *Rev. Chem. Eng.* **2020**, *39*, 127–171. [CrossRef]
2. Mechtcherine, V.; Reinhardt, H.W.; Cusson, D.; Friedrich, S.; Lura, P.; Friedemann, K.; Stallmach, F.; Mönnig, S.; Wyrzykowski, M.; Esteves, L.P.; et al. *Application of Super Absorbent Polymers (SAP) in Concrete Construction. State-of-the-Art Report Prepared by Technical Committee 225-SAP*; Mechtcherine, V., Reinhardt, H.W., Eds.; Springer Science & Business Media: Berlin, Germany, 2012; p. X-166. [CrossRef]
3. Jensen, O.M.; Hansen, P.F. Water-entrained cement-based materials I. Principles and theoretical background. *Cem. Concr. Res.* **2001**, *31*, 647–654. [CrossRef]
4. Cherel, O.C.; Wang, F.; Jin, Y.; Liu, Z. Effect of SAP on properties of high-performance concrete under Marine Wetting and Drying Cycles. *J. Wuhan Uni. Technol.-Mater. Sci. Edit.* **2019**, *34*, 1136–1142. [CrossRef]
5. Kong, X.; Zhang, Z.; Lu, Z. Effect of pre-soaked superabsorbent polymer on shrinkage of high-strength concrete. *Mater. Struct.* **2015**, *48*, 2741–2758. [CrossRef]
6. Laila, L.R.; Gurupatham, B.G.A.; Roy, K.; Lim, J.B.P. Influence of super absorbent polymer on mechanical, rheological, durability, and microstructural properties of self-compacting concrete using non-biodegradable granite pulver. *Struct. Concr.* **2020**, *22*, 1093–1116. [CrossRef]
7. Liu, J.; Farzadnia, N.; Shi, C.; Ma, X. Effects of superabsorbent polymer on shrinkage properties of ultra-high strength concrete under drying condition. *Constr. Build. Mater.* **2019**, *215*, 799–811. [CrossRef]
8. Yang, Z.; Shi, P.; Zhang, Y.; Li, Z. Effect of superabsorbent polymer introduction on properties of alkali-activated slag mortar. *Constr. Build. Mater.* **2022**, *340*, 127541. [CrossRef]
9. Wong, H.S. Concrete with Superabsorbent polymer. In *Eco-Efficient Repair and Rehabilitation of Concrete Infrastructures*; Woodhead Publishing: Sawston, UK, 2018; pp. 467–499. [CrossRef]
10. Mechtcherine, V.; Wyrzykowski, M.; Schröfl, C.; Snoeck, D.; Lura, P.; Belie, N.D.; Mignon, A.; Vlierberghe, S.V.; Klemm, A.J.; Almeida, F.C.R.; et al. Application of super absorbent polymers (SAP) in concrete construction- update of RILEM state-of-the-art-report. *Mater. Struct.* **2021**, *54*, 80. [CrossRef]
11. Qin, X.; Lin, Y.; Mao, J.; Sun, X.; Xie, Z.; Huang, Q. Research of water absorption and release mechanism of superabsorbent polymer in cement paste. *Polymers* **2023**, *15*, 3062. [CrossRef]
12. Liu, J.; Khayat, K.H.; Shi, C. Effect of superabsorbent polymer characteristics on rheology of ultra-high-performance concrete. *Cem. Concr. Compos.* **2020**, *112*, 103636. [CrossRef]
13. Mechtcherine, V.; Secrieru, E.; Schröfl, C. Effect of superabsorbent polymers (SAPs) on rheological properties of fresh cement-based mortar-development of yield stress and plastic viscosity over time. *Cem. Concr.* **2015**, *67*, 52–65. [CrossRef]
14. Jensen, O.M.; Hansen, P.F. Water-entrained cement-based materials II. Experiment observations. *Cem. Concr. Res.* **2002**, *32*, 973–978. [CrossRef]
15. Lura, P.; Guang, Y.; Cnudde, V.; Jacobs, P. Preliminary results about 3D distribution of superabsorbent polymers in mortars. In Proceedings of the International Conference on Microstructure Related Durability of Cementitious Composites, Nanjing, China, 13–15 October 2008; RILEM Proceedings. Volume 61, pp. 1341–1348.
16. Sikora, K.S.; Klemm, A.J. Effect of superabsorbent polymers on workability and hydration process in fly ash cementitious composites. *J. Mater. Civil. Eng.* **2015**, *27*, 1–13. [CrossRef]
17. Pierard, J.; Pollet, V.; Cauberg, N. Mitigating autogenous shrinkage in HPC by internal curing using superabsorbent polymers. In Proceedings of the International RILEM Conference on Volume Changes of Hardening Concrete: Testing and Mitigation, Lyngby, Denmark, 20–23 August 2006.
18. Ma, X.; Yuan, Q.; Liu, J.; Shi, C. Effect of water absorption of SAP on the rheological properties of cement-based materials with ultra-low w/b ratio. *Constr. Build. Mater.* **2019**, *195*, 66–74. [CrossRef]

19. Lee, H.X.D.; Wong, H.S.; Buenfeld, N.R. Effect of alkalinity and calcium concentration of pore solution on the swelling and ionic exchange of superbasorbent polymers in cement paste. *Cem. Concr. Compos.* **2018**, *88*, 150–164. [CrossRef]
20. Patil, A.D.; Ravande, K.; Jadhav, S.; Junead, M. Effect of superabsorbent polymer and slag cement on concrete properties. *Mater. Today Proc.* **2023**, in press. [CrossRef]
21. Kim, I.S.; Choi, S.Y.; Choi, Y.S.; Yang, E. Effect of internal pores formed by a superabsorbent polymer on durability and drying shrinkage of concrete specimens. *Materials* **2021**, *14*, 5199. [CrossRef]
22. Dang, J.; Zhao, J.; Du, Z. Effect of superabsorbent polymer on the properties of concrete. *Polymers* **2017**, *9*, 672. [CrossRef]
23. Huang, X.; Liu, X.; Rong, H.; Yang, X.; Duan, Y.; Ren, T. Effect of super-absorbent polymer (SAP) incorporation method on mechanical and shrinkage properties of internally cured concrete. *Materials* **2022**, *15*, 7854. [CrossRef]
24. Snoeck, D.; Schröfl, C.; Mechtcherine, V. Recommendation of RILEM TC 260-RSC: Testing sorption by superabsorbent polymers (SAP) prior to implementation in cement-based materials. *Mater. Struct.* **2018**, *51*, 116. [CrossRef]
25. Mechtcherine, V.; Snoeck, D.; Schröfl, C.; Belie, N.D.; Klemm, A.J.; Ichiniya, K.; Moon, J.; Wyrzykowski, M.; Lura, P.; Toropovs, N.; et al. Testing superabsorbent polymer (SAP) sorption properties prior to implementation in concrete: Results of a RILEM Round-Robin Test. *Mater. Struct.* **2018**, *51*, 28. [CrossRef]
26. Kim, S.; Choi, S.Y.; Choi, Y.S.; Yang, E.I. An experimental study on absorptivity measurement of superabsorbent polymers (SAP) and effect of SAP on freeze-thaw resistance in mortar specimen. *Constr. Build. Mater.* **2021**, *267*, 120974. [CrossRef]
27. Rajagopalan, S.; Lee, B.; Kang, S. Prediction of the rheological properties of fresh cementitious suspensions considering microstructural parameters. *Materials* **2022**, *15*, 7044. [CrossRef]
28. Kazemian, M.; Shafei, B. Investigation of type, size, and dosage effects of superabsorbent polymers on the hydration development of high-performance cementitious materials. *Constr. Build. Mater.* **2024**, *422*, 135801. [CrossRef]
29. Zhu, Q. Effect of Multivalent Ions on the Swelling and Mechanical Behavior of Superabsorbent Polymers (SAPs) for Mitigation of Mortar Autogenous Shrinkage. Master's Thesis, Pardue University, West Lafayette, IN, USA, 2014.
30. Adams, C.; Bose, B.; Olek, J.; Erk, K.A. Evaluation of mix design strategies to optimize flow and strength of mortar internally cured with superabsorbent polymers. *Constr. Build. Mater.* **2022**, *324*, 126664. [CrossRef]
31. Zhong, P.; Wang, J.; Wang, X.; Liu, J.; Li, Z.; Zhou, Y. Comparison of different approaches for testing sorption by a superabsorbent polymer to be used in cement-based materials. *Materials* **2020**, *13*, 5015. [CrossRef]
32. Fort, J.; Migas, P.; Cerny, R. Effect of absorptivity of superabsorbent polymers on design of cement mortars. *Materials* **2020**, *13*, 5503. [CrossRef]
33. Kang, S.H.; Hong, S.G.; Moon, J. Absorption kinetics of superabsorbent polymers (SAP) in various cement-based solutions. *Cem. Concr. Res.* **2017**, *97*, 73–83. [CrossRef]
34. Zhu, Q.; Barney, C.W.; Erk, K.A. Effect of ionic crosslinking on the swelling and mechanical response of model superabsorbent polymer hydrogels for internally cured concrete. *Mater. Struct.* **2014**, *48*, 2261–2276. [CrossRef]
35. Yun, K.K.; Kim, K.K.; Choi, W.; Yeon, J.H. Hygral behavior of superabsorbent polymers with various particles sizes and cross-linking densities. *Polymers* **2017**, *9*, 600. [CrossRef]
36. Secrieru, E.; Mechtcherine, V.; Schröfl, C.; Borin, D. Rheological characterization and prediction of pumpability of strain-hardening cement- based composites (SHCC) with and without addition of superabsorbent polymers (SAP) at various temperatures. *Constr. Build. Mater.* **2016**, *112*, 581–594. [CrossRef]
37. Aryanfar, A.; Sanal, I.; Marian, J. Percolation-based image processing for the plastic viscosity for cementitious mortar with super absorbent polymer. *Int. J. Concr. Struct. Mater.* **2021**, *15*, 25. [CrossRef]
38. Oh, S.; Choi, S. Effects of superabsorbent polymers (SAP) on the rheological behavior of cement mortars: A rheological study on performance requirements for 3D printable cementitious materials. *Constr. Build. Mater.* **2023**, *392*, 131856. [CrossRef]
39. Senff, L.; Modolo, R.C.E.; Ascensao, G.; Hotza, D.; Ferreira, V.M.; Labrincha, J.A. Development of mortars containing superabsorbent polymer. *Constr. Build. Mater.* **2015**, *95*, 575–584. [CrossRef]

Disclaimer/Publisher's Note: The statements, opinions and data contained in all publications are solely those of the individual author(s) and contributor(s) and not of MDPI and/or the editor(s). MDPI and/or the editor(s) disclaim responsibility for any injury to people or property resulting from any ideas, methods, instructions or products referred to in the content.

Article

Tension Capacity of Crushed Limestone–Cement Grout

Muawia Dafalla *, Ahmed M. Al-Mahbashi and Ahmed Alnuaim

Civil Engineering Department, College of Engineering, King Saud University, P.O. Box 800, Riyadh 11421, Saudi Arabia; aalmahbashi@ksu.edu.sa (A.M.A.-M.); alnuaim@ksu.edu.sa (A.A.)
* Correspondence: mdafalla@ksu.edu.sa

Abstract: The feasibility of using crushed limestone instead of sand in cement grout is examined in this work. This study entails performing several tests, including the Brazilian test, the compressive strength test, and the stress–strain correlation test. The curing times used were 7, 14, and 28 days for mixtures with various proportions of cement to limestone (1:1, 1:2, and 1:4). The conventional sand–cement grout laboratory tests were prepared using a similar methodology to examine the effectiveness of the suggested substitute. The findings show that the limestone-based grout has sufficient strength, but that it is less than that of the typical sand material. The values of the tensile strength and elastic modulus were determined. A focus was made on the tensile strength and stress–strain relationship. A special laboratory set-up was used to look at the progress of failure using strain gauges fitted to the cylindrical samples both vertically and horizontally. The angular shape of the particles' ability to interlock is responsible for the material's increase in strength. According to this study, crushed limestone can be used as a substitute for sand in circumstances where sand supply is constrained. The suggested grout can be used in the shotcrete of tunnels and rock surfaces.

Keywords: limestone; grout; stress–strain relationship; compressive strength

Citation: Dafalla, M.; Al-Mahbashi, A.M.; Alnuaim, A. Tension Capacity of Crushed Limestone–Cement Grout. Materials 2024, 17, 3860. https://doi.org/10.3390/ma17153860

Academic Editors: Miguel Ángel Sanjuán, Yuan Gao, Junlin Lin and Xupei Yao

Received: 7 May 2024
Revised: 17 July 2024
Accepted: 30 July 2024
Published: 4 August 2024

Copyright: © 2024 by the authors. Licensee MDPI, Basel, Switzerland. This article is an open access article distributed under the terms and conditions of the Creative Commons Attribution (CC BY) license (https://creativecommons.org/licenses/by/4.0/).

1. Introduction

The Brazilian test, which is also known as the indirect tensile test, is often used to determine the ability of materials to resist tension. The evaluation results of the stress–strain relationship and the failure properties obtained by this test are of great significance in the assessment of construction materials. This study is aimed at investigating the use of crushed limestone in cement grout for the stabilization and shotcrete treatment of tunnels. Previous research has shown that grouting techniques based on cement could effectively protect concrete structures from deterioration due to corrosion in steel bars [1] or the development of cracks in tension zones [2]. In addition, this material could be used to mitigate or repair the developed cracks in infrastructure construction exposed to static or dynamic loads [3]. In special applications, cement–water grout may be required for several reasons such as enhancement of the basic properties such as strength and tension resistance. The introduction of the metro transportation system to Riyadh City during the last five years has resulted in large amounts of excavated materials [4]. Tunnel boring machine technology was employed to construct the proposed tracks. The spoil material is dominantly crushed limestone. The majority of these components are variously graded crushed limestone. These materials have been employed in several geotechnical applications, including fluid management in clay liners [5], green structural concrete aggregation [4], and a reduction in the usage of commercial bentonite to be ecologically friendly [6].

Cement-based grout is primarily used in tunnel lining, crack repair, soil enhancement (jet grouting), and concrete applications. It is created by combining cement, sand, and water, with the addition of admixtures on some occasions [7,8].

Because limestone aggregates are efficient and reasonably priced, their usage in the concrete industry has been growing [9]. Aquino et al. [10] investigated the effects on the characteristics of concrete of varying ratios of fine limestone to sand at varying water-to-cement ratios. The obtained results demonstrate that adding more finely ground limestone to the concrete mixture increases the concrete's flexural strength, compressive strength, and elasticity modulus. Burhan Alshahwany [11] assessed how some characteristics of regular concrete were affected when sand was substituted with limestone filler. Additives to cement–water grout have been used to improve workability or to achieve the required properties of strength and resistance to tension cracks [12–14]. Shannag [13] reported that the use of natural materials such as pozzolan with silica fume has a significant role in producing a high-performance cement-based grout for special use in the concrete industry.

The findings indicate that adding up to 20% more limestone to the sand does not influence the strength of the concrete. Conversely, the scarcity of natural sand in many regions of the world has the potential to impede the expansion of the concrete industry's demand. For this reason, crushed limestone has been proven to be a successful substitute for natural sand. Compared to concrete–sand-based combinations, the concrete made with crushed limestone requires more water for mixing [15]. The mix design of the grout for the tunnel lining requires an understanding of the environment and stresses within the close boundaries of the tunnel opening. Tension stresses may be created within the rock mass or the broken parts along the rim or edges of the excavation. The gradation, particle size distribution, density, compaction, and flowability are very important factors in establishing a grout mix design. Knowledge of these parameters will help in supplying durable materials. The mix design of the grout should always consider its ability to resist tension to achieve stable, sound, and durable grout.

The limestone filler was proved to increase the degree of hydration. Bonaveet et al. [16] stated that the use of limestone filler is a rational option for reduced energy consumption and emission reduction. They found that the optimum level of limestone addition depends on the concrete mixture proportions. Wang et al. [17] conducted a review on the effects of the addition of limestone powder on the properties of concrete and confirmed that the use of limestone powder to replace fine aggregate improves the properties of concrete. Grout, known as highly fluid concrete that flows under its own weight or pressure, can be simulated to self-consolidated concrete (SCC). Daoud and Mahgoub [18] studied the addition of limestone powder to self-consolidated concrete and found that the addition of 30% of the cement weight can reduce the cost and enhance the performance of SCC in terms of workability and flowability. A similar outcome is presented by Valcuende et al. [19] who confirmed the speeding up of the hydration reaction and that the higher the fine limestone content in the mix, the shorter the initial and final concrete setting times. Properties including compression strength, splitting tensile, and workability were apparently improved due to the addition of crushed limestone dust to concrete [20].

The goal of this study is to look into how a finely ground, crushed limestone powder that is taken out of a TBM could be used as a cement grout material for shotcrete and tunnel lining repair. The present study also provides an assessment of the engineering properties and tension capacity of the grout. The recommended mixture may serve as a cost-effective substitute for lean concrete or sand–cement grouts, which are employed in tunnel lining. With a reduced need for cement and crushed sand, this material should be both economical and environmentally benign.

The Brazilian test and tension capacity are examined in this study for suggested mixtures where sand is entirely replaced with crushed limestone material. The elasticity and compression behavior are presented along with the obtained failure mode.

2. Materials and Methods

2.1. Materials

The primary ingredients in the mixtures employed in this investigation include cement (c), sand (s), and crushed limestone. The limestone (LS) was obtained from Riyadh, Saudi Arabia, and is intended as a substitute or replacement for sand in grout mixtures. This limestone is a spoil crushed material produced by tunnel boring machines. The percentages of the limestone material used in the mixtures compared to cement are shown in Table 1. In this investigation, three percentages of limestone aggregates, 1, 1.2, and 1.4 of the dry weight of cement content, were taken into consideration. The effectiveness of employing the crushed limestone material in place of sand was compared and evaluated using four different mixtures, the fourth of which is a control mixture of cement and sand (1:1). For blending and combining cement with limestone or sand, distilled water was used.

In this study, Yamama Portland cement, which was produced locally in KSA (Alkharj Plant Sation), was used. The sand used in this study is common construction local sand and is classified as poorly graded (SP) in accordance with ASTM D2487-17 [21,22]. The sand origin is the Al-Thumamah area east of Riyadh, which is overlain by a huge reserve of windblown and subsurface sand deposits. The coefficient of uniformity and the coefficient of concavity are 1.713 and 0.945, respectively. The sand was substituted with different amounts of crushed limestone (obtained from the tunnel boring equipment) that passed through sieve number 4. Figure 1 presents the grain size distribution for the crushed limestone, including the sample passing through #4, used in the mixtures.

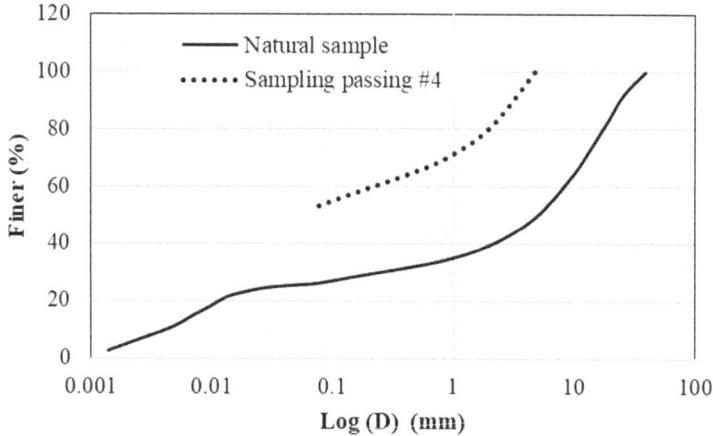

Figure 1. The particle size distribution of the crushed limestone.

The natural sample refers to the sample obtained from the site. Screening using a 3/4" sieve gave a material referred to as natural soil passing through sieve 3/4". The material passing through sieve #4 is the material used in the grout mixture.

Table 1. The cement, limestone, and sand mixtures adopted in this study.

	Mixture No.	1	2	3	4
	Cement, C	1	1	1	1
Material ratio	Limestone, LS	1	1.2	1.4	-
	Sand, S	-	-	-	1
	Water/cement ratio	0.6	0.6	0.6	0.6 *
Specimens	Curing period (days)	7	7	7	7
		14	14	14	14
		28	28	28	28

* Adjusted for workability issues.

2.2. Mixture Preparation

The materials of cement, sand, and crushed limestone were prepared based on the mixture percentages shown in Table 1. The sand and limestone materials were allowed to dry in an oven for a day. All mixtures were mixed in a dry condition, and then, a predetermined amount of distilled water was added to the mixture, which was then thoroughly mixed in a suitable pan. First of all, the control mixture with a cement–sand composition was prepared. Three different percentages of the limestone material as a proportion of the cement were added to compose the targeted mixtures. The amount of distilled water added to the mixtures was estimated to be within the ratio of 0.6 for the water-to-cement content. One exception is that for the cement–sand mixture, this ratio was reduced by 10 percent for workability considerations, and the final results were adjusted using the appropriate curves to be the equivalent of the same water/cement ratio.

2.3. Specimen Preparation

The prepared mixture in the previous section was cast into appropriate molds. As shown in Figure 2, molds measuring 100 mm in height and 50 mm in diameter were built. A longitudinal groove was created to make it easier to extrude the specimen, and it was filled with silicon grease to stop water leakage after casting. A thin layer of oil was applied to coat the interior wall of the mold; this step was necessary to eliminate friction during extruding the specimen. Three to five layers of mixture were cast onto the mold, and each layer was compacted with a steel rod to ensure a homogeneous specimen without air voids. A smooth-edged knife was used to finish each specimen's upper surface.

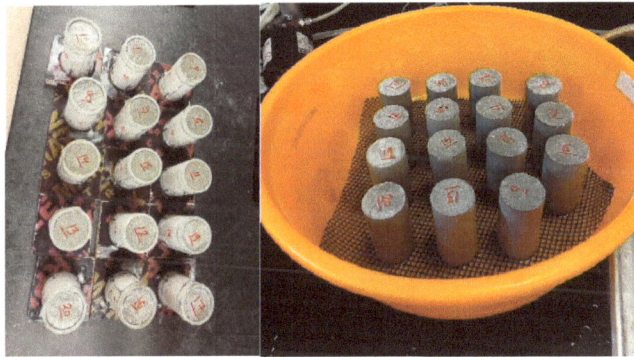

Figure 2. The prepared specimens after casting.

The specimens were extruded from the casting molds after being kept at room temperature throughout the night. A special water basin was prepared, and the specimens were transferred and placed above plastic mesh to ensure water accessibility (Figure 2). The basin was filled with distilled water and the specimens were kept for curing periods of 7, 14, or 28 days.

2.4. Testing Procedure

By the end of the scheduled curing periods of 7, 14, or 28 days, the specimens were taken out of the water basin, and the excess water was removed from the specimens. The dimensions for each specimen, including height and diameter, were measured at different levels (i.e., top, middle, and bottom) and the average values for two heights and three diameters were considered, with the weight of each specimen also recorded. The smoothness of the tested specimens was checked before performing the Brazilian test. Rough sandpapers were used to ensure that irregularities across the surfaces were less than 0.25 mm. Figure 3 shows the 3 MN Toni/Technik compression machine that was employed in this investigation. An inbuilt transducer allows the machine to measure the displacement and control the rate of vertical deformation. An external data logger equipped with additional strain gauges (5 mm) was used to register the strain in two directions along the exterior surface of the cylindrical specimen during the test. The stress–strain curve for each specimen can be drawn using this system.

The specimens were then placed in the compression machine, as shown in Figure 3.

A low rate was applied at 0.09 kN/s. The nominal range of the loading rate according to each specimen's size and specifications ranged between 0.09 kN/s and 0.18 kN/s. Alokili et al. [23] present in more detail the testing procedures followed in this study and a brief description of the compression and crushing strength of the limestone–cement grout.

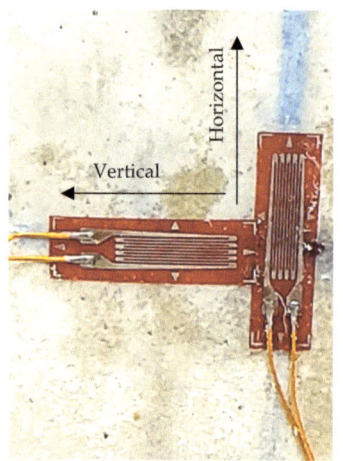 

Figure 3. Fitted strain gauges and compression machine set-up.

3. Results and Discussion

3.1. Compressive Strength over Examined Curing Periods

Table 2 presents the test results of the compressive strength for the specimens tested at specified curing times.

Table 2. Compressive strength in MPa for three curing periods.

	Specimens	1	2	3	4
Material portions	Cement, C	1	1	1	1
	Limestone, LS	1	1.2	1.4	-
	Sand, S	-	-	-	1
Curing in days	7	16.620	23.710	18.320	24.940
	14	21.980	24.090	26.870	32.440
	28	25.970	28.130	30.050	39.140

3.2. Plots of Load vs. Deformation

The vertical load versus deformation curves for the compression tests are presented here to help us understand the progress and development of cracks and the failure of the newly introduced limestone–cement grout mixture and the standard sand–cement grout (Figure 4).

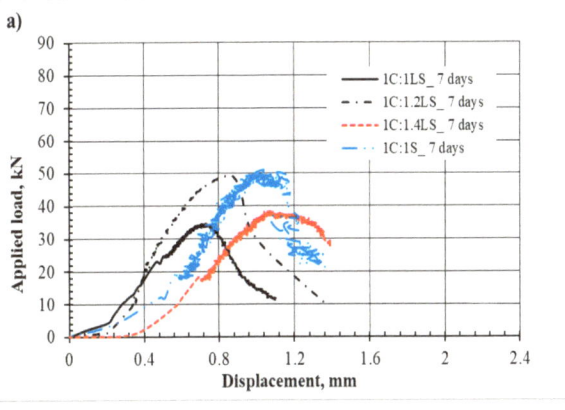

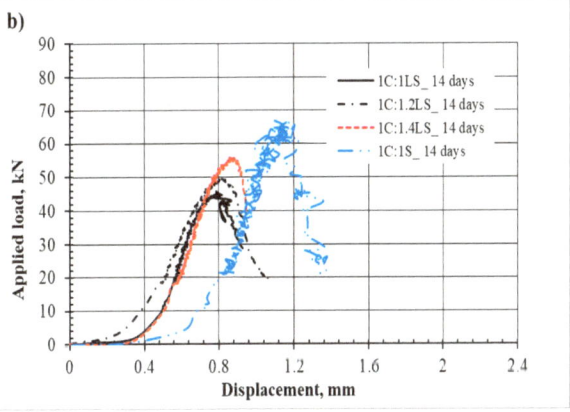

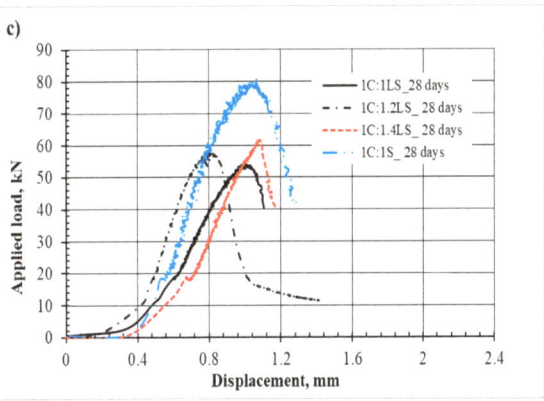

Figure 4. Load versus deformation at curing times of (**a**) 7 days, (**b**) 14 days, and (**c**) 28 days.

The load–displacement data and charts presented here are aimed at showing the general trends of the load–displacement relationships and may not be taken as a quantitative measure.

3.3. Modulus of Elasticity (E)

The calculation of the modulus of elasticity was performed in accordance with ASTM D3148-96 [24]; it represents the ratio between the variation in stress ($\Delta\sigma$) and the variation in strain (Δε) from the stress–strain curve. The slope of the straight line portion of each specimen's stress–strain curve is known as the average Young's modulus or E_{avg}. The Secant Young's modulus (E_{secant}) is the slope of the stress–strain line from the origin to a specific percentage of the ultimate strength, typically estimated to be between 50 and 75% of the ultimate strength. Table 3 presents the calculated moduli of elasticity (i.e., E_{secant} and E_{avg}).

Table 3. Secant modulus of elasticity (E_{secant} and E_{avg}) for all mixtures at variable curing times.

Mixture	Secant Modulus of Elasticity (E_{secant}) GPa			Average Modulus of Elasticity (E_{avg}) GPa		
	Curing Days			Curing Days		
	7	14	28	7	14	28
1C:1LS	22.59	31.34	27.36	32.50	27.27	29.45
1C:1.2LS	47.72	23.73	33.70	30.56	33.09	21.09
1C:1.4LS	22.22	43.75	33.59	22.43	32.50	27.73
1C:1S	20.31	32.44	57.20	17.43	44.44	57.14

3.4. Modes of Failure

The type of failure and deformation that took place during the testing of the specimens for the compression tests as well as the Brazilian tensile test for the limestone–cement grout is shown in Figure 5. The rupture took place along defined, clear vertical planes. Multiple shear failure planes are observed for the compression test, and a single central plane is noted for the Brazilian tensile test. Negative lateral stresses acting perpendicular to the loading vertical axis are responsible for splitting the cylindrical specimen into two equal parts, with a plane of failure that is not uniform or smooth due to the presence of the non-uniform crushed aggregate of limestone. Vertical displacement versus load in the Brazilian tests is shown in Figure 6. Vertical and lateral strains during split tests (strain gauges) are shown in Figures 7–10.

The compressive strength obtained at three different curing times indicates that 60 to 84% strength is gained in the first 7 days, which is not different from the levels obtained for concrete. The compressive strength of limestone–cement grout is in the order of 72% compared to the sand–cement grout. This is due to the reduced water/cement ratio used in the sand–cement grout mix. This can indicate that the use of crushed limestone instead of sand may not reduce the compressive strength.

(**a**) Brazilian tensile test failure mode

(**b**) Axial compression test failure mode

Figure 5. Typical failure mode for tensile and compression tests of limestone–cement grout.

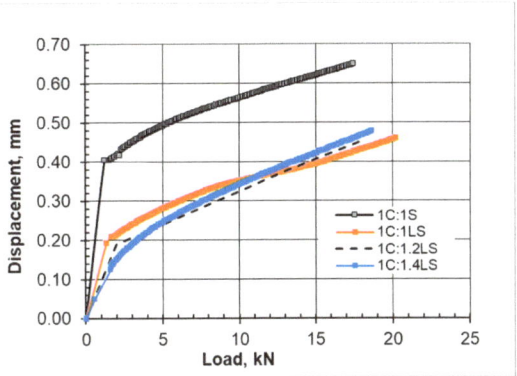

Figure 6. Vertical displacement versus load in Brazilian tests.

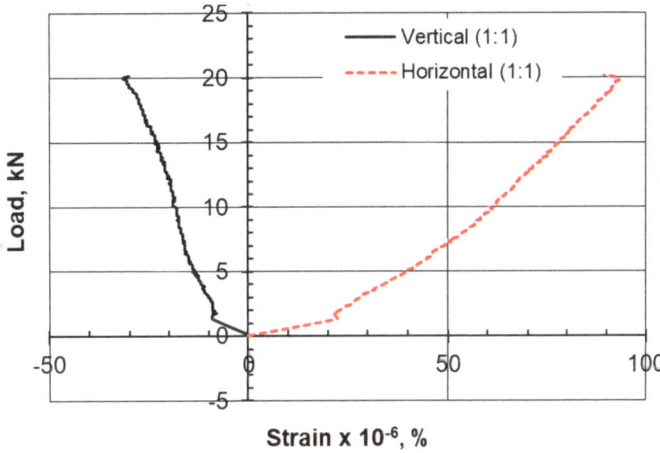

Figure 7. Vertical and lateral strains during split tests (strain gauges), 1C:1LS.

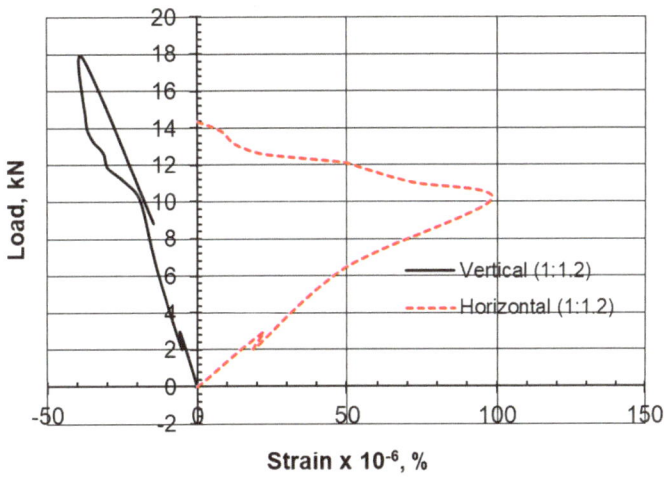

Figure 8. Vertical and lateral strains during split tests (strain gauges), 1C:1.2LS.

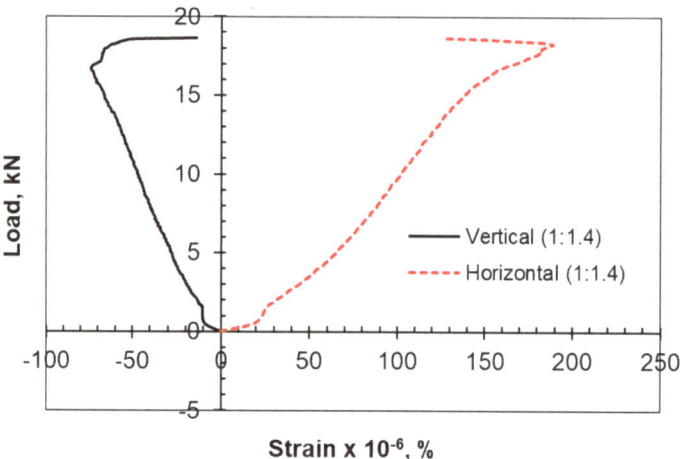

Figure 9. Vertical and lateral strains during split tests (strain gauges), 1C:1.4LS.

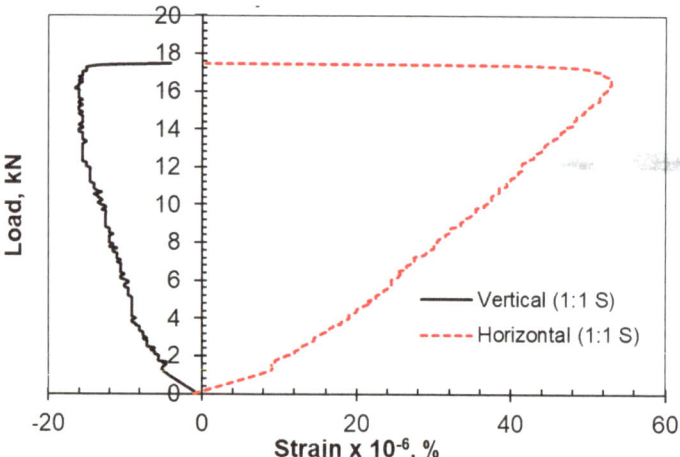

Figure 10. Vertical and lateral strains during split tests (strain gauges), 1:1 sand/cement.

4. Discussion

The Brazilian test conducted in accordance with ASTM C496/C496M-17 [25] or ASTM D3967-16 [26] is generally computed using the formula in Equation (1):

$$\sigma_t = 2* P/(\pi* D*t). \tag{1}$$

D is the diameter of the specimen, t is the width of the specimen, and P is the recorded load.

This is reduced to

$$\sigma_t = 0.636 * P/D*, \tag{2}$$

where D is the diameter of the tested sample and t is the width of the sample (Equation (2)). The computed tensile strength obtained at failure is given in Table 4.

Table 4. Split test results.

Mixture	(1C:1 LS)	(1C:1.2 LS)	(1C:1.4 LS)	(1C:1 Sand)
Max load kN	20.15	17.81	18.63	17.46
S_{pt} N/mm^2	2.54	2.27	2.38	2.29

It can be observed that the tensile strength does not improve with a greater addition of crushed limestone, unlike the compressive strength, where an increase from 25.97 kN/m^2 to 30.05 kN/m^2 was observed when the cement/limestone ratio was increased from 1 to 1.4.

The majority of the elasticity modulus variation is negligible and centers on the values of the specimens that belong to the cement–sand combination.

An increase in the limestone material results in an increase in strength when analyzing the evolution of compressive strength over the curing period. The growing compressive strength of 90% of the cement–sand control mixture is indicated by the percentage of the limestone material (1.1). The angular shape of the particles interlocks, which is responsible for the strength increase in the proposed limestone material mixes.

This shows that in situations where sand is scarce, using this limestone material can effectively replace the sand. For combinations with a greater percentage of limestone (1.4) and for the control mixture, this range has exceeded 1.5%. The strain at the ultimate load and failure is generally found between 0.5% and 1%.

The procedure for split testing is enhanced with strain gauges at the exterior surface of the cylindrical samples. An analytical approach using the Kirish solution (Figure 11) for a biaxially loaded plate of homogeneous, isotropic, and elastic material with a circular opening is the most widely used method for determining the induced stresses [27].

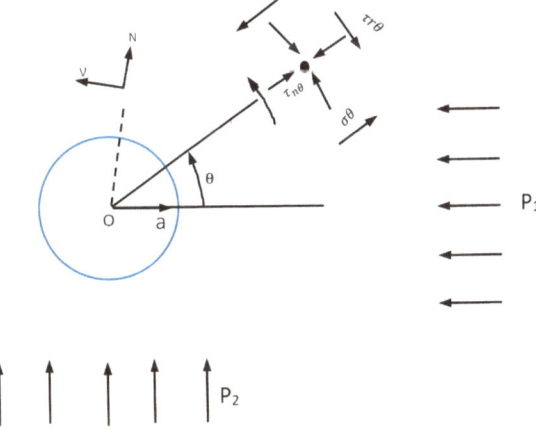

Figure 11. The radial and tangential stresses for a circular hole in an infinite disk [27].

The concept of plain strain can estimate the radial and tangential stresses using Equations (3)–(5):

$$\sigma_r = 0.5 \, (P_1 + P_2) \, (1 - a^2/r^2) + 0.5 \, (P_1 - P_2)(1 - 4a^2/r^2 + 3a^4/r^4) \cos 2\theta, \qquad (3)$$

$$\sigma_\theta = 0.5 \, (P_1 + P_2) \, (1 + a^2/r^2) - 0.5 \, (P_1 - P_2) \, (1 + 3a^4/r^4) \cos 2\theta, \qquad (4)$$

$$\tau_{r\,\theta} = -0.5 \, (P_1 - P_2) \, (1 + 4a^2/r^2 - 3a^4/r^4) \sin 2\theta. \qquad (5)$$

Considering that the inner opening circle is an extremely small part of the stresses where the strain gauges are fitted, $\theta = 0$, $P_1 = 0$, and $a = 0$.

Equations (3)–(5) lead to Equations (6)–(8):

$$\sigma_r = 0.5\,(P_2) + 0.5\,(-P_2) \tag{6}$$

$$\sigma_\theta = 0.5\,(P_2) - 0.5\,(-P_2) \tag{7}$$

$$\tau_{r\theta} = 0 \tag{8}$$

when a is equal to r, and σ_θ can increase to 3 times the axial pressure ($3P_2$). This solution applies to tunnels with a circular opening rather than a solid core subjected to stresses.

The axial stresses are always associated with bulging, and a negative lateral stress causes the cylindrical specimen to increase in the lateral dimension until failure occurs. The computational approaches using finite elements are very useful, despite the fact that the assumed shape keeps changing to an ellipsoidal geometry of varying definition. The exact profile of changes can be obtained experimentally for solid cylindrical specimen materials, excluding all approximations made in mathematical solutions.

Previous studies on the crack initiation of rock material conducted by Nicksiar and Martin [28], Tao et al. [29], and Mutaz [30] defined four zones of rock core cracking under stress. The first zone is defined as crack closure, followed by a crack initiation zone where distress is caused by shear forces and distortion in planes. The third zone is an area of instability and the growth of cracks, and the fourth zone is the peak resistance and failure zone. The crack initiation zone was claimed to start at 30% to 60% of the uniaxial compressive strength. In this study, it was decided to compare two perpendicular strains at the outer side of the cylindrical specimen diameter and observe the sample integrity at each stage.

Examining the horizontal-to-vertical strain ratio at the outer surface of the core, it can be observed that the growth of the $\varepsilon_h/\varepsilon_v$ ratio from the stable elastic zone by 10% can move the status of the stress–strain relationship to the crack initiation zone or failure. The stable elastic zone is assumed when the bilinear stress–strain response ceases. The choice of 25% failure load is believed to be within the stable elastic zone.

Lajtai [31] suggested that lateral strain is the prime indicator for crack initiation. According to this reference, the lateral strain on the stress–strain response is more sensitive to the development of cracks than the axial strain is, even before unstable crack propagation begins. As a result, the point at which the lateral strain diverges from the linearity of the axial stress (σ_1) vs. the lateral strain curve is identified as crack initiation (σ_{ci}).

The initiation of cracks, according to the lateral strain model, takes place at a higher level for the limestone–cement grout compared to the sand–cement grout. The 1:1.4 ratio of the limestone–cement grout scored a stable lateral strain of 150, which is three times the crack initiation level (CI) of that established for the 1:1 sand–cement grout (50).

However, the authors suggest considering the strain in the perpendicular direction as a factor. It was observed that when the $\varepsilon_h/\varepsilon_v$ ratio is increased by 10%, tension failure is reached (Tables 5–8).

Table 5. Horizontal/vertical strain ratio at four stages of loading, 1C: 1LS.

Stage of Loading Relative to Failure Load (20.15 kN)	Vertical Strain ε_v (10^{-6})	Horizontal Strain ε_h (10^{-6})	Horizontal/Vertical Strain Ratio $\varepsilon_h/\varepsilon_v$
25%	−15	40	2.67
50%	−20	62	3.1
75%	−25	80	3.2
100%	−30	90	3.75

Table 6. Horizontal/vertical strain ratio at four stages of loading, 1C:2LS.

Stage of Loading Relative to Failure Load (17.81 kN)	Vertical Strain ε_v (10^{-6})	Horizontal Strain ε_h (10^{-6})	Horizontal/Vertical Strain Ratio $\varepsilon_h/\varepsilon_v$
25%	−10	38	3.8
50%	−20	85	4.25
75%	−35	-	Failure range
100%	−37	-	Failure range

Table 7. Horizontal/vertical strain ratio at four stages of loading, 1C:1.4 LS.

Stage of Loading Relative to Failure Load (18.63 kN)	Vertical Strain ε_v (10^{-6})	Horizontal Strain ε_h (10^{-6})	Horizontal/Vertical Strain Ratio $\varepsilon_h/\varepsilon_v$
25%	−25	58	2.32
50%	−43	100	2.32
75%	−65	140	2.5
100%	-	170	Failure zone

Table 8. Horizontal/vertical strain ratio at four stages of loading, 1C:1S.

Stage of Loading Relative to Failure Load (17.46 kN)	Vertical Strain ε_v (10^{-6})	Horizontal Strain ε_h (10^{-6})	Horizontal/Vertical Strain Ratio $\varepsilon_h/\varepsilon_v$
25%	−10	24	2.40
50%	−14	34	2.42
75%	−18	46	2.55
100%	−17	54	3.20

The axial strain at failure in the sand–cement grout is found to be 1.2 for the compression test, while this strain is reported to be from 0.8 to 1.0 for the limestone–cement grout replacement. In the Brazilian tensile test, it was also observed that the axial strain along the loading axis is much higher than the strain for the limestone–cement grout material.

The tensile stress is highest along the loading axis, and failure occurs by splitting along the same axis, unlike compression test failures where multiple vertical shear plane failures occur (Figure 12).

The lateral strain model combined with the $\varepsilon_h/\varepsilon_v$ growth are good predictors for crack initiation in shotcrete or grout material. The sand can be successfully replaced by crushed limestone materials for shotcrete grout material in tunnels and underground utilities or other applications.

When comparing the compressive strength of the limestone–cement grout to that of the sand–cement grout, the low moisture/cement ratio used in the conventional sand–cement material must be corrected for.

A water/cement ratio of 0.48 can achieve a compressive strength of 27.6 MPa, while a water/cement ratio of 0.59 can provide a compressive strength of 20.7 MPa, as quoted by Sidney M. Levy [32]. This implies an increase in the measured strength for the sand–cement grout by about 25% when compared with other grout groups mixed at a 0.60 moisture/cement ratio. The percentage of the limestone material of 1.1 showed a compressive strength in the range of 90% of the control mixture of cement–sand. Comparing other mixtures in which the aggregate/cement ratio is different from 1:1 needs to be corrected for if reliable numbers are required.

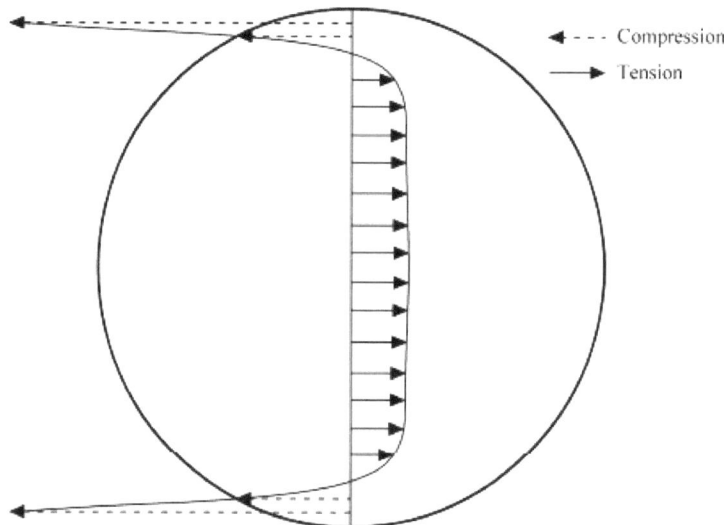

Figure 12. Stress distribution along the loading axis of a Brazilian core test [25].

Abdelgader and Elgalhud [33] presented many formulas to express relationships between the compressive and tensile strengths of grout materials, but these were not found to be representative of grouts made of crushed limestone. The experiments conducted in this study yielded values of split tension capacities ranging from 2.27 to 2.54 N/mm^2. This can be attributed to the type of cement, testing conditions, and other factors. The figures obtained in this study are in agreement with the typical split tension capacity of grout and low-strength concrete.

5. Conclusions

This study is focused on the initiation of cracks and failure in a limestone–cement grout material proposed as an alternative to sand–cement grout. It was found that crushed limestone can be utilized successfully as a shotcrete material with satisfactory performance with regard to compressive strength and tensile strength capacity. The percentage of the limestone material of 1.1 showed a compressive strength in the range of 90% of the control mixture of cement–sand. The 1:1.2 and 1:1.4 mixtures demonstrated high compressive strength values exceeding that of the control mixture. The crack initiation model based on lateral strain can be examined using the ratio of $\varepsilon_h/\varepsilon_v$ to assess crack initiation and failure zones. It was observed that when the $\varepsilon_v/\varepsilon_h$ ratio is increased by 10%, tension failure is reached. In the Brazilian tensile test, it was also observed that the axial strain along the loading axis is much higher than the strain for the limestone–cement grout material. The 1:1.4 ratio of limestone–cement grout scored a stable lateral strain of 150, which is three times the crack initiation level (CI) that is established for the 1:1 sand–cement grout (50). The split tensile strength of the limestone–cement grout was found to be equal to or higher than that of the conventional sand–cement grout.

It can be concluded that the crushed limestone material can replace sand in grout mixtures and perform well within the range of compressive and tensile stresses when reported within the design range specified. Crack initiation occurs at higher stresses when the limestone/cement ratio is 1.4. This study is limited to the common limestone rock material found in the Riyadh area and other sedimentary rocks. Rocks of different origin need to be investigated using a similar approach. Impacts on long-term performance and environmental effects are suggested for future studies. Considering the use of a plasticizer in future studies can be useful to manage the water/cement ratio of the grout mixture.

Author Contributions: Conceptualization, M.D.; data curation, A.M.A.-M.; funding acquisition, M.D.; materials, A.A; investigation, A.M.A.-M.; methodology, A.M.A.-M. and M.D.; writing—original draft, M.D.; writing—review and editing, A.M.A.-M., M.D., and A.A. All authors have read and agreed to the published version of the manuscript.

Funding: This article is funded by the Researchers Supporting Project of King Saud University, Riyadh, Saudi Arabia. Project number RSPD2024R1059.

Institutional Review Board Statement: Not applicable.

Informed Consent Statement: Not applicable.

Data Availability Statement: The data are contained within the article.

Acknowledgments: The authors gratefully acknowledge the Researchers Supporting Project number RSPD2024R1059, King Saud University, Riyadh, Saudi Arabia, for the financial support of the research work reported in this article.

Conflicts of Interest: The authors declare no conflicts of interest.

References

1. Lankard, D.R.; Thompson, N.; Sprinkel, M.M.; Virmani, Y.P. Grouts for bonded post-tensioned concrete construction: Protecting prestressing steel from corrosion. *Mater. J.* **1993**, *90*, 406–414.
2. Dong, H.-L.; Zhou, W.; Wang, Z. Flexural performance of concrete beams reinforced with FRP bars grouted in corrugated sleeves. *Compos. Struct.* **2019**, *215*, 49–59. [CrossRef]
3. Müller, U.; Miccoli, L.; Fontana, P. Development of a lime based grout for cracks repair in earthen constructions. *Constr. Build. Mater.* **2016**, *110*, 323–332. [CrossRef]
4. Alnuaim, A.; Abbas, Y.M.; Khan, M.I. Sustainable Application of Processed TBM Excavated Rock Material as Green Structural Concrete Aggregate. *Constr. Build. Mater.* **2021**, *274*, 121245. [CrossRef]
5. Alnuaim, A.; Dafalla, M.; Al-Mahbashi, A. Enhancement of Clay–Sand Liners Using Crushed Limestone Powder for Better Fluid Control. *Arab. J. Sci. Eng.* **2019**, *45*, 367–380. [CrossRef]
6. Dafalla, M.; Al-Mahbashi, A.; Alnuaim, A. Characterization and assessment of crushed limestone powder and its environmental applications. *IOP Conf. Ser. Earth Environ. Sci.* **2021**, *727*, 012013. [CrossRef]
7. El-Kelesh, A.M.; Matsui, T. Compaction Grouting and Soil Compressibility. In Proceedings of the Twelfth International Offshore and Polar Engineering Conference, Kitakyushu, Japan, 26–31 May 2002.
8. Bayer, I.R. Use of Preplaced Aggregate Concrete for Mass Concrete Application. Master's Thesis, The Graduate School of Natural and Applied Sciences of Middle East Technical University, Ankara, Türkiye, 2004.
9. Lothenbach, B.; Le Saout, G.; Gallucci, E.; Scrivener, K. Influence of limestone on the hydration of Portland cements. *Cem. Concr. Res.* **2008**, *38*, 848–860. [CrossRef]
10. Aquino, C.; Inoue, M.; Miura, H.; Mizuta, M.; Okamoto, T. The effects of limestone aggregate on concrete properties. *Constr. Build. Mater.* **2010**, *24*, 2363–2368. [CrossRef]
11. Alshahwany, R.B.A. Effect of partial replacement of sand with limestone filler on some properties of normal concrete. *AL-Rafdain Eng. J. (AREJ)* **2011**, *19*, 37–48. [CrossRef]
12. Woodward, R.J.; Miller, E.J. Grouting post-tensioned concrete bridges: The prevention of voids. *Highw. Transp.* **1990**, *37*, 9–17.
13. Shannag, M. High-performance cementitious grouts for structural repair. *Cem. Concr. Res.* **2002**, *32*, 803–808. [CrossRef]
14. Zhang, B.; Gao, F.; Zhang, X.; Zhou, Y.; Hu, B.; Song, H. Modified cement-sodium silicate material and grouting technology for repairing underground concrete structure cracks. *Arab. J. Geosci.* **2019**, *12*, 680. [CrossRef]
15. Makhloufi, Z.; Bouziani, T.; Bederina, M.; Hadjoudja, M. Mix proportioning and performance of a crushed limestone sand-concrete. *J. Build. Mater. Struct.* **2014**, *1*, 10–22. [CrossRef]

16. Bonavetti, V.; Donza, H.; Menéndez, G.; Cabrera, O.; Irassar, E. Limestone filler cement in low w/c concrete: A rational use of energy. *Cem. Concr. Res.* **2003**, *33*, 865–871. [CrossRef]
17. Wang, D.; Shi, C.; Farzadnia, N.; Shi, Z.; Jia, H. A review on effects of limestone powder on the properties of concrete. *Constr. Build. Mater.* **2018**, *192*, 153–166. [CrossRef]
18. Daoud, O.M.A.; Mahgoub, O.S. Effect of limestone powder on self-compacting concrete. *FES J. Eng. Sci.* **2021**, *9*, 71–78. [CrossRef]
19. Valcuende, M.; Marco, E.; Parra, C.; Serna, P. Influence of limestone filler and viscosity-modifying admixture on the shrinkage of self-compacting concrete. *Cem. Concr. Res.* **2012**, *42*, 583–592. [CrossRef]
20. Singh, J.; Mukherjee, A.; Dhiman, V.K. Deepmala Impact of crushed limestone dust on concrete's properties. *Mater. Today Proc.* **2021**, *43 Pt 1*, 341–347. [CrossRef]
21. *ASTM D2487-17*; Standard Practice for Classification of Soils for Engineering Purposes (Unified Soil Classification System). ASTM International: West Conshohocken, PA, USA, 2017. [CrossRef]
22. Al-Mahbashi, A.M.; Dafalla, M. Shear strength prediction for an unsaturated Sand Clay Liner. *Int. J. Geotech. Eng.* **2022**, *16*, 282–292. [CrossRef]
23. Alokili, T.; Dafalla, M.; Al-Mahbashi, A.; Alnuaim, A.; Mutaz, E. Tunnel Boring Machine Crushed Limestone as a Cement Grout. In Proceedings of the Geo-Congress 2023, Los Angeles, CA, USA, 26–29 March 2023.
24. *ASTM D3148-96*; Standard test for elastic moduli of intact rock core specimens in uniaxial compression. ASTM International: West Conshohocken, PA, USA, 1996.
25. *ASTM C496/C496M-17*; Standard Test Method for Splitting Tensile Strength of Cylindrical Concrete Specimens. ASTM International: West Conshohocken, PA, USA, 2017.
26. *ASTM D3967-16*; Standard Test Method for Splitting Tensile Strength of Intact Rock Core Specimens. ASTM International: West Conshohocken, PA, USA, 2016.
27. Hazrati Aghchai, M.; Moarefvand, P.; Salari Rad, H. On Analytic Solutions of Elastic Net Displacements around a Circular Tunnel. *J. Min. Environ. (JME)* **2020**, *11*, 419–432. [CrossRef]
28. Nicksiar, M.; Martin, C.D. Evaluation of Methods for Determining Crack Initiation in Compression Tests on Low-Porosity Rocks. *Rock Mech. Rock Eng.* **2012**, *45*, 607–617. [CrossRef]
29. Wen, T.; Tang, H.; Ma, J.; Wang, Y. Evaluation of methods for determining crack initiation stress under Compression. *Eng. Geol.* **2018**, *235*, 81–97. [CrossRef]
30. Mutaz, E.; Serati, M.; Williams, D.J. Crack Initiation Evolution Under Triaxial Loading Conditions. *IOP Conf. Ser. Earth Environ. Sci.* **2021**, *833*, 012012. [CrossRef]
31. Lajtai, E. Brittle fracture in compression. *Int. J. Fract.* **1974**, *10*, 525–536. [CrossRef]
32. Levy, S.M. *Construction Calculations Manual*; Elsevier: Amsterdam, The Netherlands, 2012; ISBN 978-0-12-382243-7. [CrossRef]
33. Abdelgader, H.S.; Elgalhud, A.A. Effect of grout proportions on strength of two-stage concrete. *Struct. Concr.* **2008**, *9*, 163–170. [CrossRef]

Disclaimer/Publisher's Note: The statements, opinions and data contained in all publications are solely those of the individual author(s) and contributor(s) and not of MDPI and/or the editor(s). MDPI and/or the editor(s) disclaim responsibility for any injury to people or property resulting from any ideas, methods, instructions or products referred to in the content.

MDPI AG
Grosspeteranlage 5
4052 Basel
Switzerland
Tel.: +41 61 683 77 34

Materials Editorial Office
E-mail: materials@mdpi.com
www.mdpi.com/journal/materials

Disclaimer/Publisher's Note: The statements, opinions and data contained in all publications are solely those of the individual author(s) and contributor(s) and not of MDPI and/or the editor(s). MDPI and/or the editor(s) disclaim responsibility for any injury to people or property resulting from any ideas, methods, instructions or products referred to in the content.

www.ingramcontent.com/pod-product-compliance
Lightning Source LLC
LaVergne TN
LVHW072351090526
838202LV00019B/2522